Other Titles in This Series

571 **Henry L. Kurland,** Intersection pairings on Conley indices, 1996

570 **Bernold Fiedler and Jürgen Scheurle,** Discretization of homoclinic orbits, rapid forcing and "invisible" chaos, 1996

569 **Eldar Straume,** Compact connected Lie transformation groups on spheres with low cohomogeneity, I, 1996

568 **Raúl E. Curto and Lawrence A. Fialkow,** Solution of the truncated complex moment problem for flat data, 1996

567 **Ran Levi,** On finite groups and homotopy theory, 1995

566 **Neil Robertson, Paul Seymour, and Robin Thomas,** Excluding infinite clique minors, 1995

565 **Huaxin Lin and N. Christopher Phillips,** Classification of direct limits of even Cuntz-circle algebras, 1995

564 **Wensheng Liu and Héctor J. Sussmann,** Shortest paths for sub-Riemannian metrics on rank-two distributions, 1995

563 **Fritz Gesztesy and Roman Svirsky,** (m)KdV solitons on the background of quasi-periodic finite-gap solutions, 1995

562 **John Lindsay Orr,** Triangular algebras and ideals of nest algebras, 1995

561 **Jane Gilman,** Two-generator discrete subgroups of $PSL(2,R)$, 1995

560 **F. Tomi and A. J. Tromba,** The index theorem for minimal surfaces of higher genus, 1995

559 **Paul S. Muhly and Baruch Solel,** Hilbert modules over operator algebras, 1995

558 **R. Gordon, A. J. Power, and Ross Street,** Coherence for tricategories, 1995

557 **Kenji Matsuki,** Weyl groups and birational transformations among minimal models, 1995

556 **G. Nebe and W. Plesken,** Finite rational matrix groups, 1995

555 **Tomás Feder,** Stable networks and product graphs, 1995

554 **Mauro C. Beltrametti, Michael Schneider, and Andrew J. Sommese,** Some special properties of the adjunction theory for 3-folds in $\mathbb{P}^5$, 1995

553 **Carlos Andradas and Jesús M. Ruiz,** Algebraic and analytic geometry of fans, 1995

552 **C. Krattenthaler,** The major counting of nonintersecting lattice paths and generating functions for tableaux, 1995

551 **Christian Ballot,** Density of prime divisors of linear recurrences, 1995

550 **Huaxin Lin,** C^*-algebra extensions of $C(X)$, 1995

549 **Edwin Perkins,** On the martingale problem for interactive measure-valued branching diffusions, 1995

548 **I-Chiau Huang,** Pseudofunctors on modules with zero dimensional support, 1995

547 **Hongbing Su,** On the classification of C*-algebras of real rank zero: Inductive limits of matrix algebras over non-Hausdorff graphs, 1995

546 **Masakazu Nasu,** Textile systems for endomorphisms and automorphisms of the shift, 1995

545 **John L. Lewis and Margaret A. M. Murray,** The method of layer potentials for the heat equation on time-varying domains, 1995

544 **Hans-Otto Walther,** The 2-dimensional attractor of $x'(t) = -\mu x(t) + f(x(t-1))$, 1995

543 **J. P. C. Greenlees and J. P. May,** Generalized Tate cohomology, 1995

542 **Alouf Jirari,** Second-order Sturm-Liouville difference equations and orthogonal polynomials, 1995

541 **Peter Cholak,** Automorphisms of the lattice of recursively enumerable sets, 1995

(Continued in the back of this publication)

MEMOIRS
of the
American Mathematical Society

Number 571

Intersection Pairings on Conley Indices

Henry L. Kurland

January 1996 • Volume 119 • Number 571 (fourth of 5 numbers) • ISSN 0065-9266

American Mathematical Society
Providence, Rhode Island

1991 *Mathematics Subject Classification.*
Primary 55N45, 58F25, 58F35, 34C35; Secondary 34B15, 34E15.

Library of Congress Cataloging-in-Publication Data

Kurland, Henry L., 1947–
Intersection pairings on Conley indices / Henry L. Kurland.
p. cm. – (Memoirs of the American Mathematical Society, ISSN 0065-9266; no. 571)
"January 1996, volume 119, number 571 (fourth of 5 numbers)."
Includes bibliographical references.
ISBN 0-8218-0440-5 (alk. paper)
1. Flows (Differentiable dynamical systems) 2. Topological dynamics. 3. Intersection theory.
I. Title. II. Series.
QA3.A57 no. 571
[QA614.82]
510 s–dc20
[514′.74] 95-39138
CIP

Memoirs of the American Mathematical Society

This journal is devoted entirely to research in pure and applied mathematics.

Subscription information. The 1996 subscription begins with Number 568 and consists of six mailings, each containing one or more numbers. Subscription prices for 1996 are $391 list, $313 institutional member. A late charge of 10% of the subscription price will be imposed on orders received from nonmembers after January 1 of the subscription year. Subscribers outside the United States and India must pay a postage surcharge of $25; subscribers in India must pay a postage surcharge of $43. Expedited delivery to destinations in North America $30; elsewhere $92. Each number may be ordered separately; *please specify number* when ordering an individual number. For prices and titles of recently released numbers, see the New Publications sections of the *Notices of the American Mathematical Society.*

Back number information. For back issues see the *AMS Catalog of Publications.*

Subscriptions and orders should be addressed to the American Mathematical Society, P. O. Box 5904, Boston, MA 02206-5904. *All orders must be accompanied by payment.* Other correspondence should be addressed to Box 6248, Providence, RI 02940-6248.

Memoirs of the American Mathematical Society is published bimonthly (each volume consisting usually of more than one number) by the American Mathematical Society at 201 Charles Street, Providence, RI 02904-2213. Second-class postage paid at Providence, Rhode Island. Postmaster: Send address changes to Memoirs, American Mathematical Society, P. O. Box 6248, Providence, RI 02940-6248.

This publication is indexed in *Science Citation Index®*, *SciSearch®*, *Research Alert®*, *CompuMath Citation Index®*, *Current Contents®/Physical, Chemical & Earth Sciences.*
Printed in the United States of America.
♾ The paper used in this book is acid-free and falls within the guidelines established to ensure permanence and durability.
♻ Printed on recycled paper.
10 9 8 7 6 5 3 2 1 01 00 99 98 97 96

CONTENTS

ABSTRACT

Given an isolated invariant set of a flow on a manifold of dimension m oriented over a PID R, an intersection class pairing of degree $-m$ on the tensor product of the singular homology modules of the forward and reverse time Conley indices of the isolated invariant set with values in the Čech homology of the invariant set is defined. Restriction of the pairing to elements of degree m results in an intersection number pairing that is invariant under continuation along a continuous path of flows and isolated invariant sets. More generally, the unrestricted pairing defines continuous lifts to a space of Čech homology classes along such a path. Further, when the homology modules of the Conley indices are torsion free, the intersection number pairing is non-singular. Also, the pairing associated to an isolated invariant set of a product flow is, modulo torsion, the product (up to sign) of the pairings associated to the factor isolated invariant sets. Intersection classes of lowest and highest dimension are computed for any R-orientable, normally hyperbolic invariant submanifold whose expanding and contracting normal subbundles are also R-orientable. These computations yield, due to dimensional considerations, a complete computation of the intersection class and number pairings for hyperbolic critical points and hyperbolic closed orbits. Application is made in an appendix to the existence of solution of a class of singularly perturbed two-point boundary value problems such problems having provided strong motivation for the present study.

1991 *Mathematics Subject Classification.* Primary: 58F25, 58F35, 34C35, 55N45. Secondary: 34B15, 34E15.

Key words and phrases. isolated invariant set, Conley index, reverse time Conley index, homology of a Conley index, intersection numbers, intersection number pairing, intersection class pairing, homology intersection number pairing, invariance under continuation, connected simple system, normally hyperbolic invariant submanifold, singularly perturbed non-linear two-point boundary value problem.

FOR HER ENDURING SUPPORT,
TO MY WIFE *Lynn*

INTRODUCTION

A compact invariant set of a flow in a manifold is an isolated invariant set if it is the largest invariant set in some compact neighborhood of itself called an isolating neighborhood of the invariant set. The present work arose from the author's study of two-point boundary value problems when in the course of proving some existence theorems it became necessary to ask how closely does the nature of intersections of local stable and unstable sets of an isolated invariant set and the behavior of such intersections under continuation along a path of flows resemble at the level of homology those found for the subclass of isolated invariant sets consisting of normally hyperbolic invariant submanifolds of flows generated by continuous vectorfields. The answer is that the resemblance is remarkably close with regard to intersection classes of degree zero; the resemblance is much weaker for intersection classes of positive degree. The definition of the intersection classes is based on the intersection product of singular homology classes as described in Dold's book [D] and, requiring no assumption of additional smoothness, is naturally formulated in the context of continuous flows on a C^0 manifold. However, the presence of a minimal amount of smoothness, viz., that the manifold is at least C^1 and that the flow is generated by a continuous vectorfield, hereinafter referred to as the minimal smoothness conditions, does simplify the proofs of a few results as will be pointed out in the main body of the text and may be necessary to the proof of non-singularity of the intersection number pairing on Conley indices developed here. A more detailed description of the contents of this work follows.

Let S be an isolated invariant set of a flow in M, an m-dimensional, second countable, topological manifold without boundary oriented over a PID R. For G any R-module, associated to S are the (reduced) singular homology modules with coefficients in G of its Conley indices in forward and reverse time, respectively denoted by $\widetilde{H}_*\mathcal{C}(S;G)$ and $\widetilde{H}_*\mathcal{C}^*(S;G)$. The graded module $\widetilde{H}_*\mathcal{C}(S;G)$ is a homological approximation to the reduced singular homology of the one-point compactification of any local unstable set of S with the approximation exact when S is a normally hyperbolic invariant submanifold of a smooth flow; the analogous remarks apply to $\widetilde{H}_*\mathcal{C}^*(S;G)$ and any local stable set of S. The precise definitions of these modules as well as other basic notions of the Conley index theory are reviewed in Chapter 1. To simplify the presentation, all homology modules henceforth mentioned in this introduction are taken with coefficients in R and the coefficient module is therefore suppressed from the notation. Thus, $H_*(M)$ denotes the singular homology of M and $\check{H}_*(S)$ denotes the Čech homology of S, both with coefficients in R.

Received by the editor August 30, 1990, and in revised form June 15, 1994.

The main objects of study in this work are defined in Chapter 2 and are a bilinear pairing of degree $-m$

$$\mathfrak{L}\colon \widetilde{H}_*\mathcal{C}(S) \otimes \widetilde{H}_*\mathcal{C}^*(S) \to \check{H}_*(S),$$

called the intersection class pairing on the Conley indices of S, and the closely related pairing

$$^{\#}\mathfrak{L}\colon \left[\widetilde{H}_*\mathcal{C}(S) \otimes \widetilde{H}_*\mathcal{C}^*(S)\right]_m \to R,$$

called the intersection number pairing on the Conley indices of S, obtained from $\mathfrak{L}$ by first restricting $\mathfrak{L}$ to the indicated submodule of elements of degree m, then composing the natural projection of $\check{H}_0(S)$ into $H_0(M)$ with the restricted $\mathfrak{L}$, and finally superposing on this composite the Kronecker index.

When $M = \mathbf{R}^m$ a third pairing

$$\mathsf{L}\colon \left[\widetilde{H}_*\mathcal{C}(S) \otimes \widetilde{H}_*\mathcal{C}^*(S)\right]_m \to H_m(\mathbf{R}^m, \mathbf{R}^m \setminus \{0\}),$$

called the homology intersection number pairing on the Conley indices of S, is defined whose definition is independent of those of $\mathfrak{L}$ and $^{\#}\mathfrak{L}$, but is related to $^{\#}\mathfrak{L}$ by the identity

$$\mathsf{L} = \pm {}^{\#}\mathfrak{L} \cdot o_m$$

where $o_m \in H_m(\mathbf{R}^m, \mathbf{R}^m \setminus \{0\})$ is the given orientation of $\mathbf{R}^m$. (See Proposition 2.9 below for a resolution of the sign ambiguity; also see part (D) of Definition 2.8 immediately preceding it.) Given the identity, the obvious question is why bother with L at all. There are two reasons for doing so. First, it is much more difficult to define intersection numbers homologically in the general manifold setting than it is in the Euclidean setting and the definition of L is based on the simpler definition available in the Euclidean setting. Since those properties of $\mathfrak{L}$ most likely to find wide use in applications to the study of differential equations are in fact properties of $^{\#}\mathfrak{L}$ necessarily shared with L, the author hopes that including the definition of L will make these results available to a wider audience. Second, that L is well-defined requires proof, and that proof when suitably abstracted has as corollary the result needed to show $\mathfrak{L}$ well-defined. Thus, in Chapter 2 first L is defined, then $\mathfrak{L}$, and finally $^{\#}\mathfrak{L}$. However, the proofs of the theorems showing L and $\mathfrak{L}$ well-defined are delayed until Chapter 4.

The pairings $\mathfrak{L}$ and $^{\#}\mathfrak{L}$ are of interest chiefly because of the properties they enjoy with respect to continuation of the flow and isolated invariant set. Stated precisely in Chapter 3, but with the proofs delayed until Chapter 5, these are, roughly, (1) the pairing $\mathfrak{L}$ defines continuous lifts of a path of flows and isolated invariant sets to a space of Čech homology classes associated to the isolated invariant sets of the family of flows under consideration; (2) the pairing $^{\#}\mathfrak{L}$ is invariant along any path of flows and isolated invariant sets; i.e, the Kronecker index is constant along a lift by $\mathfrak{L}$ to a path of intersection classes in dimension zero. Here, a family of isolated invariant sets, each member determined by a flow along a continuous path of flows, defines a continuous path of isolated invariant sets if at each parameter value an isolating neighborhood of the invariant set at that parameter value is an isolating neighborhood of the invariant set at nearby parameter values. Continuity of the lift of such a path to the space of Čech classes means that at each parameter value if N is an isolating neighborhood of the invariant set at that parameter value, then the projection to $H_*(N)$ of the Čech class on the lift at that point is the common value of the projections to $H_*(N)$ of the Čech classes on the lift at parameter values sufficiently close to the original one. Statement (1) is weak, for as illustrated by Example 3.9 below, it is possible to have a non-zero intersection class of positive dimension continue to a zero intersection class in that dimension. This merely reflects the fact that continuity in the context of a space of isolated invariant sets associated to a continuous family of flows ensures upper, but not lower, semi-continuity of the isolated invariant set relative to the topology on compact subsets of M induced by the Hausdorff metric so that the homeomorphism type of isolated invariant sets along a path can vary drastically in dramatic contrast to what could happen if the path of invariant sets were confined to a path of normally hyperbolic invariant submanifolds where as a consequence of transversality theory the diffeomorphism type must remain constant. However, the continuity of the lifts determined by $\mathfrak{L}$ over a path of isolated invariant sets does provide more information about the isolated invariant sets on the path than is available with the Conley index alone. On the other hand, the invariance under continuation of the intersection number pairing ${}^{\#}\mathfrak{L}$, a previously unsuspected algebraic manifestation of the invariance of the Conley index under continuation, is easily the most important property of $\mathfrak{L}$ proved here because it is the one most likely to have wide use in applications of the pairing to the study of differential equations. Specific application of these results to a problem arising in population genetics is discussed further on in this introduction.

The first part of Chapter 6 contains several technical results describing computation of $\mathfrak{L}$ and ${}^{\#}\mathfrak{L}$ in terms of singular chains representing classes in $\widetilde{H}_*\mathcal{C}(S)$ and $\widetilde{H}_*\mathcal{C}^*(S)$ that will not be elaborated upon here. The second part of Chapter 6 shows that $\mathfrak{L}$ is invariant, i.e., transforms naturally, under orientation preserving, orbit preserving homeomorphisms. Such homeomorphisms are not assumed to preserve time parameterizations. This result extends to transformation by continuous, proper, orbit preserving surjections if the pairing $\mathfrak{L}$ is replaced by its analogue on the Čech homology of the Conley indices denoted by

$$\check{\mathfrak{L}}\colon \check{\widetilde{H}}_*\mathcal{C}(S) \otimes \check{\widetilde{H}}_*\mathcal{C}^*(S) \to \check{H}_*(S)\,.$$

In general, $\mathfrak{L}$ always factors through $\check{\mathfrak{L}}$ by the natural homomorphism of singular theory into Čech theory. When the minimal smoothness conditions are met the Čech and singular modules of the Conley indices of S are naturally identified, and modulo this identification $\check{\mathfrak{L}}$ and $\mathfrak{L}$ coincide.

To describe the computation of $\mathfrak{L}$ and ${}^{\#}\mathfrak{L}$ on R-orientable, normally hyperbolic invariant submanifolds made in Chapter 7 (which in part relies on the technical results of Chapter 6) requires introducing the notion of an isolating block. For the purposes of this introduction, an isolating neighborhood B is an isolating block neighborhood (sometimes abbreviated to "block") if its boundary is the union of two closed and possibly overlapping sets B^+ and B^-, called respectively the exit and entrance sets of B, so that a point lies in B^+ (respectively, B^-) if it exits B immediately in positive (respectively, negative) time under the action of the flow.

The flow is therefore externally "tangent" to a block at points in the intersection of its exit and entrance sets. When the minimal smoothness conditions are met, a result of Wilson and Yorke [WY, Theorem 2.1] shows that there exists a cofinal family of isolating block neighborhoods of any isolated invariant set with the property that each block in the family is a C^0 manifold with boundary whose exit and entrance sets lie in codimension one, C^1 submanifolds without boundary to which the vectorfield generating the flow is transverse. In the present context, isolating blocks are of interest because for any block B there are canonical isomorphisms

$$H_*(B, B^+) \simeq \widetilde{H}_*\mathcal{C}(S) \qquad \text{and} \qquad H_*(B, B^-) \simeq \widetilde{H}_*\mathcal{C}^*(S)\ .$$

A result of Pugh and Shub [PS, Theorem 2] shows that when S is a normally hyperbolic invariant submanifold of a smooth flow in M, the flow is topologically conjugate on a tubular neighborhood of S to the linearization of the flow restricted to the normal bundle. It is fairly clear that the technique of their proof which defines expanding-contracting product coordinates in a neighborhood of S and which is based in part on earlier work of Palis and Smale [Pa], [PaS] can be used to show that S admits a cofinal family of isolating blocks with each member block B homeomorphic to the total space of the Whitney sum of two disk bundles, one the unit disk bundle of the expanding subbundle of the normal bundle over S, the other the unit disk bundle of the contracting subbundle of the normal bundle over S with B^+ corresponding to the Whitney sum of the unit sphere bundle of the expanding normal subbundle and the unit disk bundle of the contracting normal subbundle and with B^- described similarly by interchanging the roles of the two disk subbundles. Because $[\mathcal{C}(S)]$, the Conley homotopy index of S in forward time, must equal the homotopy type of the quotient B/B^+, it follows that $[\mathcal{C}(S)]$ equals the homotopy type of the Thom space of the expanding subbundle; similarly for the reverse time homotopy index $[\mathcal{C}^*(S)]$ and the contracting subbundle. It follows with the aid of the Thom isomorphism theorem that for S and the expanding and contracting normal subbundles R-orientable and where k equals either (i) the dimension of the unstable (stable) manifold of S or (ii) the rank of the expanding (contracting) subbundle of the normal bundle over S, a generator $\boldsymbol{\alpha}$ of $\widetilde{H}_k\mathcal{C}(S)$ (respectively, $\boldsymbol{\gamma}$ of $\widetilde{H}_k\mathcal{C}^*(S)$) can be geometrically realized: in case (i) as the local unstable (stable) manifold of S within B regarded as a relative cycle in B modulo its exit (entrance) set; in case (ii) as the local strong unstable (strong stable) manifold within B of any point in S regarded as a relative cycle in B modulo its exit (entrance) set. Hyperbolicity ensures that the intersection of such a representative for $\boldsymbol{\alpha}$ with one for $\boldsymbol{\gamma}$ is a submanifold of S, and this intersection submanifold geometrically represents $\mathfrak{L}(\boldsymbol{\alpha} \otimes \boldsymbol{\gamma})$. When $\boldsymbol{\alpha}$ represents a local unstable manifold of S and $\boldsymbol{\gamma}$ a local stable, it follows that $\mathfrak{L}(\boldsymbol{\alpha} \otimes \boldsymbol{\gamma}) = \pm o_S$ where o_S is the fundamental class of S; when $\boldsymbol{\alpha} \otimes \boldsymbol{\gamma}$ has degree m (e.g., a strong local unstable manifold represents $\boldsymbol{\alpha}$ and a local stable manifold represents $\boldsymbol{\gamma}$) it follows by definition of the Kronecker index that ${}^{\#}\mathfrak{L}(\boldsymbol{\alpha} \otimes \boldsymbol{\gamma})$ gives an algebraic count of the number of points in the 0-dimensional intersection submanifold representing $\mathfrak{L}(\boldsymbol{\alpha} \otimes \boldsymbol{\gamma})$—in the cases at hand, ${}^{\#}\mathfrak{L}(\boldsymbol{\alpha} \otimes \boldsymbol{\gamma}) = \pm 1$. In all these cases, the ambiguous sign is determined by the choice of orientation classes for the expanding and contracting subbundles of the normal bundle. In particular, the generators described in cases (i) and (ii) can be characterized in terms of

Thom classes of the expanding and contracting subbundles of the normal bundle of S, a fundamental class of S, and a cofinal family of isolating blocks of S with the requisite Whitney sum structure in which generators of the homology modules of the Conley indices can be geometrically realized as described above by cases (i) and (ii) compatibly over the cofinal family.

Below, although a partial computation of $\mathfrak{L}$ and $^{\#}\mathfrak{L}$ as just described will be made for the general normally hyperbolic invariant submanifold modulo the detailed construction of the isolating block described above and assuming that S and the expanding and contracting subbundles of the normal bundle of S are R-orientable, a complete proof of the statements of the previous paragraph will be made only for S a hyperbolic critical point or a hyperbolic closed orbit. In these two cases, the partial computation is in fact complete due to dimensional considerations. Also, the description of $[\mathcal{C}(S)]$ in terms of a Thom space results, to the author's knowledge, in the first complete computation of the Conley indices of a hyperbolic closed orbit: In [C] the correct values of the indices as pointed homotopy types were given for the two possibilities of the local unstable manifold being orientable or non-orientable, viz., $S^{P+1} \vee S^P$ in the orientable case and $S^{P-1} \wedge \mathbf{RP}^2$ in the non-orientable when there are (counting multiplicities) $P \geq 1$ Floquet multipliers of modulus greater than one; however, no computation is made in the non-orientable case, and the heuristic computation made in the orientable case assumes that the closed orbit is an isolated invariant set of a product flow, viz., the product of S^1 regarded as the unique non-empty isolated invariant set of a uniform rotation on the circle and a hyperbolic critical point of a flow on a Euclidean space of suitable dimension, and therefore has a product block. In general this assumption is invalid: for example, if M is non-orientable, the periodic orbit might have an orientable unstable manifold but a non-orientable stable manifold precluding the possibility that a tubular neighborhood of the orbit is trivializable. However, existence of a product block is provable via Floquet normal form when the linearization of the Poincaré map at one (hence every) point on the closed orbit has a real square root.[1] Thus, Conley's computation is salvageable by lifting the vectorfield to a connected double cover of a tubular neighborhood of the closed orbit and then using Floquet normal form to trivialize the problem in the double cover. In particular, a product block for the doubled closed orbit having the Whitney sum structure described earlier is easily constructed which when "averaged" over the $\mathbf{Z}_2$-action of the double cover yields an invariant block whose projection to the base tubular neighborhood is a block for the original closed orbit also having the desired Whitney sum structure. Further, the construction does not depend on the orientability of the invariant manifolds and makes it easy to compute the index when the unstable manifold is non-orientable.

The main result of Chapter 8 is that modulo torsion the intersection pairing on the Conley indices of an isolated invariant set of a product flow is the product of the intersection pairings on the Conley indices of the factor isolated invariant sets; a corresponding result holds for $^{\#}\mathfrak{L}$. Also presented is a Kunneth theorem for Conley

[1] In general, no real square root exists, nor does one exist generically: a non-singular real matrix has a real square root if, and only if, each elementary Jordan block associated to a negative eigenvalue of the matrix appears an even number of times in the Jordan normal form of the matrix. A proof can be carried out using the real Jordan form of the matrix.

indices which follows straightforwardly from the standard Kunneth theorem.

The last chapter, Chapter 9, presents an alternative characterization of $\mathfrak{L}$ using an analogue of the classical cap product representation of intersection products and a cofinal family of isolating block neighborhoods of S. An immediate consequence of this cap product representation is that ${}^{\#}\mathfrak{L}$ is a non-singular (hence, non-degenerate) pairing of $\widetilde{H}_{m-k}\mathcal{C}(S)$ and $\widetilde{H}_k\mathcal{C}^*(S)$ if both modules are torsion-free. The result is strongly suggested by the results of Chapter 7 and the invariance of ${}^{\#}\mathfrak{L}$ under continuation. The development of the cap product representation, and therefore the proof of non-singularity, currently requires that the minimal smoothness conditions be met by M and the flow. It is unclear if these results can be extended to the case where M and the flow are at best C^0. When the minimal smoothness conditions are met, there is a Poincaré duality of forward and reverse time Conley indices which in terms of a block B that is a C^0 manifold with boundary and which has its exit and entrance sets as submanifolds with boundary is expressed by $H^{m-k}(B, B^{\mp}) \simeq H_k(B, B^{\pm})$. This type of duality was first noted by Montgomery [M] using field coefficients and with all modules Alexander-Spanier cohomology modules, but he gave no proof. Recently, McCord [Mc2] has given a short proof of these duality relations. A second, longer proof comes out in the wash in the course of proving the cap product representation of $\mathfrak{L}$ which is

$$\mathfrak{L}(\boldsymbol{\alpha} \otimes \boldsymbol{\gamma}) = \varprojlim(x_B \frown \gamma_B)$$

where $x_B \in H^*(B, B^-)$ is the cohomology class dual to the homology class in $H_*(B, B^+)$ whose canonical image is $\boldsymbol{\alpha} \in \widetilde{H}_*\mathcal{C}(S)$ and where $\gamma_B \in H_*(B, B^-)$ has canonical image $\boldsymbol{\gamma} \in \widetilde{H}_*\mathcal{C}^*(S)$. The inverse limit is with respect to the inverse system of homology modules and inclusion induced homomorphisms of the cofinal family of isolating block neighborhoods of S directed by reverse inclusion; however, it is far from obvious that the inverse limit of the cap products is well-defined.

As mentioned in the first paragraph of this Introduction, largely motivating the present study is the application of L and ${}^{\#}\mathfrak{L}$ and their invariance under continuation to the statement and proof of existence results ensuring the presence of endpoint or interior transition layer behavior for singularly perturbed, vector two-point boundary value problems on the finite interval $[a, b]$ of the form

$$\varepsilon \mathbf{u}' = \mathbf{F}(\mathbf{u}, x, \varepsilon) \tag{0.1a}$$

$$\mathbf{u}(a) \in \mathbf{E}(a, \varepsilon), \quad \mathbf{u}(b) \in \mathbf{E}(b, \varepsilon) \tag{0.1b}$$

where differentiation is with respect to $x \in \mathbf{R}$, $\mathbf{u} \in \mathbf{R}^m$, and $\mathbf{E}(\lambda, \varepsilon)$ ($\lambda = a, b$, $\varepsilon > 0$) is the locus of points in the phase space $\mathbf{R}^m$ satisfying the endpoint condition at λ and where existence of solution is to be proved for $0 < \varepsilon \ll 1$. Because of the large motivational role of problem (0.1), in Appendix A to the current study the weakest form of such an existence theorem is stated and proved and then applied to the analysis of Example 3.5 (which illustrates the invariance of intersection numbers under continuation) to yield existence of a non-trivial stationary solution satisfying Neumann boundary conditions of a Fisher equation for the variation in allelic density over space and time of the two alleles present at a single gene locus in a population that is genetically isolated, has sufficiently low genetic diffusion, and is confined to a finite, one-dimensional habitat where heterozygote individuals are assumed to have lower genetic fitness than homozygote individuals of either type. In fact, this problem from population genetics as well as certain defects in the results of [Bg] initiated the author's general study of problem (0.1).

As there has been a long delay in the publication of this work since its original submission due, on the one hand, to a long refereeing process and, on the other, to the author's glacial speed in rewriting the originally submitted papers, and since at least one paper closely related to this work has appeared in the interim, viz., McCord's [Mc2], a few comments on the history of this work seem in order. The two papers originally comprising this work were submitted in April and July of 1990. However, the author first proved the existence of the pairing L in the unpublished paper [K7] which is cited in [Mc2] and possibly played some part in motivating that paper. In any event, the author received a preprint of McCord's article too late for it to have any substantial influence on the original submissions although it did result in the addition to the first paper submitted of a few remarks on how the method used to construct the pairing L could be used to construct an isomorphism giving the Poincaré duality of Conley indices. In the work at hand those remarks are represented by Proposition 9.12 and Theorem 9.15. Also, in the original submission, $\mathfrak{L}$ was computed only for a hyperbolic critical point of a vectorfield on M whose Hessian at the critical point had m distinct eigenvalues. Thus, as far as content is concerned, the main difference between the originally submitted papers and the present work is that in the former nearly all of the material in Chapters 7 and 9 was not present.

The author would like to thank John Milnor for asking several stimulating questions of the author when he presented the results of Chapters 2 and 3 and Appendix A at an April 1992 meeting in dynamical systems held at Union College, Schenectady, NY which led the author to tackle again (this time successfully) the question of whether or not $^{\#}\mathfrak{L}$ was non-degenerate. The author would also like to thank Mark Steinberger for several helpful conversations on the classification of fibre bundles and to thank the Department of Mathematics and Statistics at SUNY-Albany for permitting the author to use their computing facilities in the preparation of this work and for sometimes providing employment. Finally, the author would like to thank Kristoffer H. Rose and Ross Moore for writing and for their help in using the XY-pic diagram package used to great effect throughout this work, but especially in Chapters 8 and 9 and Appendix B, in typesetting the many non-rectangular diagrams.

CHAPTER 1

BASIC NOTATION AND BACKGROUND DEFINITIONS

A. Basic Notation. The symbols $\mathbf{N}$, $\mathbf{Z}$, $\mathbf{R}$, $\mathbf{R}^+$, and $\mathbf{R}^-$ will respectively denote the natural numbers (0 included), the integers, the reals, the non-negative reals, and the non-positive reals. For $a, b \in \mathbf{R}$, $]a, b[$ and $[a, b]$ respectively denote the open and closed intervals from a to b; similarly, $]a, b]$ and $[a, b[$ denote half-open intervals open respectively on the left and right. Throughout, M denotes a second countable C^r manifold without boundary $(0 \leq r \leq \infty)$ of dimension m. The second countability assumption ensures that M is a Euclidean neighborhood retract (hereinafter abbreviated to ENR) and that M is metrizable. Further, M is assumed oriented over the PID R.

A C^k *flow* $(0 \leq k \leq r)$ *in* M is a C^k map on an open $\mathfrak{G}$ in $M \times \mathbf{R}$ into M, generally denoted by $(\mathbf{u}, t) \mapsto \mathbf{u} \cdot t$, with the following properties: (1) $\mathfrak{G} \supset M \times \{0\}$ and $\mathbf{u} \cdot 0 = \mathbf{u}$ for each $\mathbf{u} \in M$, (2) $(\mathbf{u} \cdot t_1) \cdot t_2 = \mathbf{u} \cdot (t_1 + t_2)$ whenever $(\mathbf{u}, t_1)$, $(\mathbf{u} \cdot t_1, t_2)$, and $(\mathbf{u}, t_1 + t_2) \in \mathfrak{G}$, (3) for each $\mathbf{u} \in M$, the trajectory of $\mathbf{u}$, i.e., the map $t \mapsto \mathbf{u} \cdot t$ defined for all t satisfying $(\mathbf{u}, t) \in \mathfrak{G}$, has closed, connected graph in the topology inherited from $\mathbf{R} \times M$. It is immediate from the openness of $\mathfrak{G}$ and properties (3) and (1) that the domain of each trajectory is an open interval containing 0. Note that any locally Lipschitz vectorfield on a C^1 manifold generates a C^0 flow in the manifold assuming integral curves of the vectorfield are taken with maximally extended domain. Also, any continuous Hamiltonian vectorfield on a two-dimensional symplectic manifold generates a continuous flow in M.

Assume given a flow in M. The *orbit through* $\mathbf{u} \in M$ is the image of the trajectory of $\mathbf{u}$ and is *complete* if the trajectory has domain $\mathbf{R}$; else it is *non-complete.* The *positive* (resp. *negative*) *semi-orbit of* $\mathbf{u}$ is the image of the trajectory restricted to the non-negative (resp. non-positive) reals in its domain and is *complete* if the restricted trajectory has domain $\mathbf{R}^+$ (resp. $\mathbf{R}^-$); else it is *non-complete.* Because each trajectory has an open interval as domain, no non-complete semi-orbit can lie in a compact set as a simple consequence of property (3) of a flow. A subset of M is called *invariant* (resp. *positively invariant*, resp. *negatively invariant*) if, and only if, it is the union, the empty union not excluded, of complete orbits (resp. complete positive semi-orbits, resp. complete negative semi-orbits). Thus, the union of invariant sets is invariant; hence, each $N \subset M$ contains a unique maximal invariant set relative to the partial order induced by inclusion on the invariant sets contained in N. It is a simple consequence of property (3) of a flow that the closure of a relatively compact, positively (resp. negatively) invariant set (relative compactness is not needed if $\mathfrak{G} = M \times \mathbf{R}$) is again positively (resp. negatively) invariant. It follows that the maximal invariant set in each compact $N \subset M$ is itself compact. Call a compact $N \subset M$ an *isolating neighborhood* relative to the given flow if the maximal invariant set it contains lies in its interior, $\text{int}_M(N)$. An invariant set S of the flow is an *isolated invariant set* if it is the maximal invariant set in some isolating neighborhood. For those Y contained in some compact, positively (resp. negatively) invariant subset of M, the maximal invariant subset of $\text{cl}\,(Y \cdot \mathbf{R}^+)$ (resp. $\text{cl}\,(Y \cdot \mathbf{R}^-)$) is denoted by $\omega(Y)$ (resp. $\omega^*(Y)$) and coincides with the usual omega (resp. alpha) limit set of Y. In particular, for those $\mathbf{u} \in M$ having a complete positive semi-orbit and having $\text{cl}\,(\mathbf{u} \cdot \mathbf{R}^+)$ compact, $\omega(\mathbf{u})$ is a well-defined, non-empty, compact, connected subset of M; similarly for $\omega^*(\mathbf{u})$ for $\mathbf{u}$ having a complete negative semi-orbit. For S an isolated invariant set and U a relatively compact neighborhood of S, the notations $W^u(S;U)$ and $W^s(S;U)$ respectively denote *the local unstable set and the local stable set of S within U* relative to the given flow. Explicitly, $\mathbf{u} \in W^s(S;U)$ if, and only if, $\mathbf{u} \cdot \mathbf{R}^+ \subset U$ and $\omega(\mathbf{u}) \subset S$; replacement of $\mathbf{R}^+$ with $\mathbf{R}^-$ and of $\omega(\mathbf{u})$ with $\omega^*(\mathbf{u})$ yields $W^u(S;U)$. Also, set $A^{\pm}(U) := \{\mathbf{u} \in U : \mathbf{u} \cdot \mathbf{R}^{\pm} \subset U\}$. In general, $W^u(S;U)$ is a proper subset of $A^-(U)$, but the two coincide if S is the maximal invariant set in U.

To facilitate the investigation of the variance of the intersection pairings under continuation of the flow and isolated invariant set, the following conventions and notations are adopted in discussing parameterized families of flows. Let Λ be a locally compact Hausdorff space and assume given a C^k flow φ_λ in M with domain $\mathfrak{G}_\lambda$ for each $\lambda \in \Lambda$. To conform to the previously introduced notation denoting a flow as a right $\mathbf{R}$-action, set

$$\mathbf{u} \overset{\lambda}{\cdot} t := \varphi_\lambda(\mathbf{u}, t) \qquad \text{for} \quad (\mathbf{u}, t) \in \mathfrak{G}_\lambda\,, \ \lambda \in \Lambda\,.$$

The family of flows $\{\varphi_\lambda\}_{\lambda \in \Lambda}$ is a *continuous family of C^k flows in M* if, and only if, the set $\mathfrak{G}_\Lambda$ that is the union over $\lambda \in \Lambda$ of the sets $\mathfrak{G}_\lambda \times \{\lambda\}$ is open in $M \times \mathbf{R} \times \Lambda$ and the map $\varphi: \mathfrak{G}_\Lambda \to M$ defined by $\varphi(\mathbf{u}, t, \lambda) := \mathbf{u} \overset{\lambda}{\cdot} t$ is continuous. By an abuse of language, given $\mathfrak{G}_\Lambda$ open in $M \times \mathbf{R} \times \Lambda$, a map $\varphi: \mathfrak{G}_\Lambda \to M$ is referred to as a continuous family of C^k flows in M whenever φ is continuous and its restriction to $\mathfrak{G}_\lambda \times \{\lambda\} := \mathfrak{G}_\Lambda \cap M \times \{\lambda\}$ defines a C^k flow $\varphi_\lambda: \mathfrak{G}_\lambda \to M$ for each $\lambda \in \Lambda$. Note that if M is at least C^1 and $\lambda \mapsto X_\lambda$ is a continuous map of Λ into the space of locally Lipschitz vectorfields on M with any topology at least as fine as the compact-open topology, then the family of vectorfields $\{X_\lambda\}_{\lambda \in \Lambda}$ generates a C^0 family of flows in M.

For $\varphi: \mathfrak{G}_\Lambda \to M$ a continuous family of flows in M, the following glosses are used in referring to isolated invariant sets and isolating neighborhoods of a particular φ_λ. An invariant set of φ_λ is called a λ-invariant set, and if S is an isolated λ-invariant set, then S is said to be λ-isolated. Also, if N is an isolating neighborhood relative to φ_λ, then N is called a λ-isolating neighborhood, and if S is λ-isolated with λ-isolating neighborhood N, then N is said to λ-isolate S.

1.1 DEFINITION. (A) The *space of isolated invariant sets associated to the continuous family of flows* $\varphi: \mathfrak{G}_\Lambda \to M$, denoted $\mathcal{S}(\varphi)$, is the set of ordered pairs (S, λ) where $\lambda \in \Lambda$ and S is λ-isolated topologized as follows. For each compact $N \subset M$, let $\Lambda(N)$ be the set of $\lambda \in \Lambda$ for which N is a λ-isolating neighborhood, and for $\lambda \in \Lambda(N)$, let $S(N, \lambda)$ denote the isolated λ-invariant set λ-isolated by N. For each compact $N \subset M$, define $\sigma_N: \Lambda(N) \to \mathcal{S}(\varphi)$ by $\sigma_N(\lambda) := (S(N, \lambda), \lambda)$. The topology on $\mathcal{S}(\varphi)$ is declared to be the finest topology consistent with each $\sigma_N: \Lambda(N) \to \mathcal{S}(\varphi)$

being continuous where N ranges over compact subsets of M and $\Lambda(N)$ inherits its topology from Λ. In fact, $\Lambda(N)$ is open in Λ for each compact $N \subset M$. The collection of sets $\sigma_N(U)$ where U ranges over open subsets of $\Lambda(N)$ and N ranges over compact subsets of M is a basis for this topology. Also, the natural projection $\pi_\varphi: \mathcal{S}(\varphi) \to \Lambda$ is a surjective local homeomorphism with σ_N a local inverse on $\Lambda(N)$. Details of the above can be found in [RAII], [C], or [Sl].

(B) Throughout this work, whenever $g: Y \to \mathcal{S}(\varphi)$ is a continuous map, the components of g are denoted by S_g and λ_g, i.e., $g(y) =: (S_g(y), \lambda_g(y))$ where $S_g(y)$ is an isolated invariant set of the flow $\varphi_{\lambda_g(y)}$ for each $y \in Y$.

B. The Conley Index. The Conley index will be described below as a certain type of category called a connected simple system. The following notation is used in describing categories. For any category $\mathcal{K}$, its class of objects is denoted below by $\mathrm{ob}(\mathcal{K})$ and for $X, Y \in \mathrm{ob}(\mathcal{K})$, the set of morphisms from X to Y is denoted by $\mathcal{K}(X, Y)$. Recall that a category $\mathcal{K}$ is a *small category* if $\mathrm{ob}(\mathcal{K})$ is a set.

1.2 DEFINITION. A small category $\mathcal{K}$ is a *connected simple system* if $\mathcal{K}(X, Y)$ is a singleton for every $X, Y \in \mathrm{ob}(\mathcal{K})$. The unique element of $\mathcal{K}(X, Y)$ will be denoted by $h_{\mathcal{K}}^{X\,Y}$. Consequently, morphisms in $\mathcal{K}$ satisfy the following identities: for $X, X', X'' \in \mathrm{ob}(\mathcal{K})$,

$$(1.1) \qquad h_{\mathcal{K}}^{X\,X} = 1_X\,, \qquad h_{\mathcal{K}}^{X'\,X''} \circ h_{\mathcal{K}}^{X\,X'} = h_{\mathcal{K}}^{X\,X''}\,, \qquad h_{\mathcal{K}}^{X'\,X} = (h_{\mathcal{K}}^{X\,X'})^{-1}\,.$$

The middle and left identities hold because $\mathcal{K}$ is a category and because $\mathcal{K}(X, Y) = \{h_{\mathcal{K}}^{X\,Y}\}$; the identity on the right holds as a simple consequence of the other two and uniqueness of inverses. Consequently, any two objects of $\mathcal{K}$ are equivalent.

For $\mathcal{K}$ a connected simple system, the facts that $\mathrm{ob}(\mathcal{K})$ is a set, that for every $X, X' \in \mathrm{ob}(\mathcal{K})$, $\mathcal{K}(X, X') = \{h_{\mathcal{K}}^{X\,X'}\}$, and that the identities (1.1) hold are collectively referred to as *the connected simple system properties of* $\mathcal{K}$.

As shown in [K3], for any category $\mathcal{H}$, there is a corresponding category, denoted $\mathcal{CSS}(\mathcal{H})$, of connected simple systems that are subcategories of $\mathcal{H}$. The notion of morphism in $\mathcal{CSS}(\mathcal{H})$ is reviewed below in Chapter 5 and is of particular interest to us in the case $\mathcal{H} := \mathcal{T}^{*\prime}$, the homotopy category of topological spaces with basepoint, because a Conley index is by definition an object of $\mathcal{CSS}(\mathcal{T}^{*\prime})$ and because this notion of morphism undergirds the proof of invariance of intersection numbers under continuation. Other substitutions for $\mathcal{H}$ of interest to us in this context are $\mathcal{G}_R\mathcal{M}$ and $\partial_R\mathcal{M}'$, the former the category of graded left R-modules, the latter the homotopy category of chain complexes over R.

1.3 DEFINITION. Let S be an isolated invariant set of a C^k flow ψ in M, set $\mathbf{u} \cdot t := \psi(\mathbf{u}, t)$, and let $\mathcal{O}^+(\mathbf{u})$ denote the positive semi-orbit of $\mathbf{u}$. A description of the Conley index of S relative to the flow ψ follows.

(A) Compact subsets N_1 and N_0 of M form an *index pair* $\langle N_1, N_0\rangle$ *for* S relative to ψ if, and only if, the following three conditions are satisfied:

(1) $S \subset \mathrm{int}_M(N_1 \setminus N_0)$ and $\mathrm{cl}\,(N_1 \setminus N_0)$ is an isolating neighborhood of S;
(2) if $\mathbf{u} \in N_0$, $t > 0$, and $\mathbf{u} \cdot [0, t] \subset N_1$, then $\mathbf{u} \cdot [0, t] \subset N_0$;
(3) if $\mathbf{u} \in N_1$ and $\mathcal{O}^+(\mathbf{u}) \not\subset N_1$, then there exists $t \geq 0$ so that $\mathbf{u} \cdot [0, t] \subset N_1$ and $\mathbf{u} \cdot t \in N_0$.

These three properties are called respectively the isolating, the relative positive invariance, and the exit properties of an index pair.

If an index pair $\langle N_1, N_0 \rangle$ in fact forms a topological pair, i.e., if $N_0 \subset N_1$, then call it a *nested index pair* and use the notation (N_1, N_0) rather than $\langle N_1, N_0 \rangle$ in writing the index pair. Note that if $\langle N_1, N_0 \rangle$ is an index pair for S, then $(N_1, N_1 \cap N_0)$ is a nested index pair for S, and $(N_1, \cap N_0)$ is used as a shorthand for the latter. The notion of non-nested index pair is needed to prove invariance of intersection numbers under continuation in the non-smooth case, i.e., when either the ambient manifold is only C^0 or when some flow in the continuation is only absolutely continuous but not C^1. Readers interested only in the smooth case ($r \geq k \geq 1$) can assume that all index pairs are nested. Existence of index pairs is proven in [C] and [Sl].

(B) The *index space associated to an index pair* $\langle N_1, N_0 \rangle$ is the quotient space N_1/N_0; i.e., when $N_1 \cap N_0$ is non-empty, N_1/N_0 is the pointed space obtained from N_1 by collapsing $N_1 \cap N_0$ to a point denoted by $[N_0]$ and taken as the basepoint, and when $N_1 \cap N_0$ is empty, with $\{*\}$ any one point space disjoint from M, N_1/N_0 is the disjoint union $N_1 \cup \{*\}$ with basepoint $*$ which for convenience will also be denoted $[N_0]$. In both cases $[\mathbf{u}]$ denotes the image of a point $\mathbf{u} \in N_1$ under the quotient map $N_1 \to N_1/N_0$. Too, note that $(N_1, \cap N_0)$ and $\langle N_1, N_0 \rangle$ have the same index space. Unless stated explicitly to the contrary, an index space is always regarded as a topological space with basepoint, i.e., as a pair with the space as first member and the basepoint as second; however, the notation is usually abbreviated to just writing the space.

(C) The *Conley index of* S relative to the flow ψ, denoted by $\mathcal{C}(S)$ if ψ is clear from context but by $\mathcal{C}(S, \psi)$ if necessary to resolve possible ambiguity, is the connected simple system whose set of objects is the set of index spaces for S relative to ψ and for two such index spaces $X := N_1/N_0$ and $X' := N_1'/N_0'$, the unique morphism from X to X' is the homotopy class of the map $F^t\colon X \to X'$ defined for $t \geq 0$ sufficiently large by

$$F^t([\mathbf{u}]) := F([\mathbf{u}], t) := \begin{cases} [\mathbf{u} \cdot 3t] & \text{if } \mathbf{u} \cdot [0, 2t] \subset N_1 \setminus N_0 \\ & \qquad \text{and } \mathbf{u} \cdot [t, 3t] \subset N_1' \setminus N_0'; \\ [N_0'] & \text{otherwise.} \end{cases}$$

Here, the phrase "$t \geq 0$ sufficiently large" means that there exists $T \geq 0$, called a *common squeeze time* for $\langle N_1, N_0 \rangle$ and $\langle N_1', N_0' \rangle$, so that for $t \geq T$,

(1) $\mathbf{u} \cdot [-t, t] \subset N_1 \setminus N_0$ implies $\mathbf{u} \in N_1' \setminus N_0'$;
(2) $\mathbf{u} \cdot [-t, t] \subset N_1' \setminus N_0'$ implies $\mathbf{u} \in N_1 \setminus N_0$.

This definition of the unique morphism in $\mathcal{C}(S)$ from N_1/N_0 to N_1'/N_0' and the proof of continuity of F for $t \geq 0$ sufficiently large is due to Salamon [Sl]. See [K5] for a proof that $\mathcal{C}(S)$ with this definition of its morphisms is a connected simple system and is the same one obtained using the original definition of Conley [C] as modified by the author in [K1].

For simplicity, when the flow is clear from context *abuse notation and write* $h_S^{X\,X'}$ *for the unique morphism in* $\mathcal{C}(S)$ *from* X *to* X' rather than $h_{\mathcal{C}(S,\psi)}^{X\,X'}$ as specified by the general notational scheme for connected simple systems introduced in Definition 1.2.

(D) The *Conley homotopy index of an isolated invariant set* S is the common pointed homotopy type of the index spaces in $\mathcal{C}(S)$ and is denoted by $[\mathcal{C}(S)]$.

(E) Each flow ψ in M gives rise to a new flow ψ^* in M defined by $\psi^*(\mathbf{u},t) := \psi(\mathbf{u},-t)$ and called the *reverse time flow* associated to ψ. Note that ψ^* has the same isolated invariant sets as ψ. However, an index pair relative to ψ^* is not an index pair relative to ψ and conversely; hence, $\mathcal{C}(S,\psi)$ and $\mathcal{C}(S,\psi^*)$ are distinct connected simple systems.

Usually, the flow ψ under consideration will be clear from context without being explicitly named (e.g., a subscript on the invariant set could specify the flow) and will be viewed as defining the preferred time direction. In such instances, call ψ the *forward time flow*, set $\mathcal{C}(S) := \mathcal{C}(S,\psi)$ and call it the *forward time Conley index of* S, and set $\mathcal{C}^*(S) := \mathcal{C}(S,\psi^*)$ and call it the *reverse time Conley index of* S.

(F) Whenever a single upper-case Latin letter is used in denoting both members of a nested index pair in forward (respectively, reverse) time, the corresponding lower case Latin letter with a plus (respectively, minus) sign as superscript is used to denote the quotient map from the pair to its index space. For example, given (P_1,P_0) and (P_1^*,P_0^*), nested index pairs in forward and reverse time respectively, the quotient maps are denoted by

$$p^+ : (P_1,P_0) \to (P_1/P_0,[P_0]) \qquad \text{and} \qquad p^- : (P_1^*,P_0^*) \to (P_1^*/P_0^*,[P_0^*]) .$$

The Conley index as described above grew out of Conley's earlier notion of homotopy index for an isolated invariant set based on the existence of isolating blocks for an isolated invariant set. Several definitions of isolating block have been made and their existence proven in various contexts, but they are all special cases of the following notion suggested by R. McGehee.

1.4 DEFINITION. (A) For N an isolating neighborhood relative to a given flow, let τ_N denote the *time until first exit map* on N into $[0,\infty]$ defined by

$$\tau_N(\mathbf{u}) := \sup\{t \geq 0 : \mathbf{u}\cdot[0,t] \subset N\},$$

and taken relative to the reverse time flow, let the time until first exit map on N be denoted by τ_N^* and call it the *time since last entrance map* on N. Call $(\tau_N)^{-1}(0)$ the exit set of N and $(\tau_N^*)^{-1}(0)$ the entrance set of N.

(B) Call an isolating neighborhood B an *isolating block* if, and only if, τ_B and τ_B^* are continuous maps. The exit and entrance sets of an isolating block B will be denoted here by B^+ and B^- respectively. The existence of an isolating block interior to a given isolating neighborhood has been proved in various contexts and with varying choices of structure given to the boundary; see [C-E], [Ch], [WY]. Also see Proposition 4.4 below.

N.B. *The usage here of a plus (minus) sign as superscript to denote the exit (entrance) set is the opposite of the historical usage introduced by Conley and Easton in* [C-E], *but coincides with the usage in* [Sm].

(C) Continuity of τ_B and τ_B^* implies that the exit and entrance sets of a block are closed. Conversely, if an isolating neighborhood has closed exit and entrance sets it is a block as follows from [Conley, Theorem II.2.3]. It is trivial to verify that (B, B^+) and (B, B^-) are nested index pairs with respect to the forward and reverse time flows respectively. In keeping with the notation introduced above, $b^{\pm}: (B, B^{\pm}) \to (B/B^{\pm}, [B^{\pm}])$ denotes the quotient map from the index pair to its index space. It is a simple consequence of the continuity of τ_B and τ_B^* that B^+ and B^- are strong deformation retracts of closed neighborhoods in B, whence it follows [Sp, Theorem 4.8.9] that the quotient maps b^+ and b^- induce isomorphisms on homology. This property is shown to generalize to certain classes of index pairs in Chapter 4.

1.4.1 Remark. One advantage of the definition of isolating block given above is that it is a triviality that the intersection of two isolating blocks for an isolated invariant set is again an isolating block for the invariant set: if B_1 and B_2 are the blocks, then certainly $B_1 \cap B_2$ is an isolating neighborhood and $\tau_{B_1 \cap B_2} = \min\{\tau_{B_1}, \tau_{B_2}\}$ so that $\tau_{B_1 \cap B_2}$ is continuous; similarly $\tau^*_{B_1 \cap B_2}$ is continuous.

C. Homology and Cohomology of Conley Indices. The definitions in this subsection yield the graded (co)homology modules of a Conley index.

1.5 DEFINITION. *(A)* Let $\mathcal{X}$ be one of the following categories: ${}_R\mathcal{M}$, the category of left R-modules, $\mathcal{G}_R\mathcal{M}$, the category of graded left R-modules, or $\partial_R\mathcal{M}'$, the category of chain complexes over R and chain homotopy classes of chain maps. Also, let F be a functor on the homotopy category of topological pairs with values in $\mathcal{X}$, let $\widetilde{\mathsf{F}}$ be the corresponding reduced functor on $\mathcal{T}^{*\prime}$, and let $\mathcal{K}$ be a connected simple system that is a subcategory of $\mathcal{T}^{*\prime}$.

Then $\operatorname{ob}(\mathcal{K})$ is a directed set with respect to the quasi-order (i.e., reflexive, transitive relation) $\preceq$ defined by $X \preceq X'$ for every $X, X' \in \operatorname{ob}(\mathcal{K})$. Hence, for $\widetilde{\mathsf{F}}$ covariant (respectively, contravariant), application of $\widetilde{\mathsf{F}}$ to the objects and morphisms of $\mathcal{K}$ yields a direct (respectively, inverse) system of objects and bonding isomorphisms in $\mathcal{X}$ indexed over the directed set $\operatorname{ob}(\mathcal{K})$, viz.,

$$\left\{\{\widetilde{\mathsf{F}}(X)\}_{X \in \operatorname{ob}(\mathcal{K})}, \{\widetilde{\mathsf{F}}(h_{\mathcal{K}}^{X\,X'})\}_{X \preceq X' \in \operatorname{ob}(\mathcal{K})}\right\}. \tag{1.2}$$

For $\widetilde{\mathsf{F}}$ covariant (respectively, contravariant), define $\widetilde{\mathsf{F}}(\mathcal{K})$ to be the direct (respectively, inverse) limit of the system in (1.2). Also, for each $X \in \operatorname{ob}(\mathcal{K})$, let $\iota^X: \widetilde{\mathsf{F}}(X) \to \widetilde{\mathsf{F}}(\mathcal{K})$ (respectively, $\pi_X: \widetilde{\mathsf{F}}(\mathcal{K}) \to \widetilde{\mathsf{F}}(X)$) denote the canonical homomorphism.

Key to much of what follows is that ι^X (respectively, π_X) is an $\mathcal{X}$-isomorphism as follows from the connected simple system properties of $\mathcal{K}$, in particular, the identities (1.1). In this regard it is helpful to note that for $\widetilde{\mathsf{F}}$ covariant the direct limit $\widetilde{\mathsf{F}}(\mathcal{K})$ can be realized as the set of equivalence classes of the equivalence relation $\sim$ defined on the union of the $\widetilde{\mathsf{F}}(X)$, $X \in \operatorname{ob}(\mathcal{K})$ as follows: for $\alpha \in \widetilde{\mathsf{F}}(X)$, $\alpha' \in \widetilde{\mathsf{F}}(X')$, $\alpha \sim \alpha'$ if, and only if, $\widetilde{\mathsf{F}}(h_{\mathcal{K}}^{X\,X'})\alpha = \alpha'$. Similarly, for $\widetilde{\mathsf{F}}$ contravariant, it is helpful to note that the inverse limit $\widetilde{\mathsf{F}}(\mathcal{K})$ can be realized as those elements (ϕ_X) in the direct product $\prod\{\widetilde{\mathsf{F}}(X) : X \in \operatorname{ob}(\mathcal{K})\}$ satisfying $\widetilde{\mathsf{F}}(h_{\mathcal{K}}^{X\,X'})\phi_{X'} = \phi_X$.

Should $\mathcal{X} = \partial_R\mathcal{M}'$, then in the case F is a covariant functor into $\partial_R\mathcal{M}'$, the boundary operator $\partial^{\mathcal{K}}$ for $\widetilde{\mathsf{F}}(\mathcal{K})$ is for any choice of $X \in \operatorname{ob}(\mathcal{K})$ unambiguously defined for each integer k by

$$\partial_k^{\mathcal{K}} := \iota_{k-1}^X \circ \partial_k^X \circ (\iota_k^X)^{-1}.$$

No ambiguity arises from the choice of X because the collection in (1.2) is a direct system of chain complexes and chain homotopy equivalences over R. An analogous definition can be made in the contravariant case.

In general, elements of $\widetilde{\mathsf{F}}(\mathcal{K})$ will be denoted with lower case, bold face Greek letters while correspondents in any $\widetilde{\mathsf{F}}(X)$, $X \in \mathrm{ob}(\mathcal{K})$, will be represented by the corresponding lower case, normal weight Greek letter.

(B) Let S be an isolated invariant set of a C^k flow in M, let F be a covariant (respectively, contravariant) functor of the type considered in part (A), and apply the construction of (A) to the connected simple systems $\mathcal{C}(S)$ and $\mathcal{C}^*(S)$. The resulting direct (respectively, inverse) limit graded R-modules are, for simplicity, denoted by $\widetilde{\mathsf{F}}\mathcal{C}(S)$ and $\widetilde{\mathsf{F}}\mathcal{C}^*(S)$ respectively. If (P_1, P_0) and (P_1^*, P_0^*) are nested index pairs for S, the former in forward time, the latter in reverse, then if F is covariant (respectively, contravariant) set $\overline{\mathsf{F}(p^{\pm})} = \iota^X \circ \mathsf{F}(p^{\pm})$ (respectively, set $\overline{\mathsf{F}(p^{\pm})} := F(p^{\pm}) \circ \pi_X$) where $X := P_1/P_0$ if the plus sign is taken, but $X := P_1^*/P_0^*$ if the minus sign is taken. Call $\overline{\mathsf{F}(p^{\pm})}$ the *canonical homomorphism*. When F is covariant (respectively contravariant), $\overline{\mathsf{F}(p^{\pm})}$ may also be called the canonical injection (projection). Note that $\overline{\mathsf{F}(p^{\pm})}$ will be an $\mathcal{X}$-isomorphism if $\mathsf{F}(p^{\pm})$ is.

With $\mathsf{F} := H_*(\cdot\,; G)$, the singular homology functor with coefficients in the R-module G, the construction of part (A) applied to $\mathcal{C}(S)$ and $\mathcal{C}^*(S)$ yields, respectively, the graded R-modules $\widetilde{H}_*\mathcal{C}(S; G)$ and $\widetilde{H}_*\mathcal{C}^*(S; G)$ with the former referred to as the *graded singular homology module (coefficients in G) of the forward time Conley index of S* and the latter referred to as the *graded singular homology module (coefficients in G) of the reverse time Conley index of S*.

If instead one takes the singular cohomology functor $\mathsf{F} = H^*(\cdot\,; G)$, one obtains the *graded singular cohomology modules (coefficients in G) of the forward and reverse time Conley indices of S* as inverse limits denoted by $\widetilde{H}^*\mathcal{C}(S; G)$ and $\widetilde{H}^*\mathcal{C}^*(S; G)$, respectively. Obviously, the homology and cohomology modules of any Conley index obtained using any other homology or cohomology theory can be similarly constructed and will be referred to as needed. Here and throughout this work, the phrases "homology theory" and "cohomology theory" are used in the broad sense to include extraordinary theories, i.e., ones that fail to satisfy the dimension axiom although the other Eilenberg-Steenrod axioms are satisfied.

The following notation and glosses are used when $\mathsf{F} = H_*(\cdot\,; G)$ and (P_1, P_0) is a nested index pair for S in forward time. For simplicity, $\bar{p}_*^+ : H_*(P_1, P_0; G) \to \widetilde{H}_*\mathcal{C}(S; G)$ is used to denote the canonical homomorphism. Also, $\alpha \in H_*(P_1, P_0; G)$ is said to represent $\boldsymbol{\alpha} \in \widetilde{H}_*\mathcal{C}(S; G)$ if, and only if, $\bar{p}_*^+\alpha = \boldsymbol{\alpha}$. Similarly, a singular chain c that is a relative cycle in P_1 modulo P_0 with homology class α is said to represent $\boldsymbol{\alpha}$. Analogous notation is used in the reverse time case.

(C) Let us note that the singular homology and cohomology modules of a Conley index as just defined are in fact isomorphic to the homology and cohomology modules of a chain complex in $\partial_R\mathcal{M}'$. In fact, for any connected simple system $\mathcal{K}$ that is a subcategory of $\mathcal{T}^{*\prime}$, apply the construction described in part (A) with $\widetilde{\mathsf{F}} = \tilde{\Delta}_*$, the reduced singular chain functor with coefficients in R. The direct limit chain complex, $\tilde{\boldsymbol{\Delta}}_*(\mathcal{K})$, is called the singular chain complex of $\mathcal{K}$ with coefficients in R. The homology and cohomology of the chain complex $\tilde{\boldsymbol{\Delta}}_*(\mathcal{K}) \otimes G$ are isomorphic to $\widetilde{H}_*(\mathcal{K};G)$ and $\widetilde{H}^*(\mathcal{K};G)$, respectively, as follows from the observation made in (A) that the canonical homomorphisms to the direct and inverse limits are isomorphisms. In the case of homology, this also follows because the direct limit of the homology modules of the constituent chain complexes in a direct system of chain complexes is isomorphic to the homology of the direct limit of the chain complexes; see [Sp, Theorem 4.1.7].

D. Sign Conventions for Products in Homology and Cohomology. The statement and verification of the evaluations of the intersection pairing in Chapter 7 and of the product formulae in §8.D depend upon the choice of sign convention in the definition of the exterior cohomology product. Here we adopt the sign conventions of [D]. Thus, with all coefficients taken in R, if $y_i \in H^*(X_i, A_i)$ and $\eta_i \in H_*(X_i, A_i)$, $i = 1, 2$, then define the exterior cohomology product so that the following duality statement relating the exterior products on cohomology and homology holds:

$$\langle y_1 \times y_2, \eta_1 \times \eta_2 \rangle = (-1)^{|\eta_1||y_2|} \langle y_1, \eta_1 \rangle \langle y_2, \eta_2 \rangle .$$

Other authors define the exterior cohomology product so that in the corresponding duality statement there is no alternating sign; e.g., see [Sp]. The choice of sign convention affects the sign that appears in the standard formulae relating exterior products and cup and cap products. Specifically, for $i = 1, 2$, if $y_i \in H^*(X_i, A_i)$ and $\eta_i \in H_*(X_i, A_i \cup B_i)$ and appropriate excision conditions are met, then in $H_*\big((X_1, B_1) \times (X_2, B_2)\big)$ as a consequence of using the sign conventions of [D]

$$(y_1 \times y_2) \frown (\eta_1 \times \eta_2) = (-1)^{|y_2||\eta_1|} (y_1 \frown \eta_1) \times (y_2 \frown \eta_2) \tag{1.3}$$

whereas if the sign conventions of [Sp] were used the exponent on minus one would be $|y_1|(|\eta_2| - |y_2|)$.

Mnemonically, for the exterior, cup, and cap products, when two adjacent elements in an expression are interchanged, then the sign of the expression changes by a factor of (-1) raised to the product of the degrees of the interchanged elements.

CHAPTER 2

THE INTERSECTION PAIRINGS L, $\mathfrak{L}$, AND $^{\#}\mathfrak{L}$

A. Pairs of Index Pairs Admissible for the Intersection Pairing.

2.1 DEFINITION. (A) Let $\Delta\colon M \to M \times M$ denote the diagonal embedding and suppose (X, A) and (Y, B) are topological pairs in M. Then the pair of pairs $\big((X, A), (Y, B)\big)$ satisfies *the off-diagonal condition* if, and only if,

$$(X \times B \cup A \times Y) \cap \Delta(M) = \emptyset\,,$$

or equivalently, if, and only if,

$$X \cap B = \emptyset = A \cap Y\,. \tag{2.1}$$

Alternatively, (X, A) and (Y, B) could be said to lie in general position.

(B) Let S be an isolated invariant set of a flow in M, let (P_1, P_0) be a nested index pair for S with respect to forward time and let (P_1^*, P_0^*) be a nested index pair for S with respect to reverse time. Then the pair of index pairs $P := \big((P_1, P_0), (P_1^*, P_0^*)\big)$ is *admissible for the intersection pairing on the Conley indices of* S (hereinafter abbreviated to AIP for S) if, and only if,

(1) the pair of pairs P satisfies the off-diagonal condition,
(2) the quotient maps

$$p^+\colon (P_1, P_0) \to (P_1/P_0, [P_0]) \quad \text{and} \quad p^-\colon (P_1^*, P_0^*) \to (P_1^*/P_0^*, [P_0^*])$$

induce isomorphisms on singular homology, coefficients in any R-module G.

Proposition 4.4 below states the existence of index pairs that are AIP for S.

From condition (2), it follows that the canonical homomorphisms

$$\bar{p}_*^+\colon H_*(P_1, P_0; G) \to \widetilde{H}_*\mathcal{C}(S; G) \quad \text{and} \quad \bar{p}_*^-\colon H_*(P_1^*, P_0^*; G') \to \widetilde{H}_*\mathcal{C}^*(S; G')$$

induced by the quotient maps p^+ and p^- are isomorphisms. It then follows that also $\bar{p}_*^+ \otimes \bar{p}_*^-$ is an isomorphism because the tensor product of two R-module isomorphisms is again an isomorphism—e.g., see [Sp, Lemmas 5.1.4–5].

Remarks analogous to those of the previous paragraph hold if instead of singular homology any functor F on the homotopy category of compact Hausdorff pairs to the category of left R-modules, R a ring with unit 1 acting like the identity on every R-module, is considered. That is, whenever $\mathsf{F}(p^+)$ is an isomorphism, $\overline{\mathsf{F}(p^+)}$ is an R-module isomorphism of $\mathsf{F}(P_1, P_0)$ onto $\widetilde{\mathsf{F}\mathcal{C}}\,(S)$ in the covariant case and of $\widetilde{\mathsf{F}\mathcal{C}}\,(S)$ onto $\mathsf{F}(P_1, P_0)$ in the contravariant with analogous properties holding for $\mathsf{F}(p^-)$ relative to the modules $\mathsf{F}(P_1^*, P_0^*)$ and $\widetilde{\mathsf{F}\mathcal{C}}^*(S)$; also $\overline{\mathsf{F}(p^+)} \otimes \overline{\mathsf{F}(p^-)}$ will be an isomorphism.

(C) For S as in (B), let $\mathcal{AIP}(S)$ be the small category whose set of objects consists of pairs of index pairs that are AIP for S. A morphism in this category from $((P_1, P_0), (P_1^*, P_0^*))$ to $((Q_1, Q_0), (Q_1^*, Q_0^*))$ is an ordered pair of inclusion maps (i, j):

$$i\colon (P_1, P_0) \hookrightarrow (Q_1, Q_0) \quad \text{and} \quad j\colon (P_1^*, P_0^*) \hookrightarrow (Q_1^*, Q_0^*)\,.$$

The following notation will be employed. If $P \in \mathrm{ob}(\mathcal{AIP}(S))$, then P is the pair of index pairs given by having $P = ((P_1, P_0), (P_1^*, P_0^*))$. Also, if $P_j \in \mathrm{ob}(\mathcal{AIP}(S))$, j varying through some index set, then $P_j = ((P_{1\,j}, P_{0\,j}), (P_{1\,j}^*, P_{0\,j}^*))$. If P and Q are objects of $\mathcal{AIP}(S)$ and if there exists a morphism in $\mathcal{AIP}(S)$ from P to Q, this fact is denoted symbolically by writing $P \hookrightarrow Q$ or $Q \hookleftarrow P$ as is convenient.

B. The Euclidean Case: the Homology Intersection Number Pairing L. In this section let S be an isolated invariant set of a flow in $\mathbf{R}^m$. All homology modules occurring in this section are assumed to have coefficients in R and the coefficient module is therefore suppressed from the notation for simplicity. The definition of the homology intersection number pairing on the Conley indices of S is based on the following notion of intersection numbers in $\mathbf{R}^m$ taken from [D].

2.2 DEFINITION. Let $\big((X, A), (Y, B)\big)$ be a pair of topological pairs in $\mathbf{R}^m$ satisfying the off-diagonal condition, set $\Delta_m := \Delta(\mathbf{R}^m)$, let

$$i\colon (X \times Y, (X \times B) \cup (A \times Y)) \hookrightarrow (\mathbf{R}^m \times \mathbf{R}^m, \mathbf{R}^m \times \mathbf{R}^m \setminus \Delta_m)$$

be inclusion, and define

$$d\colon (\mathbf{R}^m \times \mathbf{R}^m, \mathbf{R}^m \times \mathbf{R}^m \setminus \Delta_m) \to (\mathbf{R}^m, \mathbf{R}^m \setminus \{0\})$$

by

$$d(\mathbf{u}, \mathbf{v}) := \mathbf{u} - \mathbf{v}.$$

Then for $\alpha \in H_{m-k}(X, A)$ and $\gamma \in H_k(Y, B)$, *the homology intersection number of the pair* (α, γ), denoted $\alpha \# \gamma$, is defined by

$$\alpha \# \gamma := (-1)^k d_* i_* (\alpha \times \gamma).$$

As the map on $H_{m-k}(X, A) \times H_k(Y, B)$ to $H_m(\mathbf{R}^m, \mathbf{R}^m \setminus \{0\})$ given by

$$(\alpha, \gamma) \mapsto \alpha \# \gamma$$

is bilinear, it defines a linear map on the module of elements of degree m in the tensor product of the homology modules:

$$\begin{array}{ccc} [H_*(X, A) \otimes H_*(Y, B)]_m & \longrightarrow & H_m(\mathbf{R}^m, \mathbf{R}^m \setminus \{0\}) \simeq R \\ \alpha \otimes \gamma & \longmapsto & \alpha \# \gamma \end{array}.$$

Both the bilinear map on the Cartesian product and the linear map on the tensor product will be referred to as *the homology intersection number pairing in* $\mathbf{R}^m$, and in general throughout this work the word "pairing" is used as a gloss for the phrase "bilinear pairing" and refers either to a bilinear map on a Cartesian product or the corresponding linear map on the tensor product depending on context.

2.2.1 Remark. The homology intersection number pairing in $\mathbf{R}^m$ is invariant under inclusion maps. That is, if $(X', A') \subset (X, A)$ and $(Y', B') \subset (Y, B)$ are pairs in $\mathbf{R}^m$ with $((X, A), (Y, B))$ satisfying the off-diagonal condition, then so too does $((X', A'), (Y', B'))$, and if α and γ in the definition above are the images under the inclusion induced maps of $\alpha' \in H_{m-k}(X', A')$ and $\gamma' \in H_k(Y', B')$, respectively, then $\alpha' \# \gamma' = \alpha \# \gamma$. This is immediate from the naturality of the homology cross-product with respect to inclusions.

The naturality with respect to inclusions of the homology intersection number pairing in $\mathbf{R}^m$ and its generalization in the manifold setting is crucial to our development of intersection pairings on Conley indices.

2.2.2 Remark. The homology intersection number pairing is graded commutative; i.e., $\gamma \# \alpha = (-1)^{k(m-k)} \alpha \# \gamma$ with α and γ as in Definition 2.2. For a proof see [D, Proposition VII.4.12].

2.2.3 Remark. If $X \cap Y = \emptyset$, then $\alpha \# \gamma = 0$ because the diagram of inclusions of pairs

$$\begin{array}{ccc} (X \times Y, X \times B \cup A \times Y) & \xhookrightarrow{\quad i \quad} & (\mathbf{R}^m \times \mathbf{R}^m, \mathbf{R}^m \times \mathbf{R}^m \setminus \Delta_m) \\ & \searrow \qquad\qquad \nearrow & \\ & (\mathbf{R}^m \times \mathbf{R}^m \setminus \Delta_m, \mathbf{R}^m \times \mathbf{R}^m \setminus \Delta_m) & \end{array}$$

commutes, whence i induces the zero homomorphism on homology since $H_*(Z, Z) = 0$ for any non-empty space Z. Thus, if $\alpha \# \gamma \neq 0$, then $X \cap Y \neq \emptyset$, and because of (2.1) in fact $(X \setminus A) \cap (Y \setminus B) \neq \emptyset$.

2.3 Definition. For each $P \in \mathrm{ob}(\mathcal{AIP}(S))$, for $k = 0, \ldots, m$, define a pairing of R-modules

$$\mathsf{L}_{P\,k}\colon H_{m-k}(P_1, P_0) \otimes H_k(P_1^*, P_0^*) \to H_m(\mathbf{R}^m, \mathbf{R}^m \setminus \{0\})$$

by assigning $\alpha \# \gamma$ as given by Definition 2.2 to a generating element $\alpha \otimes \gamma$, i.e.,

$$\mathsf{L}_{P\,k}(\alpha \otimes \gamma) := \alpha \# \gamma\,,$$

and then extending by linearity. Because the pair of pairs P satisfies the off-diagonal condition, $\mathsf{L}_{P\,k}$ is well-defined. As

$$\big[H_*(P_1, P_0) \otimes H_*(P_1^*, P_0^*)\big]_m := \bigoplus_{k=0}^{m} H_{m-k}(P_1, P_0) \otimes H_k(P_1^*, P_0^*)$$

there is a pairing

$$\mathsf{L}_P\colon \big[H_*(P_1, P_0) \otimes H_*(P_1^*, P_0^*)\big]_m \to H_m(\mathbf{R}^m, \mathbf{R}^m \setminus \{0\})$$

defined by

$$\mathsf{L}_P := \bigoplus_{k=0}^{m} \mathsf{L}_{P\,k}\,.$$

Remembering that for each $P \in \mathcal{AIP}(S)$, $\bar{p}_*^+ \otimes \bar{p}_*^-$ is an isomorphism, consider the family of pairings $\bar{\mathsf{L}}_P := \mathsf{L}_P \circ (\bar{p}_*^+ \otimes \bar{p}_*^-)^{-1}$, $P \in \mathrm{ob}(\mathcal{AIP}(S))$, and for $P, Q \in \mathrm{ob}(\mathcal{AIP}(S))$, define P to be equivalent to Q if, and only if, $\bar{\mathsf{L}}_P = \bar{\mathsf{L}}_Q$. It is a consequence of Remark 2.2.1 (see the proof of Theorem 4.7 below) that if $P \hookrightarrow Q$ is an inclusion morphism in $\mathcal{AIP}(S)$, then P is equivalent to Q. One can then show with some work (see the proofs of Lemma 4.6 and Theorem 4.7 below) that given $P_1, P_2 \in \mathrm{ob}(\mathcal{AIP}(S))$ one can produce a string of inclusion morphisms in $\mathcal{AIP}(S)$ mediating between P_1 and P_2—in general the arrows do not all point in the same direction; hence, there is no composite from P_1 to P_2 or vice versa. The observation from Remark 2.2.1 and transitivity of the relation then yield that the equivalence relation on $\mathrm{ob}(\mathcal{AIP}(S))$ has a unique equivalence class, whence the following is immediate.

2.4 THEOREM. *The family of pairings* L_P, $P \in \mathrm{ob}(\mathcal{AIP}(S))$, *determines a pairing*

$$\mathsf{L}\colon \left[\widetilde{H}_*\mathcal{C}(S) \otimes \widetilde{H}_*\mathcal{C}^*(S)\right]_m \to H_m(\mathbf{R}^m, \mathbf{R}^m \setminus \{0\})$$

in such manner that for each $P \in \mathrm{ob}(\mathcal{AIP}(S))$,

$$\mathsf{L} = \mathsf{L}_P \circ (\bar{p}_*^+ \otimes \bar{p}_*^-)^{-1}\,.$$

2.5 DEFINITION. Call L *the homology intersection number pairing for the Conley indices of* S, and for each integer $k \in [0, m]$, for each element $\boldsymbol{\alpha} \otimes \boldsymbol{\gamma}$ of the R-module $\widetilde{H}_{m-k}\mathcal{C}(S) \otimes \widetilde{H}_k\mathcal{C}^*(S)$, set

$$\boldsymbol{\alpha} \# \boldsymbol{\gamma} := \mathsf{L}(\boldsymbol{\alpha} \otimes \boldsymbol{\gamma})\,.$$

Call $\boldsymbol{\alpha} \# \boldsymbol{\gamma}$ *the homology intersection number of* $\boldsymbol{\alpha} \otimes \boldsymbol{\gamma}$.

2.5.1 Remark. Theorem 2.4 asserts that L is completely determined by L_P for any choice of $P \in \mathrm{ob}(\mathcal{AIP}(S))$; i.e., there are unique representatives $\alpha \in H_{m-k}(P_1, P_0)$ and $\gamma \in H_k(P_1^*, P_0^*)$ representing $\boldsymbol{\alpha}$ and $\boldsymbol{\gamma}$, respectively, and $\boldsymbol{\alpha} \# \boldsymbol{\gamma} = \alpha \# \gamma$. Consequently, computing L_P for one P computes L_P for all P given that one knows the maps on homology determined by the morphisms of $\mathcal{C}(S)$ and $\mathcal{C}^*(S)$.

There are many properties of L that follow from the analogous property for homology intersection numbers in $\mathbf{R}^m$ and Remark 2.5.1. The derivations of two simple ones of interest follow.

First, letting L^* denote the pairing obtained from Theorem 2.4 by regarding S as an isolated invariant set of the reverse time flow associated to the given flow and defining $\tau_*(\boldsymbol{\gamma} \otimes \boldsymbol{\alpha}) := (-1)^{k(m-k)}\boldsymbol{\alpha} \otimes \boldsymbol{\gamma}$ for $\boldsymbol{\alpha}$ and $\boldsymbol{\gamma}$ as in Definition 2.5, we immediately obtain from Remark 2.5.1 and Remark 2.2.2 the following graded commutativity relationship between L and L^*.

2.6 PROPOSITION. $\mathsf{L}^* = \mathsf{L}\mathcal{T}_*$, *i.e.*, $\boldsymbol{\gamma} \# \boldsymbol{\alpha} = (-1)^{k(m-k)} \boldsymbol{\alpha} \# \boldsymbol{\gamma}$.

Second, noting that for any topological pair (X, A), the homology class of a relative singular cycle c in X modulo A is the image under the inclusion induced homomorphism of the homology class of c regarded as a relative cycle in its support $|c|$ modulo the support of its boundary chain $|\partial c|$, we immediately obtain from Remarks 2.2.1, 2.2.3, and 2.5.1 the following result basic to the use of L in proving existence theorems for singularly perturbed two-point boundary value problems.

2.7 PROPOSITION. *Let* $\boldsymbol{\alpha} \in \widetilde{H}_{m-k}\mathcal{C}(S)$, $\boldsymbol{\gamma} \in \widetilde{H}_k\mathcal{C}^*(S)$, *and suppose* $\boldsymbol{\alpha} \# \boldsymbol{\gamma} \neq 0$. *Then for any* $P \in \mathrm{ob}(\mathcal{AIP}(S))$, *if* $c_{\boldsymbol{\alpha}}$ *is a relative singular cycle in* P_1 *modulo* P_0 *representing* $\boldsymbol{\alpha}$ *and* $c_{\boldsymbol{\gamma}}$ *is a relative singular cycle in* P_1^* *modulo* P_0^* *representing* $\boldsymbol{\gamma}$, *then* $(|c_{\boldsymbol{\alpha}}| \setminus P_0) \cap (|c_{\boldsymbol{\gamma}}| \setminus P_0^*) \neq \emptyset$.

In the applications the most commonly and easily constructed index pairs are those that consist of an isolating block and either its exit or entrance set, the former (latter) yielding an index pair with respect to the forward (reverse) time flow. An obvious question therefore is: Given a block for S is it possible to express every homology intersection number of homology classes in the Conley indices of S directly from a pair of singular chains, one representing non-zero homology in the block modulo its exit set, the other in the block modulo its entrance set? The answer is a qualified yes with minor restrictions on how the representing chains sit inside the block. The precise statement is deferred until after the development of the intersection class pairing and the intersection number pairing for Conley indices in the setting of a continuous flow on a manifold because the result in the Euclidean setting (Corollary 6.3 below) is but a special case of more general results holding for the intersection class pairing that appear below as Theorem 6.1 and Corollary 6.2. As other properties of L of interest to us here are also subsumed by results obtainable for the more general intersection class pairing, we turn now to its development.

C. The Manifold Case: the Intersection Class and Number Pairings $\mathfrak{L}$ and $^{\#}\mathfrak{L}$. The development of the intersection class pairing on Conley indices as carried out in the rest of this work requires specific notions of orientation, Poincaré duality, Čech homology and cohomology, and the definition of certain transfer maps on homology induced by the diagonal map Δ. An exposition of the required notions follows. Throughout, all manifolds are assumed second countable.

An *R-orientation of the m-dimensional manifold* X is a continuous global section O^X of the orientation bundle of X with coefficients in R with the property that $O^X(\mathbf{u})$ generates $H_m(X, X \setminus \{\mathbf{u}\}; R) \simeq R$ for each $\mathbf{u} \in X$. For O^X an R-orientation of X, for each non-empty, compact $K \subset X$, there exists a unique class $o_K^X \in H_m(X, X \setminus K; R)$, called the fundamental class along K, with the property that for each $\mathbf{u} \in K$ the image of o_K^X under the inclusion induced homomorphism to $H_m(X, X \setminus \{\mathbf{u}\}; R)$ is $o_{\mathbf{u}}^X := O^X(\mathbf{u})$. It follows that if $K \subset K'$ are compact subsets of X then $o_{K'}^X \mapsto o_K^X$ under the inclusion induced homomorphism. See [D, §VIII.2] for a full exposition.

Because the current setting includes only manifolds that are ENR's, Čech homology and cohomology modules with coefficients in an R-module G will be identified with limits of the usual inverse and direct systems of singular homology and cohomology modules; specific notations and restrictions follow.

For X a topological space, for each $K \subset X$, let $\mathcal{N}(K)$ denote the family of neighborhoods of K directed by containment (i.e., $U \preceq U' \iff U \supset U'$), and for each pair (K, L) in X, let $\mathcal{N}(K, L)$ denote the family of neighborhood pairs of (K, L) directed by containment. When $L = \emptyset$ the set of neighborhood pairs of the form $(U, \emptyset)$ is cofinal in $\mathcal{N}(K, \emptyset)$ and as used in arguments below it is therefore convenient to abuse notation and identify $\mathcal{N}(K, \emptyset)$ with $\mathcal{N}(K)$.

For (K, L) an arbitrary pair in a manifold X, its *graded Čech homology module with coefficients in G*, denoted $\check{H}_*(K, L; G)$, is identified with

$$\check{H}_*(K, L; G) := \varprojlim H_*(U, V; G)$$

where the limit is over $(U, V) \in \mathcal{N}(K, L)$ directed by containment and where for $(U', V') \preceq (U, V) \in \mathcal{N}(K, L)$, the bonding homomorphism $i^{(U,V)}_{(U',V')*}$ is the inclusion induced homomorphism on singular homology. By the definition of an inverse limit, for each neighborhood pair $(U, V) \in \mathcal{N}(K, L)$, there is a natural projection homomorphism

$$\pi^{(K,L)}_{(U,V)} \colon \check{H}_*(K, L; G) \to H_*(U, V; G).$$

To simplify notation, a subscript (U, V) will often be appended to a Čech homology class to denote its image under this natural projection; i.e., define

$$\check{\xi}_{(U,V)} := \pi^{(K,L)}_{(U,V)} \check{\xi} \quad \text{whenever } \check{\xi} \in \check{H}_*(K, L; G) \text{ and } (U, V) \in \mathcal{N}(K, L).$$

Clearly, if $(K, L) \subset (K', L') \subset X$, then there is an inclusion induced homomorphism

$$\check{i}^{(K,L)}_{(K',L')} \colon \check{H}_*(K, L; G) \to \check{H}_*(K', L'; G)$$

determined by the property that if $(U', V') \in \mathcal{N}(K', L')$, then $\pi^{(K',L')}_{(U',V')} \circ \check{i}^{(K,L)}_{(K',L')} = \pi^{(K,L)}_{(U',V')}$. With these definitions $\check{H}_*(\,\cdot\,; G)$ is a functor on the category of pairs in X and inclusions of such pairs which suffices for our purposes.

For (K, L) a pair of locally compact subsets of a manifold X, its *graded Čech cohomology module with coefficients in G*, denoted by $\check{H}(K, L; G)$, is identified with

$$\check{H}^*(K, L; G) := \varinjlim H^*(U, V; G)$$

where the limit is over $(U, V) \in \mathcal{N}(K, L)$ and where for $(U', V') \preceq (U, V)$ the bonding map $\rho^{(U,V)}_{(U',V')}$ is the inclusion induced restriction homomorphism on singular cohomology. For a pair of such locally compact pairs $(K', L') \supset (K, L)$ there is a restriction homomorphism

$$\check{\rho}^{(K,L)}_{(K',L')} \colon \check{H}^*(K', L'; G) \to \check{H}^*(K, L; G)$$

obtained by passage to the limit over $(U', V') \in \mathcal{N}(K', L') \subset \mathcal{N}(K, L)$ through the canonical homomorphisms $H^*(U', V'; G) \to \check{H}^*(K, L; G)$. More generally, $\check{H}(\,\cdot\,; G)$

can be extended to a contravariant functor on the category of pairs of locally compact subsets of a manifold and continuous maps between such pairs; see [D, §VIII.6]. The last assertion requires X to be an ENR; that and metrizability are the two reasons for assuming that all manifolds mentioned herein are second countable.

For (K, L) a closed pair in a manifold X, its *graded Čech cohomology module with compact supports and coefficients in* G is denoted by $\check{H}^*_c(K, L; G)$ and is defined by

$$\check{H}^*_c(K, L; G) := \varinjlim \check{H}^*(K, \omega; G)$$

where the limit is over the family, directed by containment, of locally compact subsets ω satisfying $L \subset \omega \subset K$ and ω is cobounded in K (i.e., $\mathrm{cl}\,(K \setminus \omega)$ is compact) and where each bonding homomorphism is a restriction $\check{\rho}^{(K,L)}_{(K',L')}$. Čech cohomology with compact supports is a contravariant functor on the category of closed pairs in manifolds and continuous maps $f\colon (K, L) \to (K', L')$ between such pairs that are proper over $K' \setminus L'$ (i.e, $f^{-1}(C)$ is compact for each compact $C \subset K' \setminus L'$); see [D,§VIII.6.26].

In [D,§VIII.7, pp. 291–297] relative Poincaré-Alexander duality isomorphisms are described for three successively more inclusive situations; all in the context of X an m-dimensional manifold with R-orientation O^X. The first two of these situations are (i) (K, L) is a compact pair in X and (ii) (K, L) is a closed pair in X with L cobounded in K. Both result in a duality isomorphism

$$\check{H}^i(K, L; G) \xrightarrow[\simeq]{\frown o^X} H_{m-i}(X \setminus L, X \setminus K; G)$$

for any R-module G. The third situation, (iii) (K, L) is an arbitrary closed pair in X, results in a duality isomorphism

$$\check{H}^i_c(K, L; G) \xrightarrow[\simeq]{\frown o^X} H_{m-i}(X \setminus L, X \setminus K; G)$$

and is the one of immediate interest to us although all three types of duality isomorphisms occur in Chapter 9. It is not difficult to show (a) that all three duality isomorphisms are the same for a compact pair (K, L) and (b) that the second and third are the same for (K, L) a closed pair and L cobounded in K modulo the natural identification $\check{H}^*(K, L; G) \simeq \check{H}^*_c(K, L; G)$ that occurs in situations (a) and (b) via simple cofinality arguments.

Although it does not play a role in any of the arguments below, the reader may find it of interest that usage of the notation $\frown o^X$ to denote the duality isomorphisms is particularly apt because where $H^c_q(X, \dot{X}; R)$ is, in the sense of [Sp,§6.3], the module of compatible $\mathfrak{A}$-families of homology classes (of dimension q with coefficients in R) for $\mathfrak{A}$ the family of compact subsets of X $(= X \setminus \dot{X}$ as X is boundaryless) the R-orientation O^X defines the compatible family $o^X := (o^X_K)_{K \in \mathfrak{A}}$ which generates $H^c_m(X, \dot{X}; R)$ and because the method used to define the duality isomorphisms in each of situations (i)–(iii) is naturally extended to define in situations (i) and (ii) a bilinear pairing having the properties of a cap product

$$\check{H}^i(K, L; G) \otimes H^c_q(X, \dot{X}; R) \xrightarrow{\frown} H_{q-i}(X \setminus L, X \setminus K; G)$$

and similarly for situation (*iii*) where $\check{H}^i_c(K, L; G)$ replaces $\check{H}^i(K, L; G)$; i.e., with reference to the bilinear pairings just mentioned, the duality isomorphisms of situations (*i*)–(*iii*) are given by fixing the second argument in the appropriate pairing at $o^X \in H^c_m(X, \dot{X}; R)$.

It is convenient to use $M^{(2)}_{\not\Delta}$ to denote the category of pairs of topological pairs in M satisfying the off-diagonal condition with morphisms pairs of inclusions; thus $\mathcal{AIP}(S)$ becomes a full subcategory of $M^{(2)}_{\not\Delta}$. Accordingly, the notation used to denote objects and morphisms of $\mathcal{AIP}(S)$ is extended to $M^{(2)}_{\not\Delta}$; e.g, if $X \in \text{ob}(M^{(2)}_{\not\Delta})$, then $X = \big((X_1, X_0), (X_1^*, X_0^*)\big)$. The notations $M^{(2)}_{\not\Delta\, c}$ and $M^{(2)}_{\not\Delta\, o}$ respectively denote the full subcategories of $M^{(2)}_{\not\Delta}$ whose objects are respectively pairs of compact pairs and pairs of open pairs. Also, for each $X \in \text{ob}(M^{(2)}_{\not\Delta})$, let $M^{(2)}_{\not\Delta\, c}(X)$ and $M^{(2)}_{\not\Delta\, o}(X)$ denote the full subcategories of $M^{(2)}_{\not\Delta\, c}$ and $M^{(2)}_{\not\Delta\, o}$ respectively with sets of objects defined as follows: $K \in \text{ob}(M^{(2)}_{\not\Delta\, c}(X)$ if, and only if, there exists an inclusion morphism $K \hookrightarrow X$, and $V \in \text{ob}(M^{(2)}_{\not\Delta\, o}(X)$ if, and only if, there exists an inclusion morphism $X \hookrightarrow V$.

Each of the categories defined in the previous paragraph can be regarded as a full subcategory of a corresponding category of pairs of pairs in M where objects are no longer required to satisfy the off-diagonal condition; the corresponding supercategory is denoted by dropping the subscript $\not\Delta$ from the notation. For example, $M^{(2)}$ denotes the category of arbitrary pairs of topological pairs in M and $M^{(2)}_o$ its full subcategory of pairs of open pairs, etc.

For the rest of this section, fix an R-orientation O^M of M and give $M \times M$ the canonical product R-orientation: $O^{M\times M} := O^M \times O^M$. Then for any $V \in \text{ob}(M^{(2)}_o)$, for $i \in \mathbf{N}$, for coefficients taken in any R-module G but suppressed from the notation to aid in printing, there is a homology transfer homomorphism

$$\Delta_!\colon H_i(V_1 \times V_1^*, V_0 \times V_1^* \cup V_1 \times V_0^*) \to H_{i-m}(V_1 \cap V_1^*, (V_0 \cap V_1^*) \cup (V_1 \cap V_0^*))$$

(read $\Delta_!$ as Δ-shriek) defined by

$$\Delta_! := (\frown o^M) \circ \check{\Delta}_c^* \circ (\frown o^{M\times M})^{-1}$$

where

$$\check{H}_c^{2m-i}(M \times M \setminus (V_0 \times V_1^* \cup V_1 \times V_0^*, M \times M \setminus V_1 \times V_1^*) \xrightarrow[\simeq]{\frown o^{M\times M}} H_i(V_1 \times V_1^*, V_0 \times V_1^* \cup V_1 \times V_0^*)$$

and

$$\check{H}_c^{2m-i}(M \setminus ((V_0 \cap V_1^*) \cup (V_1 \cap V_0^*)), M \setminus (V_1 \cap V_1^*)) \xrightarrow[\simeq]{\frown o^{M}} H_{i-m}(V_1 \cap V_1^*, (V_0 \cap V_1^*) \cup (V_1 \cap V_0^*))$$

are relative Poincaré-Alexander duality homomorphisms and where

$$\check{H}_c^{2m-i}(M \times M \setminus (V_0 \times V_1^* \cup V_1 \times V_0^*), M \times M \setminus V_1 \times V_1^*) \xrightarrow{\check{\Delta}_c^*} \check{H}_c^{2m-i}(M \setminus ((V_0 \cap V_1^*) \cup (V_1 \cap V_0^*)), M \setminus (V_1 \cap V_1^*))$$

is the homomorphism induced by the diagonal map Δ. Note that Δ is proper because the diagonal is closed and Δ is an embedding. The transfer $\Delta_!$ should be thought of as the map on homology induced by taking inverse images under Δ of open pairs in $M \times M$. For the general definition of transfer homomorphisms on homology and cohomology see [D, §VIII.10].

As mentioned in the Introduction, the definition of the intersection class pairing on Conley indices is based on the intersection pairings defined in [D, §VIII.13] where the intersection pairing is first defined for arbitrary pairs of open pairs in M and then extended via limiting arguments to pairs of arbitrary topological pairs in M. However, in the current work, interest is focused almost entirely on intersection pairings for pairs of pairs satisfying the off-diagonal condition; all exceptions to this occur in the proof of Lemma 7.7 wherein the intersection pairing on Conley indices is computed for normally hyperbolic invariant sets of a smooth flow. Thus, to avoid repeated mention of cofinality arguments to reduce to the off-diagonal situation, it is technically convenient to construct intersection pairings for off-diagonal pairs of pairs entirely within the category $M_{\not\Delta}^{(2)}$. For the sake of Lemma 7.7 and the reader's convenience, however, the definition of the pairing for a general pair of open pairs is made below as the first step since the definition and notation is the same for the general and off-diagonal cases for pairs of open pairs. Then, after the extension of the definition to arbitrary off-diagonal pairs of pairs is made, the minor changes needed to make the extension to the general case are pointed out.

2.8 DEFINITION. (*A*) For $V \in \mathrm{ob}(M_o^{(2)})$ and G and G' R-modules, define a pairing $\frown\!\!\!\bullet^V$ on the tensor product of $H_*(V_1, V_0; G)$ and $H_*(V_1^*, V_0^*; G')$ by defining it first on generating elements and then extending by linearity where the formula on generating elements is

(2.3)

$$\begin{aligned} H_i(V_1, V_0; G) \otimes H_j(V_1^*, V_0^*; G') &\to H_{i+j-m}(V_1 \cap V_1^*, V_0 \cap V_1^* \cup V_1 \cap V_0^*; G \otimes G') \\ \alpha \otimes \gamma &\mapsto \alpha \frown\!\!\!\bullet^V \gamma := (-1)^{m(m-j)} \Delta_!(\alpha \times \gamma) \end{aligned}$$

The pairing $\frown\!\!\!\bullet^V$ is natural with respect to inclusions of open pairs [D, Proposition VIII.13.4]; i.e., if $W := \big((W_1, W_0), (W_1^*, W_0^*)\big)$ is another pair of open pairs in M and if there is an inclusion morphism $V \hookrightarrow W$, then where the vertical arrows are inclusion induced homomorphisms and with coefficients suppressed the following is a commutative diagram:

(2.4)

$$\begin{array}{ccc} H_i(V_1, V_0) \otimes H_j(V_1^*, V_0^*) & \xrightarrow{\frown\!\!\bullet^V} & H_{i+j-m}(V_1 \cap V_1^*, (V_0 \cap V_1^*) \cup (V_1 \cap V_0^*)) \\ \downarrow & & \downarrow \\ H_i(W_1, W_0) \otimes H_j(W_1^*, W_0^*) & \xrightarrow{\frown\!\!\bullet^W} & H_{i+j-m}(W_1 \cap W_1^*, (W_0 \cap W_1^*) \cup (W_1 \cap W_0^*)) . \end{array}$$

This naturality follows from the corresponding naturality of homology cross-products and of transfer homomorphisms, and the naturality of the latter in turn follows from the naturality of the duality isomorphisms and the functorial properties of Čech cohomology with compact supports.

An alternative definition of the pairing defined by (2.3) can be given in terms of cup and cap products. Specifically, if $x \in \check{H}_c^{m-i}(M \setminus V_0, M \setminus V_1)$ and $y \in \check{H}_c^{m-j}(M \setminus V_0^*, M \setminus V_1^*)$ are Poincaré dual in M to α and γ respectively, then (suppressing certain inclusion induced maps)

$$\alpha \overset{\frown}{\bullet} \gamma = (x \smile y) \frown o^M \tag{2.5a}$$

$$= x \frown (y \frown o^M) = x \frown \gamma . \tag{2.5b}$$

Note that some interpretation is required to make sense of these formulae, especially for (2.5*b*). The interpretation and verification of these formulae is outlined in [D] on p. 337 and in Exercise 1 of §VIII.13.30 on p. 345 proved below as part of Proposition 9.18. Analogs of the formulae (2.5) in the context of the Conley index theory are obtained in Chapter 9 below and used to prove non-degeneracy of the intersection pairing on Conley indices; see Theorem 9.4 and its corollaries.

(*B*) The extension of (2.3) to an arbitrary pair of topological pairs satisfying the off-diagonal condition is made as follows. Let $(i, j) \in \mathbf{N}^2$ and let $X \in M^{(2)}_{\not\Delta}$. For each $U \in \mathcal{N}(X_1 \cap X_1^*)$, for each $K \in M^{(2)}_{\not\Delta c}(X)$, define a pairing

$$\mathfrak{L}_U^K \colon H_i(K_1, K_0; G) \otimes H_j(K_1^*, K_0^*; G') \to H_{i+j-m}(U; G \otimes G')$$

as follows. Choose $V \in \mathrm{ob}\,(M^{(2)}_{\not\Delta o}(K))$ satisfying $V_1 \cap V_1^* \subset U$. Such V are easily seen to exist via a simple construction using normality and the closedness of the sets in the pairs of K as well as that of $K_1 \cap K_1^*$. Note too that $\mathrm{ob}\,(M^{(2)}_{\not\Delta o}(K))$ is directed by containment. That is, for $V, V' \in \mathrm{ob}\,(M^{(2)}_{\not\Delta o}(K))$, define $V \cap V'$ to be the element of $\mathrm{ob}\,(M^{(2)}_{\not\Delta o}(K))$ obtained by taking the componentwise intersection of the sets in the pairs forming V and V', whence there are inclusion morphisms $V \hookleftarrow V \cap V'$ and $V' \hookleftarrow V \cap V'$. Then $\mathfrak{L}_U^K$ is that pairing making the diagram

$$\begin{array}{ccc}
H_i(K_1, K_0; G) \otimes H_j(K_1^*, K_0^*; G') & \xrightarrow{\mathfrak{L}_U^K} & H_{i+j-m}(U; G \otimes G') \\
\downarrow & & \uparrow \\
H_i(V_1, V_0; G) \otimes H_j(V_1^*, V_0^*; G') & \xrightarrow{\overset{\frown}{\bullet}{}^V} & H_{i+j-m}(V_1 \cap V_1^*; G \otimes G')
\end{array}$$

commutative where the vertical arrows are inclusion induced and the bottom horizontal arrow is given by (2.3). This definition of $\mathfrak{L}_U^K$ does not depend on the particular choice of $V \in \mathrm{ob}\,(M^{(2)}_{\not\Delta o}(K))$ satisfying $V_1 \cap V_1^* \subset U$ that is made as follows easily from our observation that $\mathrm{ob}\,(M^{(2)}_{\not\Delta o}(K))$ is directed by containment and from the naturality expressed by (2.4). It follows from this independence as to the choice of V that if $U \supset U' \in \mathcal{N}(X_1 \cap X_1^*)$, then

$$i_U^{U'} \circ \mathfrak{L}_{U'}^K = \mathfrak{L}_U^K . \tag{2.7}$$

It is immediate from (2.7) that we can pass to an inverse limit over $\mathcal{N}(X_1 \cap X_1^*)$ directed by containment and thereby obtain a pairing of R-modules

$$(2.8) \quad \mathfrak{L}^K_{X\,(i,j)}\colon H_i(K_1, K_0; G) \otimes H_j(K_1^*, K_0^*; G') \to \check{H}_{i+j-m}(X_1 \cap X_1^*; G \otimes G').$$

Note that if in fact $X \in \mathrm{ob}(M^{(2)}_{\mathbb{A}\,o})$, then the singleton $\{X_1 \cap X_1^*\}$ is cofinal in $\mathcal{N}(X_1 \cap X_1^*)$, whence the Čech and singular homology modules of $X_1 \cap X_1^*$ are naturally isomorphic. Hence, $\mathfrak{L}^K_X$ as given by (2.8) is naturally identified with $\mathfrak{L}^K_{X_1 \cap X_1^*}$ as defined by diagram (2.6) with $U := X_1 \cap X_1^*$ and $V := X$.

Next, observe that $\mathrm{ob}(M^{(2)}_{\mathbb{A}\,c}(X))$ can be regarded as a directed set where $K_1 \preceq K_2$ if, and only if, $K_1 \hookrightarrow K_2$ because where $K_1 \cup K_2$ is defined to be the pair of pairs obtained by taking the union of corresponding components in K_1 and K_2, $K_1 \cup K_2$ satisfies the off-diagonal condition because X does.

Thus, noting that if

$$j^K_L \equiv \left(j^{(K_1,K_0)}_{(L_1,L_0)}, j^{(K_1^*,K_0^*)}_{(L_1^*,L_0^*)}\right) : K \hookrightarrow L$$

is a morphism in $M^{(2)}_{\mathbb{A}\,c}(X)$ then

$$(2.9) \qquad \mathfrak{L}^K_{X\,(i,j)} = \mathfrak{L}^L_{X\,(i,j)} \circ \left(j^{(K_1,K_0)}_{(L_1,L_0)\,*} \otimes j^{(K_1^*,K_0^*)}_{(L_1^*,L_0^*)\,*}\right),$$

we find that we can pass to the direct limit over $M^{(2)}_{\mathbb{A}\,c}(X)$ and thereby obtain a pairing

$$\mathfrak{L}_{X\,(i,j)}\colon H_i(X_1, X_0; G) \otimes H_j(X_1^*, X_0^*; G') \to \check{H}_{i+j-m}(X_1 \cap X_1^*; G \otimes G')$$

because direct limits commute with tensor products and because the singular homology of any pair is the direct limit of the singular homology of the compact pairs it contains. Further, should it be the case that $X \in \mathrm{ob}(M^{(2)}_{\mathbb{A}\,c})$, then the pairing just defined coincides with $\mathfrak{L}^X_{X\,(i,j)}$ because the singleton $\{X\}$ would in this case be cofinal in $\mathrm{ob}(M^{(2)}_{\mathbb{A}\,c}(X))$. Similarly, should it be the case that $X \in \mathrm{ob}(M^{(2)}_{\mathbb{A}\,o})$, then $\mathfrak{L}_{X\,(i,j)}$ coincides with the pairing $\frown^X$ defined in the manner of (2.3).

The intersection class pairing on $X \in \mathrm{ob}(M^{(2)}_{\mathbb{A}})$ is the R-module homomorphism

$$\mathfrak{L}_X\colon H_*(X_1, X_0; G) \otimes H_*(X_1^*, X_0^*; G') \to \check{H}_*(X_1 \cap X_1^*; G \otimes G')$$

defined by $\mathfrak{L}_X := \bigoplus \mathfrak{L}_{X\,(i,j)}$ where the sum is over all pairs of non-negative integers. Also, for $\alpha \in H_*(X_1, X_0; G)$ and $\gamma \in H_*(X_1^*, X_0^*; G')$, set

$$\alpha \frown^X \gamma := \mathfrak{L}_X(\alpha \otimes \gamma)$$

and call $\alpha \frown^X \gamma$ *the intersection class of* $\alpha \otimes \gamma$.

(*C*) To construct an analogous *intersection class pairing* $\mathfrak{L}_X$ *for* $X \in \mathrm{ob}(M^{(2)})$, i.e., X fails to satisfy the off-diagonal condition hereinafter referred to as the general case, the following minor changes need to be made. First, for any $Y \in \mathrm{ob}(M^{(2)})$ set

$$Y^{\cap} := (Y_1 \cap Y_1^*, (Y_0 \cap Y_1^*) \cup (Y_1 \cap Y_0^*)).$$

Then $\mathfrak{L}_U^K$ is now defined for $K \in \mathrm{ob}(M_c^{(2)}(X)$ and $U \equiv (U_1, U_0) \in \mathcal{N}(X^{\cap})$ via the diagram obtained from diagram (2.6) by replacing $V_1 \cap V_1^*$ with $V^{\cap}$ where $V \in \mathrm{ob}(M_o^{(2)}(X)$ and satisfies $V^{\cap} \hookrightarrow U$ such V existing by an argument similar to that used in the off-diagonal case. Then equality (2.7) holds in the context of the general case and passage to the inverse limit yields a pairing

$$\mathfrak{L}^K_{X\,(i,j)}\colon H_i(K_1, K_0; G) \otimes H_j(K_1^*, K_0^*; G') \to \check{H}_{i+j-m}(X^{\cap}; G \otimes G')$$

so that (2.9) remains satisfied in the general case. Therefore, passage to the direct limit over $M_c^{(2)}(X)$ through the pairings $\mathfrak{L}^K_{X\,(i,j)}$ yields a pairing

$$\mathfrak{L}_{X\,(i,j)}\colon H_i(X_1, X_0; G) \otimes H_j(X_1^*, X_0^*; G') \to \check{H}_{i+j-m}(X^{\cap}; G \otimes G'),$$

and as before $\mathfrak{L}_X := \bigoplus \mathfrak{L}_{X\,(i,j)}$.

(*D*) From (2.8), for each $X \in M_{\Delta}^{(2)}$, the restriction of $\mathfrak{L}_X$ to the module of elements of degree m, viz., $\left[H_*(X_1, X_0) \otimes H_*(X_1^*, X_0^*)\right]_m$ with all coefficients taken in R, has image in the module $\check{H}_0(X_1 \cap X_1^*; R)$ allowing us to generalize to the manifold setting the notion of intersection number pairing given in Definition 2.1 for the Euclidean setting as follows.

For $X \in M_{\Delta}^{(2)}$, define *the intersection number pairing on* X to be the pairing

$$^{\#}\mathfrak{L}_X\colon \left[H_*(X_1, X_0) \otimes H_*(X_1^*, X_0^*)\right]_m \to R$$

given on generators by

$$^{\#}\mathfrak{L}_X(\alpha \otimes \gamma) := \langle 1, (\mathfrak{L}_X(\alpha \otimes \gamma))_M \rangle$$

and extended by linearity. Here $\langle\,\cdot\,,\,\cdot\,\rangle$ denotes the Kronecker pairing of singular cohomology and homology and 1 denotes the unit cohomology class in $H^0(M)$, i.e., the class of the usual augmentation (i.e., the Kronecker Index) on singular chains in M to R so that $^{\#}\mathfrak{L}_X(\alpha \otimes \gamma)$ gives a signed count (modulo the characteristic of R) of the number of points in the 0-chain $(\mathfrak{L}_X(\alpha \otimes \gamma))_M$ assuming $|\alpha| + |\gamma| = m$. As a matter of convenience, we also set

$$^{\#}(\alpha \,\hat{\bullet}^X\, \gamma) := {}^{\#}\mathfrak{L}_X(\alpha \otimes \gamma)$$

and call it *the intersection number of* $\alpha \otimes \gamma$. Too, the superscript X in the notation $\hat{\bullet}^X$ will often be dropped when clear from context.

(*E*) To see that the notion of intersection number just defined is indeed a generalization to the manifold setting of definition 2.1, let $o_m \in H_m(\mathbf{R}^m, \mathbf{R}^m \setminus \{0\})$ be the generator specified by a given orientation of $\mathbf{R}^m$, let $\nu^m \in H^m(\mathbf{R}^m, \mathbf{R}^m \setminus \{0\})$ be the element dual to o_m, i.e., $\langle \nu^m, o_m \rangle = 1$, let $X \in \mathrm{ob}\big((\mathbf{R}^m)^{(2)}_{\Delta}\big)$, and define

$$^{\#}\mathsf{L}_X \colon \big[H_*(X_1, X_0) \otimes H_*(X_1^*, X_0^*)\big]_m \to R$$

by setting $^{\#}\mathsf{L}_X(\alpha \otimes \gamma) := (-1)^{|\alpha||\gamma|}\langle \nu^m, \alpha \# \gamma \rangle$ on generators and extending by linearity. Thus

$$^{\#}\mathsf{L}_X(\alpha \otimes \gamma) o_m = (-1)^{|\alpha||\gamma|} \alpha \# \gamma \tag{2.10}$$

because the isomorphism $H_m(\mathbf{R}^m, \mathbf{R}^m \setminus \{0\}) \simeq R$ specified by sending o_m to $1 \in R$ is given by the map

$$\xi \mapsto \langle \nu^m, \xi \rangle = \langle 1 \smile \nu^m, \xi \rangle = \langle 1, \nu^m \frown \xi \rangle\,. \tag{2.11}$$

On the other hand with α and γ as in the last paragraph,

$$\nu^m \frown (\gamma \# \alpha) = (\alpha \overset{X}{\widehat{\bullet}} \gamma)_{\mathbf{R}^m}\,. \tag{2.12}$$

as an immediate consequence of [D, Proposition VIII.13.29].

The equalities (2.10)–(2.12) together with Remark 2.2.2 immediately yield the following result which states, in essence, that the two methods available for defining intersection numbers in the Euclidean case are the same.

2.9 Proposition. *Let $X \in \mathrm{ob}\big((\mathbf{R}^m)^{(2)}\big)$. Then $^{\#}\mathsf{L}_X = {}^{\#}\mathfrak{L}_X$; that is, for $\alpha \in H_{m-k}(X_1, X_0)$ and $\gamma \in H_k(X_1^*, X_0^*)$,*

$$\nu_m \frown (\gamma \# \alpha) = (\alpha \,\widehat{\bullet}\, \gamma)_{\mathbf{R}^m}\,.$$

Hence

$$(\gamma \# \alpha) = {}^{\#}(\alpha \,\widehat{\bullet}\, \gamma) o_m\,.$$

The construction of the intersection pairing on the homology modules of the Conley indices of an isolated invariant set as carried out by the proof of Theorem 2.11 below has as key element the following statement of naturality following from the double limit construction used in defining $\mathfrak{L}_X$; see §4.10 for the proof.

2.10 Lemma. *The families of pairings $\{\mathfrak{L}_X : X \in \mathrm{ob}(M^{(2)}_{\Delta})\}$ and $\{^{\#}\mathfrak{L}_X : X \in \mathrm{ob}(M^{(2)}_{\Delta})\}$ are natural relative to the morphisms of $M^{(2)}_{\Delta}$. The family of pairings $\{\mathfrak{L}_X : X \in \mathrm{ob}(M^{(2)}\}$ is natural relative to the morphisms of $M^{(2)}$.*

For the remainder of this section, fix S to be an isolated invariant set of some flow in the manifold M.

Let $\mathcal{N}(S; \mathcal{AIP})$ consist of those neighborhoods in $\mathcal{N}(S)$ expressible in the form $P_1 \cap P_1^*$ for some $P \in \mathrm{ob}(\mathcal{AIP}(S))$. Recall too that for $P \in \mathrm{ob}(\mathcal{AIP}(S))$ and coefficient R-modules G and G', the tensor product of the canonical maps

$$\bar{p}_*^+ \otimes \bar{p}_*^- \colon H_*(P_1, P_0; G) \otimes H_*(P_1^*, P_0^*; G') \to \widetilde{H}_*\mathcal{C}(S; G) \otimes \widetilde{H}_*\mathcal{C}^*(S; G')$$

is an isomorphism.

2.11 THEOREM. *$\mathcal{N}(S;\mathcal{AIP})$ is cofinal in $\mathcal{N}(S)$, and if $P, Q \in \mathrm{ob}(\mathcal{AIP}(S))$ with $P_1 \cap P_1^* \supset Q_1 \cap Q_1^*$, then*

$$i_{P_1\cap P_1^*}^{Q_1\cap Q_1^*} \circ \mathfrak{L}_Q \circ (\bar{q}_*^+ \otimes \bar{q}_*^-)^{-1} = \mathfrak{L}_P \circ (\bar{p}_*^+ \otimes \bar{p}_*^-)^{-1}, \tag{2.13}$$

whence passage to the inverse limit yields a pairing of degree $-m$

$$\mathfrak{L}\colon \widetilde{H}_*\mathcal{C}(S;G) \otimes \widetilde{H}_*\mathcal{C}^*(S;G') \to \check{H}_*(S;G\otimes G')$$

upon identification of $\varprojlim \check{H}_*(P_1 \cap P_1^*)$ and $\check{H}_*(S)$.

Remark 2.11.1. Note that for each $V \in \mathcal{N}(S)$, if $P \in \mathrm{ob}(\mathcal{AIP}(S))$ and $P_1 \cap P_1^* \subset V$, then where π_V^S and $\pi_V^{P_1\cap P_1^*}$ are the projections from the Čech homology of the superscript to the singular homology of the subscript,

$$\pi_V^S \circ \mathfrak{L} = \pi_V^{P_1\cap P_1^*} \circ \mathfrak{L}_P \circ (\bar{p}_*^+ \otimes \bar{p}^-)^{-1}$$

and that this relation completely characterizes $\mathfrak{L}$ by (2.13) and the universality properties of inverse limits. In particular, the right-hand side is independent of the choice of P. The proof of the theorem consists in showing this via an application of the method used to prove Theorem 2.4 which requires the naturality expressed by Lemma 2.10. See Proposition 4.4, Theorem 4.7, Corollary 4.8, and §4.11 for the proof.

2.12 DEFINITION. (A) Call $\mathfrak{L}$ *the intersection class pairing for the Conley indices of* S, and for $(i,j) \in \mathbf{N}^2$ and element $\boldsymbol{\alpha} \otimes \boldsymbol{\gamma} \in \widetilde{H}_i\mathcal{C}(S;G) \otimes \widetilde{H}_j\mathcal{C}^*(S;G')$, set

$$\boldsymbol{\alpha} \,\widehat{\bullet}\, \boldsymbol{\gamma} := \mathfrak{L}(\boldsymbol{\alpha} \otimes \boldsymbol{\gamma})$$

and call the Čech homology class $\boldsymbol{\alpha} \,\widehat{\bullet}\, \boldsymbol{\gamma}$ *the intersection class of* $\boldsymbol{\alpha} \otimes \boldsymbol{\gamma}$.

If it is necessary to distinguish between intersection class pairings for different flows, the flow or an index corresponding to the flow is appended to $\mathfrak{L}$ as a superscript. For simplicity if ψ^* is the reverse time flow associated to a flow ψ understood from context, set $\mathfrak{L}^* := \mathfrak{L}^{\psi^*}$, and for $\boldsymbol{\gamma} \in \widetilde{H}_j\mathcal{C}^*(S)$ and $\boldsymbol{\alpha} \in \widetilde{H}_i\mathcal{C}(S)$, set $\boldsymbol{\gamma} \,\widehat{\bullet}\, \boldsymbol{\alpha} := \mathfrak{L}^*(\boldsymbol{\gamma} \otimes \boldsymbol{\alpha})$.

As is the case for the homology intersection number pairing L, there is a simple graded commutativity relationship between $\mathfrak{L}$ and $\mathfrak{L}^*$. Namely with $\boldsymbol{\alpha}$ and $\boldsymbol{\gamma}$ as above,

$$\boldsymbol{\gamma} \,\widehat{\bullet}\, \boldsymbol{\alpha} = (-1)^{(m-i)(m-j)} t_*(\boldsymbol{\alpha} \,\widehat{\bullet}\, \boldsymbol{\gamma})$$

where t_* is the change in coefficient isomorphism induced by $G \otimes G' \simeq G' \otimes G$. This identity follows easily from the definition of $\mathfrak{L}$, from equality $(2.5a)$ above, and from the well-known graded commutativity of cup products. Details are left to the reader.

(B) In analogy with Definition 2.8(D), for (i,j), $\boldsymbol{\alpha}$, and $\boldsymbol{\gamma}$ as in (A) but assuming that $i+j=m$ and $G=G'=R$ and making the identification $R\otimes R \simeq R$, define *the intersection number of* $\boldsymbol{\alpha} \otimes \boldsymbol{\gamma}$, denoted by ${}^{\#}\mathfrak{L}(\boldsymbol{\alpha} \otimes \boldsymbol{\gamma})$ or, more often, by ${}^{\#}(\boldsymbol{\alpha} \,\widehat{\bullet}\, \boldsymbol{\gamma})$, via the formula

$${}^{\#}(\boldsymbol{\alpha} \,\widehat{\bullet}\, \boldsymbol{\gamma}) := \langle 1, (\boldsymbol{\alpha} \,\widehat{\bullet}\, \boldsymbol{\gamma})_M \rangle\,.$$

Extend ${}^{\#}\mathfrak{L}$ by linearity to a pairing

$$
{}^{\#}\mathfrak{L}\colon \left[\widetilde{H}_*\mathcal{C}(S) \otimes \widetilde{H}_*\mathcal{C}^*(S)\right]_m \to R
$$

called *the intersection number pairing* on the singular homology of the Conley indices of S, coefficients in R.

(C) In analogy with Definition 2.8(E), when $M := \mathbf{R}^m$ and $\boldsymbol{\alpha}$ and $\boldsymbol{\gamma}$ are as in (B), set ${}^{\#}\mathsf{L}(\boldsymbol{\alpha} \otimes \boldsymbol{\gamma}) := (-1)^{|\boldsymbol{\alpha}||\boldsymbol{\gamma}|}\langle \nu^m, \boldsymbol{\alpha} \# \boldsymbol{\gamma}\rangle$ and extend ${}^{\#}\mathsf{L}$ by linearity to an R-module homomorphism

$$
{}^{\#}\mathsf{L}\colon \left[\widetilde{H}_*\mathcal{C}(S;R) \otimes \widetilde{H}_*\mathcal{C}^*(S;R)\right]_m \to R\,.
$$

If $P \in \mathcal{AIP}(S)$ and $\alpha \in H_{m-k}(P_1, P_0; R)$ represents $\boldsymbol{\alpha}$ and $\gamma \in H_k(P_1^*, P_0^*; R)$ represents $\boldsymbol{\gamma}$, then $(\alpha \mathbin{\widehat{\bullet}}^P \gamma)_M = (\boldsymbol{\alpha} \mathbin{\widehat{\bullet}} \boldsymbol{\gamma})_M$ as can be read off from Theorem 2.11 and the fact that $\pi_M^S = \pi_M^{P_1 \cap P_1^*} \circ \tilde{\imath}^S_{P_1 \cap P_1^*}$. Also, when $M := \mathbf{R}^m$ Remark 2.5.1 states that $\alpha \# \gamma = \boldsymbol{\alpha} \# \boldsymbol{\gamma}$. Thus from Propositions 2.6 and 2.9 we get the following.

2.13 PROPOSITION. *For $M = \mathbf{R}^m$, ${}^{\#}\mathsf{L} = {}^{\#}\mathfrak{L}$; i.e.,*

$$
\nu_m \frown (-1)^{k(m-k)}(\boldsymbol{\alpha} \# \boldsymbol{\gamma}) = (\boldsymbol{\alpha} \mathbin{\widehat{\bullet}} \boldsymbol{\gamma})_{\mathbf{R}^m}\,.
$$

Hence,

$$
(-1)^{k(m-k)}\boldsymbol{\alpha} \# \boldsymbol{\gamma} = {}^{\#}(\boldsymbol{\alpha} \mathbin{\widehat{\bullet}} \boldsymbol{\gamma})o_m\,.
$$

CHAPTER 3

STATEMENT OF THE CONTINUATION RESULTS AND EXAMPLES

Throughout this chapter $\{\varphi_\lambda\}_{\lambda\in\Lambda}$ is a continuous family of C^s flows on the C^r manifold M, $0 \le s \le r \le \infty$, and $\mathcal{S}(\varphi)$ is the corresponding space of isolated invariant sets described in Definition 1.1(A) above. By a path β in $\mathcal{S}(\varphi)$ is meant a continuous map $\beta\colon[0,1] \to \mathcal{S}(\varphi)$. As shown in [K3] and as described below in Chapter 5, functorially associated to the homotopy class (endpoints fixed) of a path β in $\mathcal{S}(\varphi)$ are continuation isomorphisms on homology (coefficients in any R-module)

$$\widetilde{H}_*\mathcal{C}(S_\beta(0)) \xrightarrow{\mathcal{C}(\beta)_*} \widetilde{H}_*\mathcal{C}(S_\beta(1)) \quad\text{and}\quad \widetilde{H}_*\mathcal{C}^*(S_\beta(0)) \xrightarrow{\mathcal{C}^*(\beta)_*} \widetilde{H}_*\mathcal{C}^*(S_\beta(1))\ .$$

(By "functorially associated" we mean that $\widetilde{H}_*\mathcal{C}$ and $\widetilde{H}_*\mathcal{C}^*$ can be regarded as functors on the fundamental groupoid of $\mathcal{S}(\varphi)$ to the category of graded R-modules.)

A. Invariance of Intersection Numbers under Continuation. The general result for the invariance of intersection numbers under continuation is the following result whose proof is given in §5.11.

3.1 THEOREM. *Let β be a path (or a path class with endpoints fixed) in $\mathcal{S}(\varphi)$. Then for $k = 0, 1, \ldots, m$, for $\boldsymbol{\alpha} \in \widetilde{H}_{m-k}\mathcal{C}(S_\beta(0); R)$ and $\gamma \in \widetilde{H}_k\mathcal{C}^*(S_\beta(0); R)$,*

$$\text{(3.1)}\qquad {}^{\#}(\boldsymbol{\alpha} \mathbin{\widehat{\bullet}} \gamma) = {}^{\#}\big(\mathcal{C}(\beta)_*\boldsymbol{\alpha} \mathbin{\widehat{\bullet}} \mathcal{C}^*(\beta)_*\gamma\big)\ .$$

When $M = \mathbf{R}^m$, in $H_m(\mathbf{R}^m, \mathbf{R}^m \setminus \{0\}; R)$,

$$\text{(3.2)}\qquad \boldsymbol{\alpha} \# \gamma = \mathcal{C}(\beta)_*\boldsymbol{\alpha} \# \mathcal{C}^*(\beta)_*\gamma\,.$$

3.1.1 Remark. For $s \in [0,1]$ and β as in the statement of Theorem 3.1, define the path β^s in $\mathcal{S}(\varphi)$ by $\beta^s(t) := \beta(st)$ for $0 \le t \le 1$. Application of Theorem 3.1 to β^s for each $s \in [0,1]$ shows that ${}^{\#}\big(\mathcal{C}(\beta^s)_*\boldsymbol{\alpha} \mathbin{\widehat{\bullet}} \mathcal{C}^*(\beta^s)_*\gamma\big)$ does not vary with s.

From the construction of $\mathcal{C}(\beta)$ reviewed below, it will be seen that Theorem 3.1 is a simple consequence of the following result proved in §5.10.

3.2 THEOREM. *Assume $[0,1] \subset W \subset \Lambda \subset \mathbf{R}$ with W open in $\mathbf{R}$, let $\sigma\colon W \to \mathcal{S}(\varphi)$ be a local section of $\pi_\varphi\colon \mathcal{S}(\varphi) \to \Lambda$, and let β be the restriction of σ to $[0,1]$. Then equality (3.1) holds for each $\boldsymbol{\alpha} \in \widetilde{H}_{m-k}\mathcal{C}(S_\sigma(0); R)$ and $\gamma \in \widetilde{H}_k\mathcal{C}^*(S_\sigma(0); R)$. If $M = \mathbf{R}^m$, then equality (3.2) also holds.*

Theorem 3.2 will first be proved under additional assumptions of smoothness where the main idea of the proof is not obscured by technical difficulties. Specifically, the first continuation result proved is the following. The proof appears in §5.8.

3.3 THEOREM. *In addition to the hypotheses of Theorem 3.2, assume M is C^r $(r \geq 1)$, give the space of C^0 vectorfields on M the compact-open topology, and for each $\lambda \in \Lambda$, assume φ_λ is generated by a continuous vectorfield X_λ with $\lambda \mapsto X_\lambda$ continuous. Then equality (3.1) holds for each $\boldsymbol{\alpha} \in \widetilde{H}_{m-k}\mathcal{C}(S_\sigma(0); R)$ and $\gamma \in \widetilde{H}_k\mathcal{C}^*(S_\sigma(0); R)$. If $M = \mathbf{R}^m$, then equality (3.2) also holds.*

The following discussion and proposition give a useful criterion for establishing continuity of a map $g: Y \to \mathcal{S}(\varphi)$ that will ease the reader's understanding of some examples presented afterward illustrating the above continuation theorems and the analogue of these results for the intersection class pairing.

It follows straightforwardly from the definitions that a map $g: Y \to \mathcal{S}(\varphi)$ is continuous at y_0 if, and only if, λ_g is continuous at y_0 and for each N that $\lambda_g(y_0)$-isolates $S_g(y_0)$, there exists W an open neighborhood of y_0 so that for each $y \in W$, N also $\lambda_g(y)$-isolates $S_g(y)$. In fact it suffices to consider a single isolating neighborhood of $S_g(y_0)$ as a consequence of the following:

3.4 PROPOSITION. *A map $g: Y \to \mathcal{S}(\varphi)$ is continuous at y_0 if, and only if, λ_g is continuous at y_0 and there exist W and N that together satisfy W is an open neighborhood of y_0 and N is a $\lambda_g(y)$-isolating neighborhood of $S_g(y)$ for each $y \in W$.*

Proof. Proof of necessity of the continuity condition follows straightforwardly from the definitions and is left to the reader. To prove sufficiency, assume N and W exist and suppose $g(y_0) \in \sigma_{N'}(U)$ where U is open in $\Lambda(N')$ and N' is a compact subset of M. Then $S_g(y_0) = S(N', \lambda_g(y_0))$; i.e., N and N' λ_0-isolate the same invariant set where $\lambda_0 := \lambda_g(y_0)$. Set $K := N \cap N'$ so that K also λ_0-isolates $S_g(y_0)$ and $K \subset N$ and $K \subset N'$. Now if two compact sets one containing the other λ_0-isolate the same invariant set, then by [K3, Lemma 1.5] they also λ-isolate the same invariant set for λ in a neighborhood of λ_0. Thus, there exists U' an open neighborhood of λ_0 satisfying $U' \subset \Lambda(N) \cap \Lambda(K) \cap U$ so that K, N and N' λ-isolate the same invariant set for each $\lambda \in U'$. Then continuity of λ_g at y_0 implies that we can choose W' an open neighborhood of y_0 with $W' \subset W$ so that $\lambda_g(W') \subset U'$. Then on the one hand, by choice of U' and W', the compact sets N and N' both $\lambda_g(y)$-isolate the same invariant set for each $y \in W'$, and on the other, by definition of N and W and as $W' \subset W$, necessarily N $\lambda_g(y)$-isolates $S_g(y)$ for each $y \in W'$. Thus $S_g(y) = S(N', \lambda_g(y))$ for each $y \in W'$ and $\lambda_g(W') \subset U'$, i.e., $g(W') \subset \sigma_{N'}(U)$, which yields continuity of g at y_0. $\square$

3.5 EXAMPLE. Let a be a continuous map of some open interval $W \supset [0,1]$ into the open interval $]0,1[$. For each $\lambda \in \Lambda := W$, call the planar system (differentiation is with respect to a variable τ and is denoted by a dot over a variable)

$$\begin{aligned} \dot{v} &= w \\ \dot{w} &= -v(1-v)(v-a(\lambda)) \end{aligned} \tag{3.3$_\lambda$}$$

the fast system at λ associated to the scalar second-order equation

$$\varepsilon^2 v'' + v(1-v)(v-a(x)) = 0\,.$$

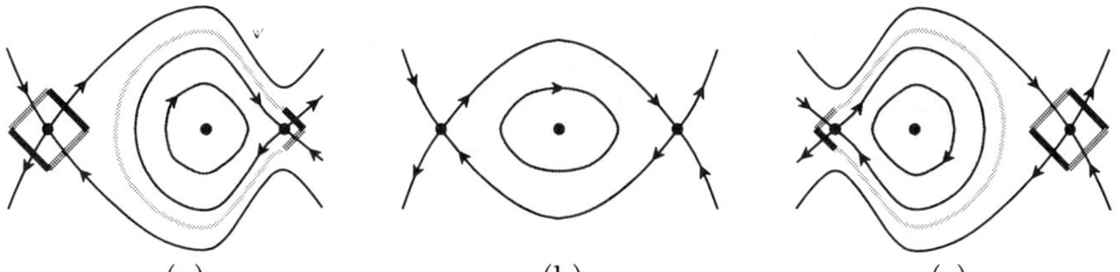

(a) (b) (c)

FIGURE 1

This scalar equation considered on the interval $0 \le x \le 1$ with Neumann endpoint conditions is the equilibrium equation of the Fisher equation

$$v_t = \varepsilon^2 v_{xx} + v(1-v)(v-a(x)) \tag{3.4}$$

on the infinite half-strip $[0,1] \times \mathbf{R}^+$ satisfying Neumann conditions on the infinite edges that is a model for the evolution of a diploid population in a genetically isolated non-homogeneous finite one-dimensional habitat in the heterozygote inferior case where v is the frequency of one of the alleles and $x \in [0,1]$ represents position in the habitat; i.e., with a and A corresponding to the two alleles, the heterozygote genotype Aa is observed to be inferior to either of the homozygote genotypes aa or AA and v is the ratio of the number of alleles A to the total number of alleles of both types in the population at a given point in space and time [Fisher,Ha].

Sketched in Figure 1 are three phase portraits which qualitatively exhaust the possibilities for system (3.3_λ). These were sketched by drawing, for fixed λ, level curves of the Hamiltonian

$$H(v,w,\lambda) := \frac{1}{2}w^2 + V(v,\lambda) \quad \text{where} \quad V(v,\lambda) := \int_0^v u(1-u)(u-a(\lambda))\,du\,.$$

The technique for sketching such level curves from the graph of V for fixed λ via graphical integration is described, for example, in [A]. The phase portraits in Figures 1a, b, c correspond, respectively to the cases (i) $1/2 < a(\lambda) < 1$, (ii) $a(\lambda) = 1/2$, (iii) $0 < a(\lambda) < 1/2$. In each case, there is a largest compact invariant subset $S_a(\lambda)$ of the flow φ_λ generated by system (3.3_λ), which in Figures 1a, b, c is the union of the shaded region (a closed 2-cell) together with any critical point outside that region (either $(0,0)$ or $(1,0)$, each represented by a heavy dot), whose analytic description is:

$$S_a(\lambda) := \begin{cases} \{(v,w) : (v,w) = (0,0) \text{ or } v \in \,]0,1] \text{ and } H(v,w,\lambda) \le V(1,\lambda)\} \\ \qquad\qquad\qquad\qquad\qquad\qquad\qquad\qquad \text{if } 1 > a(\lambda) > 1/2; \\ \{(v,w) : v \in [0,1] \text{ and } H(v,w,\lambda) \le 0\} \quad \text{if } a(\lambda) = 1/2; \\ \{(v,w) : (v,w) = (1,0) \text{ or } v \in [0,1[\, \text{ and } H(v,w,\lambda) \le V(0,\lambda)\} \\ \qquad\qquad\qquad\qquad\qquad\qquad\qquad\qquad \text{if } 0 < a(\lambda) < 1/2. \end{cases}$$

Using Proposition 3.4, the reader should have little difficulty showing that $\sigma_a : W \to \mathcal{S}(\varphi)$ defined by $\sigma_a(\lambda) := (S_a(\lambda), \lambda)$ is a continuous local section of π_φ. It may help the reader to note that $S_a(\lambda)$ varies upper semi-continuously with λ and is the largest compact invariant subset of the flow φ_λ.

Let β be the path in $\mathcal{S}(\varphi)$ that is the restriction of σ_a to $[0,1]$. Our goal in this example is to analyze Theorem 3.1 as it applies to the path β and to discern what it tells us about the intersection of the local invariant manifolds that are geometric representatives vis-à-vis a suitable isolating block of generating classes of the homology of the forward and reverse time Conley indices of the isolated invariant sets at the beginning and end of β. The point to be made is best illustrated if we assume henceforth that $(a(0)-1/2)(a(1)-1/2)<0$, and for definiteness let us assume that $a(0)>1/2>a(1)$. Thus, $S_a(0)$ qualitatively corresponds to the $S_a(\lambda)$ of Figure 1a and $S_a(1)$ to the $S_a(\lambda)$ of Figure 1c.

Drawn in Figures 1a, c, respectively covering the cases $a(\lambda)>1/2$ and $a(\lambda)<1/2$, are the bounding curves of an isolating block for $S_a(\lambda)$ with two components, both closed 2-cells and each a block in its own right. The exit set consists of the thick black line segments of slope minus one, the entrance set consists of the thick dark grey line segments of slope plus one, and the rest of the boundary of the block consists of the thinner light grey curve segments that are in fact orbit segments of the flow φ_λ. One component block, call it $C(\lambda)$, contains that part of $S_a(\lambda)$ that is an invariant closed 2-cell bounded by a homoclinic orbit, the other, call it $B(\lambda)$, contains the saddle point in $S_a(\lambda)$ not in the invariant 2-cell. Our computation of the intersection number pairing on the Conley indices of $S_a(\lambda)$ will be based on using geometric representatives of homology classes of the index pairs formed from these blocks and their exit and entrance sets, and for this purpose the graphical description of these blocks provided by Figure 1 will suffice. However, explicit analytic descriptions of these blocks and their exit and entrance sets is given below in the course of proving Theorem A.1 in Appendix A, an existence result for the Fisher equation (3.4) about which more is said in the last paragraph of this example. It is left as an easy exercise for the interested reader as it is not needed here that from these blocks one can compute that for $a(\lambda)\neq 1/2$ each index space for $S_a(\lambda)$ in both forward and reverse time has the homotopy type of a pointed one-sphere; hence by invariance of the homotopy type of index spaces under continuation, the same is true for $a(\lambda)=1/2$. Thus, taking all coefficients in R ($R=\mathbf{Z}_2$ is a good choice) we must show that the Conley indices of $S_a(\lambda)$ in both time directions have one-dimensional homology isomorphic to R while the homology modules must vanish in all other dimensions.

The necessary computations follow from the following sequence of observations throughout which it is assumed that $a(\lambda)\neq 1/2$. First, $H_*(C(\lambda),C(\lambda)^\pm)\equiv 0$ as follows from the long exact singular homology sequence of the pair since $C(\lambda)$ is a 2-cell and $C(\lambda)^\pm$ is an arc in its boundary. Second, $H_1(B(\lambda),B(\lambda)^\pm)\simeq R$, but $H_q(B(\lambda),B(\lambda)^\pm)\simeq 0$ for all $q\neq 1$. Again this follows from the long exact sequence of the pair using reduced singular homology since the pair $(B(\lambda),B(\lambda)^\pm)$ is homeomorphic to the pair consisting of the standard 2-cell and one pair of parallel edges. Further, an examination of the connecting map of the sequence from the 1-level to the 0-level shows that any path in $B(\lambda)$ starting in one component of $B(\lambda)^\pm$ and ending in the other when regarded as a relative 1-cycle in $B(\lambda)$ modulo $B(\lambda)^\pm$ has homology class that generates $H_1(B(\lambda),B(\lambda)^\pm)$. Third, an index pair for $S_a(\lambda)$ consisting of the block $K(\lambda):=B(\lambda)\cup C(\lambda)$ together with its exit or entrance set has one-dimensional singular homology isomorphic to R but vanishes in all other dimensions; hence, the indices $\mathcal{C}\,(S_a(\lambda))$ and $\mathcal{C}^*(S_a(\lambda))$ have one-dimensional homology isomorphic to R, but have vanishing homology in all other dimensions. For from the Meyer-Vietoris sequence for singular homology and the first observation it follows that

$$H_*(B(\lambda), B(\lambda)^{\pm}) \simeq H_*(K(\lambda), K(\lambda)^{\pm})$$

where the isomorphism is inclusion induced, and this isomorphism followed by the canonical homomorphism from $H_*(K(\lambda), K(\lambda)^{\pm})$ to the singular homology of the Conley index in the appropriate time direction is an isomorphism as follows from the observations made in Definitions 1.4(B) and 1.5(A)—also see Proposition 4.2 below. The third observation then follows from the second.

Let $\boldsymbol{\alpha}_\lambda$ generate $\widetilde{H}_1\mathcal{C}(S_a(\lambda))$, let $\boldsymbol{\gamma}_\lambda$ generate $\widetilde{H}_1\mathcal{C}^*(S_a(\lambda))$, and let s_λ denote the saddle point in $B(\lambda)$ for $a(\lambda) \neq 1/2$. Then since as observed any path in a 2-cell starting in one edge of a parallel pair and ending in the other edge of the parallel pair is a relative 1-cycle generating the homology of the 2-cell modulo the parallel pair of edges, in particular, the local unstable manifold of s_λ within $B(\lambda)$ is an arc that can be regarded as a geometric generator of $H_1(B(\lambda), B(\lambda)^+)$, and therefore by the isomorphisms described in the previous paragraph, it can be regarded as a geometric representative of $\boldsymbol{\alpha}_\lambda$. Similarly, the local stable manifold of s_λ within $B(\lambda)$ is an arc that can be regarded as a geometric representative of $\boldsymbol{\gamma}_\lambda$. Finally, because these local invariant manifolds are transverse as is easily computed from the linearization of system (3.3_λ) about s_λ, it follows from Theorem 6.5 below—also see statements (2) and (3) of Theorem 7.5—that

$$\text{(3.5}a\text{)} \qquad {}^{\#}(\boldsymbol{\alpha}_\lambda \mathbin{\widehat{\bullet}} \boldsymbol{\gamma}_\lambda) = \pm 1$$

or, equivalently, because we are working with a planar flow that

$$\text{(3.5}b\text{)} \qquad \boldsymbol{\alpha}_\lambda \# \boldsymbol{\gamma}_\lambda = \pm o_2 \,.$$

By Theorem 3.1 (or 3.2 or 3.3),

$$\text{(3.6)} \qquad \mathcal{C}(\beta)_* \boldsymbol{\alpha}_0 \# \mathcal{C}^*(\beta)_* \boldsymbol{\gamma}_0 = \boldsymbol{\alpha}_0 \# \boldsymbol{\gamma}_0 \,.$$

This equality is certainly non-trivial in the present context, for it asserts that there is a *functorial identification* of the intersection of the local invariant manifolds of the saddle point $(0,0)$ of system (3.3_0) with the intersection of the local invariant manifolds of the saddle point $(1,0)$ of system (3.3_1) even though the first saddle arises from the $v = 0$ branch of the zero set of $v(1-v)(v-a(x))$ and the second arises from the $v = 1$ branch of the zero set. Also, see Theorem 3.7 below and the paragraph preceding it.

Before leaving this example let us make one final observation. Because $\mathcal{C}^*(\beta)_*$ is an isomorphism necessarily $\mathcal{C}^*(\beta)_* \boldsymbol{\gamma}_0 = \pm \boldsymbol{\gamma}_1$. This equality combined with (3.5) and (3.6) yields that

$$\text{(3.7)} \qquad \mathcal{C}(\beta)_* \boldsymbol{\alpha}_0 \# \boldsymbol{\gamma}_1 \neq 0 \,.$$

One way to interpret (3.7) is that any arc representing $\boldsymbol{\alpha}_0$ relative to some forward time index pair for the saddle point s_0 continues to an arc that intersects any arc that represents γ_1 relative to some reverse time index pair for the saddle point s_1. This interpretation together with the main result of [K4] (which in part was motivated by this interpretation) yields Theorem A.1 of Appendix A which states in part that for $0 < \varepsilon \ll 1$ there exists a family v_ε of non-constant equilibrium solutions for the Fisher equation (3.4) over the interval $[0,1]$ satisfying Neumann endpoint conditions.

B. Continuation of $\mathfrak{L}$ over a Path of Isolated Invariant Sets. To discuss continuation results for the intersection class pairing, a topological space of Čech homology classes of isolated invariant sets will be introduced.

3.6 Definition and Proposition. *(A)* For each $k \in \mathbf{N}$, for any R-module G'', define $\check{\mathcal{H}}_k\mathcal{S}(\varphi; G'')$ to be the set of ordered triples $(\check{\zeta}, S, \lambda)$ satisfying $(S,\lambda) \in \mathcal{S}(\varphi)$ and $\check{\zeta} \in \check{H}_k(S; G'')$ and define $\check{\mathcal{H}}_*\mathcal{S}(\varphi; G'') := \bigcup\{\check{\mathcal{H}}_k\mathcal{S}(\varphi; G'') : k \in \mathbf{N}\}$. Also, for $k \in \mathbf{N}$, let $\pi_{\check{\mathcal{H}}_k} : \check{\mathcal{H}}_k\mathcal{S}(\varphi; G'') \to \mathcal{S}(\varphi)$ be given by projection of the last two coordinates.

(B) For $k \in \mathbf{N}$, for $(\check{\zeta}, S, \lambda) \in \check{\mathcal{H}}_k\mathcal{S}(\varphi; G'')$, for N a λ-isolating neighborhood of S, and for U open in $\Lambda(N)$ with $\lambda \in U$, define $\mathcal{N}(\check{\zeta}, S, \lambda, N, U)$ to be the set of points $(\check{\zeta}', S', \lambda') \in \check{\mathcal{H}}_k\mathcal{S}(\varphi; G'')$ satisfying $\lambda' \in U$, $S' = S(N, \lambda')$, $\check{\zeta}' \in \check{H}_k(S'; G'')$ and $\pi_N^S\check{\zeta} = \pi_N^{S'}\check{\zeta}'$ where π_N^S and $\pi_N^{S'}$ are respectively the canonical projections of $\check{H}_*(S; G'')$ and $\check{H}_*(S'; G'')$ to $H_*(N; G'')$.

For $k \in \mathbf{N}$, define $\mathbf{B}_k$ to be the set consisting of sets of the form $\mathcal{N}(\check{\zeta}, S, \lambda, N, U)$.

(C) $\mathbf{B}_k$ *is a basis for a topology on* $\check{\mathcal{H}}_k\mathcal{S}(\varphi)$ *(call it the fine topology) and* $\pi_{\check{\mathcal{H}}_k} : \check{\mathcal{H}}_k\mathcal{S}(\varphi; G'') \to \mathcal{S}(\varphi)$ *is continuous.*

Proof. Clearly $\mathbf{B}_k$ covers $\check{\mathcal{H}}_k\mathcal{S}(\varphi; G'')$. If $(\check{\zeta}'', S'', \lambda'') \in \mathcal{N}(\check{\zeta}_i, S_i, \lambda_i, N_i, U_i)$ for $i = 1,2$, then by [K3, Lemma 1.5] for some open neighborhood U'' of λ'', $U'' \subset \Lambda(N_1) \cap \Lambda(N_2) \cap \Lambda(N_1 \cap N_2)$, and $\lambda' \in U''$ implies $S(N_i, \lambda') = S(N_1 \cap N_2, \lambda')$ for $i = 1, 2$. Claim

$$\mathcal{N}(\check{\zeta}'', S'', \lambda'', N_1 \cap N_2, U'') \subset \bigcap_{i=1}^{2} \mathcal{N}(\check{\zeta}_i, S_i, \lambda_i, N_i, U_i)\,.$$

For suppose $(\check{\zeta}', S', \lambda') \in \mathcal{N}(\check{\zeta}'', S'', \lambda'', N_1 \cap N_2, U'')$. Then $\pi_{N_1 \cap N_2}^{S'}\check{\zeta}' = \pi_{N_1\cap N_2}^{S''}\check{\zeta}''$, whence, for $i = 1, 2$,

$$\begin{aligned}
\pi_{N_i}^{S''}\check{\zeta}'' &= i_{N_i}^{N_1\cap N_2} \circ \pi_{N_1\cap N_2}^{S''}\check{\zeta}'' \\
&= i_{N_i}^{N_1\cap N_2} \circ \pi_{N_1\cap N_2}^{S'}\check{\zeta}' \\
&= \pi_{N_i}^{S'}\check{\zeta}'\,.
\end{aligned}$$

From this and because $(\check{\zeta}'', S'', \lambda'') \in \mathcal{N}(\check{\zeta}_i, S_i, \lambda_i, U_i)$ for $i = 1, 2$, it follows that also $(\check{\zeta}', S', \lambda') \in \mathcal{N}(\check{\zeta}_i, S_i, \lambda_i, U_i)$ for $i = 1, 2$, whence $\mathbf{B}_k$ is a basis.

Continuity of $\pi_{\check{\mathcal{H}}_k}$ at each point of $\check{\mathcal{H}}_k\mathcal{S}(\varphi; G'')$ is immediate from the inclusion $\pi_{\check{\mathcal{H}}_k}(\mathcal{N}(\check{\zeta}, S, \lambda, N, U)) \subset \sigma_N(U)$ holding for each $\mathcal{N}(\check{\zeta}, S, \lambda, N, U) \in \mathbf{B}_k$. □

Remark. Unless $\check{H}_*(S; G'') \simeq H_*(S; G'')$ for each $(S, \lambda) \in \mathcal{S}(\varphi)$ (for example, if S is always a manifold), the "projection" $\pi_{\check{\mathcal{H}}_k}$ need not be an open map. However, if the Čech and singular homologies are isomorphic at each point of $\mathcal{S}(\varphi)$, then it is an easy exercise to construct local sections of $\check{\mathcal{H}}_k\mathcal{S}(\varphi; G'')$ making $\pi_{\check{\mathcal{H}}_k}$ a local homeomorphism.

The next theorem is an algebraic statement of the continuity along a continuous path of isolated invariant sets of the intersection of the stable and unstable sets of the invariant sets on the path. For its statement, assume G and G' are R-modules. Its proof is given in §5.14.

3.7 THEOREM. *For β a path in $\mathcal{S}(\varphi)$, for $\boldsymbol{\alpha} \in \widetilde{H}_i\mathcal{C}(S_\beta(0); G)$, and for $\boldsymbol{\gamma} \in \widetilde{H}_j\mathcal{C}^*(S_\beta(0); G')$, set $k := i+j-m$ and define the path $\ell(\beta, \boldsymbol{\alpha}, \boldsymbol{\gamma})$ in $\check{\mathcal{H}}_k\mathcal{S}(\varphi; G\otimes G')$ by*

$$\ell(\beta, \boldsymbol{\alpha}, \boldsymbol{\gamma})(s) := \left(\mathcal{C}\,(\beta^s)_*\boldsymbol{\alpha} \mathbin{\widehat{\bullet}} \mathcal{C}^*(\beta^s)_*\boldsymbol{\gamma}\,, \beta(s)\right) \qquad \text{for } s \in [0,1].$$

Then relative to the fine topology on $\check{\mathcal{H}}_k\mathcal{S}(\varphi; G\otimes G')$, $\ell(\beta, \boldsymbol{\alpha}, \boldsymbol{\gamma})$ is a continuous lift of β with initial point $(\boldsymbol{\alpha} \mathbin{\widehat{\bullet}} \boldsymbol{\gamma}, \beta(0))$.

3.8 DEFINITION. (*A*) For G'' an R-module, if $(S, \lambda) \in \mathcal{S}(\varphi)$ and $\check{0}_k$ is the zero element of $\check{H}_k(S; G'')$, call $(\check{0}_k, S, \lambda)$ a *singular zero point of $\check{\mathcal{H}}_k\mathcal{S}(\varphi; G'')$* if, and only if, for each neighborhood $\mathcal{Z}$ of $(\check{0}_k, S, \lambda)$ there exists $(\check{\zeta}, S', \lambda') \in \mathcal{Z}$ with $\check{\zeta}$ not the zero element of $\check{H}_k(S'; G'')$. Call $(S, \lambda) \in \mathcal{S}(\varphi)$ a *G''-singular point of $\mathcal{S}(\varphi)$* if, and only if, for some $k \in \mathbf{N}$, $(\check{0}_k, S, \lambda)$ is a singular zero point of $\check{\mathcal{H}}_k\mathcal{S}(\varphi; G'')$.

(*B*) Call a G''-singular point $(S, \lambda) \in \mathcal{S}(\varphi)$ a *G''-bifurcation point of $\mathcal{S}(\varphi)$* if, and only if, for some N that λ-isolates S, for each open U satisfying $\lambda \in U \subset \Lambda(N)$, there exist $\lambda_1, \lambda_2 \in U \setminus \{\lambda\}$ so that

$$\check{H}_*(S_{\sigma_N}(\lambda_1); G'') \not\approx \check{H}_*(S_{\sigma_N}(\lambda_2); G'')\,.$$

The following example shows that, in general, a lift of β having the form $\ell(\beta,\boldsymbol{\alpha}, \boldsymbol{\gamma})$ with initial point not in the zero section of $\check{\mathcal{H}}_k\mathcal{S}(\varphi)$ can have a singular zero point in its image whose projection to $\mathcal{S}(\varphi)$ is a bifurcation point. Of course, as a consequence of Theorem 3.1, necessarily $k > 0$ where $\boldsymbol{\alpha} \mathbin{\widehat{\bullet}} \boldsymbol{\gamma} \in \check{H}_k(S_\beta(0))$.

3.9 EXAMPLE. Consider the one-parameter family of autonomous systems in $\mathbf{R}^3$ ($\lambda \in \Lambda := \mathbf{R}$)

$$\begin{aligned} \dot{u} &= u^2 - v^2 - w^2 - \lambda \\ \dot{v} &= -2uv + \min\{0, \lambda\}w + \max\{0, \lambda - u^2\} \\ \dot{w} &= -2uw - \min\{0, \lambda\}v \end{aligned} \tag{3.8$_\lambda$}$$

where the dots over u, v, and w indicate differentiation with respect $t \in \mathbf{R}$. This family of systems is discussed extensively in [K4, Example 5.1] in conjunction with the behavior for $0 < \varepsilon \ll 1$ of solutions to initial value problems for the system $\varepsilon\mathbf{u}' = \mathbf{F}(\mathbf{u}, x)$ where $\mathbf{u} := (u, v, w) \in \mathbf{R}^3$, differentiation is with respect to $x \in \mathbf{R}$, and $\mathbf{F}(\mathbf{u}, \lambda)$ is given by the right hand side of (3.8$_\lambda$). Here it is examined in terms of the peculiarities of the intersection number and class pairings it points out. *The reader is reminded that in this work when given a block B, its exit (entrance) set is B^+ (respectively, B^-) whereas in* [K4] *the opposite signs are used to designate exit and entrance sets.*

Let φ_λ be the flow in $\mathbf{R}^3$ generated by system (3.8_λ) so that φ_λ, $\lambda \in \mathbf{R}$ is a continuous family of flows in $\mathbf{R}^3$. For $\lambda < 0$ let $S(\lambda)$ be the circle locus $u = 0$, $v^2 + w^2 = -\lambda$, for $\lambda = 0$ set $S(0) := \{(0,0,0)\}$, and for $\lambda > 0$ let $S(\lambda)$ be the doubleton with elements $(\pm\sqrt{\lambda}, 0, 0)$. Then $S(\lambda)$ is the maximal isolated invariant set of system (3.8_λ) for each $\lambda \in \mathbf{R}$, and as discussed in [K4, Example 5.1], the section σ of $\pi_\varphi\colon \mathcal{S}(\varphi) \to \Lambda$ defined by $\sigma(\lambda) = (S(\lambda), \lambda)$ is continuous. For $\lambda < 0$, $S(\lambda)$ is a hyperbolic periodic orbit with two-dimensional stable and unstable manifolds, $S(0)$ is a degenerate critical point, and for $\lambda > 0$, $S(\lambda)$ consists of two critical points, one having two-dimensional unstable manifold and one-dimensional stable, the other having the dimensions of the invariant manifolds reversed. Typical phase portraits for the three cases including an isolating block are sketched in Figure 2. Call the blocks in Figures 2a, 2b and 2c, respectively $T(\lambda, \delta)$ (the case $\lambda < 0$), $B(\delta)$ (the case $\lambda = 0$), and $K(\lambda, \delta)$ (the case $\lambda > 0$) where δ is chosen to lie in $]0, \sqrt{-\lambda}[$ when $\lambda < 0$ and in $]0, \frac{1}{2}\sqrt{\lambda}[$ when $\lambda > 0$, and is an arbitrary positive number for $\lambda = 0$. $T(\lambda, \delta)$ is a solid torus of revolution (about the u-axis) with square cross-section of diagonal length 2δ centered on its point of intersection with the periodic orbit with one pair of opposite edges of the cross-section generating the exit set and the other pair the entrance set. Hence the exit set of $T(\lambda, \delta)$ consists of two disjoint annuli in the bounding torus as does the entrance set, the two sets of two annuli being complementary up to closure. $B(\delta)$ is a solid ball of radius δ centered at the origin with exit set consisting of a cap centered at the north pole (i.e., $u = \delta$, $v = w = 0$) and a band extending downward from the equator while the entrance set consists of the closure of the complement relative to the bounding sphere of the exit set and consists of a cap centered at the south pole together with a band extending upward from the equator. $K(\lambda, \delta)$ is the union of two disjoint isolating blocks each a bent solid cylinder of square cross-section with diagonal of length 2δ. Let $K_\pm(\lambda, \delta)$ be the cylinder containing $(\pm\sqrt{\lambda}, 0, 0)$. Then the exit set of $K(\lambda, \delta)$ consists of the top and bottom of $K_+(\lambda, \delta)$ and the lateral surface of $K_-(\lambda, \delta)$ while the entrance set consists of the top and bottom of $K_-(\lambda, \delta)$ and the lateral surface of $K_+(\lambda, \delta)$. The reader is referred to [K4, Example 5.1] for a precise analytic description of the blocks and their exit and entrance sets.

It is easy to compute that the index spaces $K(\lambda, \delta)/K(\lambda, \delta)^\pm$ have the homotopy type of $S^1 \vee S^2$; hence by invariance of the Conley homotopy index under continuation along any continuous curve in $\mathcal{S}(\varphi)$ the same is true of $B(\delta)/B(\delta)^\pm$ and $T(\lambda, \delta)/T(\lambda, \delta)^\pm$. It follows that $\widetilde{H}_j\mathcal{C}(S(\lambda); \mathbf{Z})$ and $\widetilde{H}_j\mathcal{C}^*(S(\lambda); \mathbf{Z})$ are isomorphic to $\mathbf{Z}$ when $j = 1, 2$ but are otherwise the trivial group.

A compatible choice of generators for the non-zero groups is made as follows where we continue to assume $\mathbf{Z}$ coefficients for the rest of this example. Choose a generator $\boldsymbol{\alpha}_{j,-1}$ of $\widetilde{H}_j\mathcal{C}(S(-1))$ and a generator $\boldsymbol{\gamma}_{j,-1}$ of $\widetilde{H}_j\mathcal{C}^*(S(-1))$ for $j = 1, 2$. Set $\lambda_\beta(t) := 2t - 1$ for $t \in [0, 1]$, define the path β in $\mathcal{S}(\varphi)$ by $\beta(t) := \sigma \circ \lambda_\beta(t)$, and for $0 < s \leq 1$ define the auxiliary path β^s in $\mathcal{S}(\varphi)$ as in Remark 3.1.1. For each $\lambda \in [-1, 1]$ set $s_\lambda := \frac{1}{2}(\lambda + 1)$ and for $j = 1, 2$ set

$$\boldsymbol{\alpha}_{j\,\lambda} := \mathcal{C}\,(\beta^{s_\lambda})_*\boldsymbol{\alpha}_{j,-1}\,, \qquad \boldsymbol{\gamma}_{j\,\lambda} := \mathcal{C}^*(\beta^{s_\lambda})_*\boldsymbol{\gamma}_{j,-1}\,.$$

Then $\boldsymbol{\alpha}_{j\,\lambda}$ generates $\widetilde{H}_j\mathcal{C}(S(\lambda))$ and $\boldsymbol{\gamma}_{j\,\lambda}$ generates $\widetilde{H}_j\mathcal{C}^*(S(\lambda))$ $(j = 1, 2)$. In [K4] there is a lengthy discussion on the various possible choices of geometric representatives for $\boldsymbol{\alpha}_{j\,\lambda}$ and an analogous discussion applies to the reverse time indices

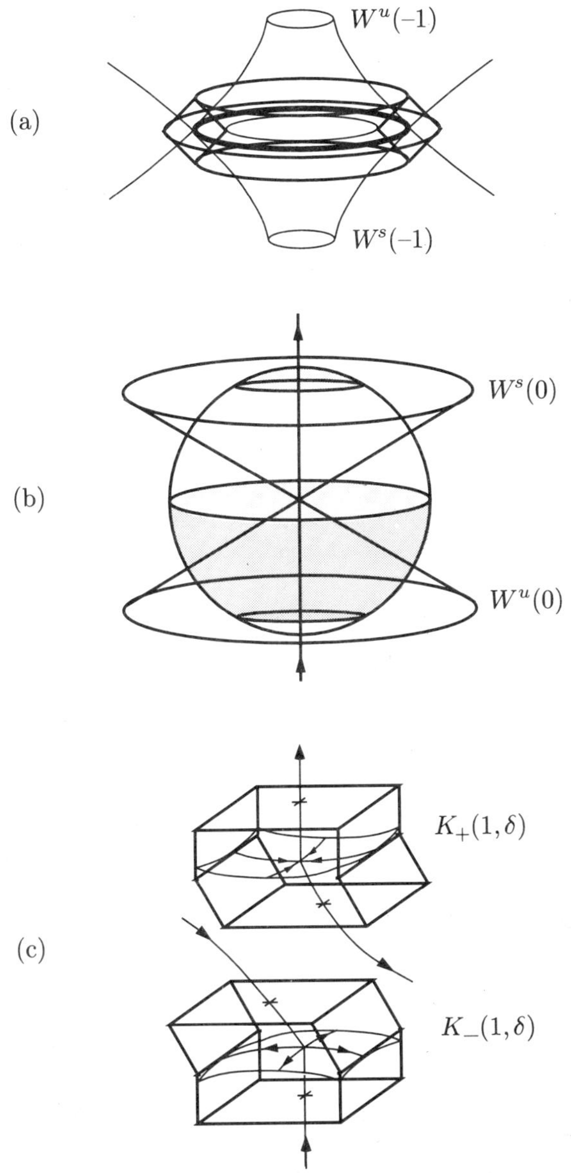

FIGURE 2. (*a*) The block $T(-1,\delta)$ and local stable and unstable manifolds of $S(-1)$. (*b*) The block $B(\delta)$ (shaded portions of the bounding sphere indicate the exit set) and local stable and unstable sets of $S(0)$. Note that the positive (negative) u-axis is part of the unstable (stable) set of $S(0)$. (*c*) The block $K(1,\delta) := K_-(1,\delta) \cup K_+(1,\delta)$ and local stable and unstable manifolds of $S(1)$.

and the choice of geometric representatives for $\gamma_{j\,\lambda}$. In particular, for $\lambda \in [-1, 0[$, the local unstable (stable) manifold of $S(\lambda)$ within $T(\lambda, \delta)$ is an annulus with one bounding circle in each annular component of the exit (entrance) set and is a geometric representative of $\boldsymbol{\alpha}_{2\,\lambda}$ (respectively, $\gamma_{2\,\lambda}$). Also, the annulus that is the local unstable or stable manifold within $T(\lambda, \delta)$ can be fibered by arcs with each arc having an endpoint in each of the bounding circles of the annulus, and such an arc in the local unstable (stable) manifold represents $\boldsymbol{\alpha}_{1\,\lambda}$ (respectively, $\gamma_{1\,\lambda}$).

Because, when $-1 \leq \lambda < 0$, any local stable and unstable manifolds of $S(\lambda)$ intersect transversely in the periodic orbit $S(\lambda)$, it is immediate from Theorem 6.5 that $\boldsymbol{\alpha}_{2\,\lambda} \,\widehat{\bullet}\, \gamma_{2\,\lambda}$ generates $\check{H}_1(S(\lambda)) \simeq \mathbf{Z}$ and also that

$$\text{(3.9)} \qquad {}^{\#}(\boldsymbol{\alpha}_{1\,\lambda} \,\widehat{\bullet}\, \gamma_{2\,\lambda}) = \pm 1 \quad \text{and} \quad {}^{\#}(\boldsymbol{\alpha}_{2\,\lambda} \,\widehat{\bullet}\, \gamma_{1\,\lambda}) = \pm 1\,.$$

Further, by the invariance of intersection numbers under continuation, it follows that (3.9) holds for all $\lambda \in [-1, 1]$. This is interesting as regards the example under discussion because $S(\lambda)$ undergoes bifurcation as λ increases through zero from a degenerate critical point to two non-degenerate ones (here, "non-degenerate" means that the local stable and unstable sets are topological cells that are topologically transverse) and for $\lambda > 0$ the two intersection numbers in (3.9) are associated to different critical points. Specifically, from the analysis of the family of systems (3.8_λ) undertaken in [K4, Example 5.1] and from a similar analysis of these systems with time reversed, it follows for $\lambda > 0$ and $0 < \delta \leq \frac{1}{2}\sqrt{\lambda}$ that $\boldsymbol{\alpha}_{2\,\lambda} \,\widehat{\bullet}\, \gamma_{1\,\lambda}$ is represented by the intersection of the local unstable and stable manifolds of $(-\sqrt{\lambda}, 0, 0)$ within $K_-(\lambda, \delta)$ while $\boldsymbol{\alpha}_{1\,\lambda} \,\widehat{\bullet}\, \gamma_{2\,\lambda}$ is represented by the intersection of the local stable and unstable manifolds of $(\sqrt{\lambda}, 0, 0)$ within $K_+(\lambda, \delta)$.

As $S(0)$ is a single point and $S(\lambda)$ for $\lambda > 0$ is two points, $\check{H}_1(S(\lambda)) = 0$ for $\lambda \geq 0$. Thus

$$\mathcal{C}\,(\beta^{s\lambda})_* \boldsymbol{\alpha}_{2,-1} \,\widehat{\bullet}\, \mathcal{C}^*(\beta^{s\lambda})_* \gamma_{2,-1} = \check{0}_1 \in \check{H}_1(S(\lambda)) \qquad \text{for } \lambda \geq 0.$$

Hence, $(\check{0}_1, S(0), 0)$ is a $\mathbf{Z}$-singular zero point in the image of $\ell(\beta, \boldsymbol{\alpha}_{2,-1}, \gamma_{2,-1})$, and also $(S(0), 0)$ is a $\mathbf{Z}$-bifurcation point of $\mathcal{S}(\varphi)$.

Definition 3.8 and Example 3.9 raise the question as to what are sufficient conditions on the manifold M, the family of flows φ, and the path $\beta \in \mathcal{S}(\varphi)$ to ensure that β has no singular points or bifurcation points in its image. For example if M and φ are smooth and $S_\beta(\lambda)$ is a non-degenerate hyperbolic set for $\lambda \in \Lambda$ a smooth manifold, then transversality theory yields that in general there will be no bifurcation points on β. A more difficult question is under what circumstances are bifurcation points unavoidable; i.e., if for a given family of flows φ and path β into $\mathcal{S}(\varphi)$ there is a bifurcation point on β, then is it the case that whenever φ is embedded in a larger family of flows Φ and β is regarded as a path in $\mathcal{S}(\Phi)$, then any path β' sufficiently close to β also has a bifurcation point in its image.

Neither of the questions just raised shall be explored here. However, we would like to point out that homotopy theoretic methods and obstruction theory probably have a role to play in the answers. To see this let us note that in some circumstances, it is possible to put a different topology on $\check{\mathcal{H}}_k\mathcal{S}(\varphi)$ so that $\ell(\beta, \boldsymbol{\alpha}, \gamma)$ is a continuous lift of β implies its first component is always a zero class or never a zero class.

For example, with coefficients in the PID R, it is possible to define an alternate topology on $\check{\mathcal{H}}_k\mathcal{S}(\varphi; R)$ if we assume that for each $(S, \lambda) \in \mathcal{S}(\varphi)$, S has arbitrarily small λ-isolating neighborhoods N satisfying $H_k(N; R)$ is a free R-module. Call this assumption Hypothesis $\tau_k(R)$.

The point of Hypothesis $\tau_k(R)$ is that for each λ-isolated invariant set S it ensures the existence of arbitrarily small λ-isolating neighborhoods N of S with the property that for $0 \neq \zeta \in H_k(N; R)$, there exists $\psi_\zeta \in \mathrm{Hom}(H_k(N; R), R)$ satisfying $\psi_\zeta(\zeta) \neq 0$—just choose ψ_ζ from a suitable dual basis. The Universal Coefficient Theorem for Cohomology then yields the existence of $z^{\zeta\, N} \in H^k(N; R)$ so that $\langle z^{\zeta\, N}, \zeta \rangle = 1$.

A sub-basis for a topology on $\check{\mathcal{H}}_k\mathcal{S}(\varphi; R)$ can then be defined as follows. For each compact $N \subset M$, for each $0 \neq z \in H^k(N; R)$, and for each open $V \subset \Lambda(N)$, define $\langle z, N, V \rangle$ to be the set of triples $(\check{\zeta}, S, \lambda)$ satisfying (i) $\lambda \in V$ and $(S, \lambda) \in \mathcal{S}(\varphi)$, (ii) $S = S(N, \lambda)$, (iii) $\check{\zeta} \in \check{H}_k(S; R)$ and $\langle z, \zeta_N \rangle \neq 0$ where ζ_N is the image of $\check{\zeta}$ under the canonical projection of $\check{H}_*(S; R)$ to $H_*(N; R)$. Also for the zero element $0^{(k)} \in H^k(N; R)$ define $\langle 0^{(k)}, N, V \rangle$ to be the set of triples $(\check{0}, S, \lambda)$ where λ and S satisfy (i) and (ii) and $\check{0}$ is the zero element of $\check{H}_k(S; R)$. To see that the collection $\check{\mathcal{B}}_k$ of sets of the form $\langle z, N, V \rangle$ covers $\check{\mathcal{H}}_k\mathcal{S}(\varphi; R)$ and can therefore be taken as a sub-basis for a topology make the following observation: If $(S, \lambda) \in \mathcal{S}(\varphi)$ and $\check{0}_k \neq \check{\zeta} \in \check{H}_k(S; R)$, then $\zeta_U \neq 0$ for some neighborhood U of S, whence necessarily $\zeta_{U'} \neq 0$ for each neighborhood U' of S with $U' \subset U$. In particular, by Hypothesis (τ_k), there exists a λ- isolating neighborhood N of S with $N \subset U$ so that $\zeta_N \neq 0$ and $H_k(N; R)$ is a free R-module. Hence, Hypothesis (τ_k) implies that $(\check{\zeta}, S, \lambda) \in \langle z^{\zeta\, N}, N, \Lambda(N) \rangle$.

Call the topology on $\check{\mathcal{H}}_k\mathcal{S}(\varphi; R)$ generated by the just described sub-basis the weak topology. Each sub-basic open set with non-zero elements is open in the fine topology since $\mathcal{N}(\check{\zeta}, S, \lambda, N, V) \subset \langle z, N, V \rangle$ if $(\check{\zeta}, S, \lambda) \in \langle z, N, V \rangle$ and $\check{\zeta} \neq 0$.

By fiat, $\check{\mathcal{H}}_k\mathcal{S}(\varphi; R)$ has the zero section as one component relative to this weak topology, whence with the weak topology it is impossible to continue a non-zero class to a zero class. However, this topology is very weak—it is not even Hausdorff on stalks of $\pi_{\check{\mathcal{H}}_k}$ although this is easily remedied if R is an ordered ring or admits a non-trivial absolute value—and it does not seem readily possible to prove useful results of the type that would give sufficient conditions on when a path β in $\mathcal{S}(\varphi)$ has a continuous lift to $\check{\mathcal{H}}_*\mathcal{S}(\varphi)$ with non-zero initial point. From the presentation of this alternate topology on $\check{\mathcal{H}}_k\mathcal{S}(\varphi; R)$, the reader might surmise that it would be better to deal directly with cohomology classes. That is, rather than examining continuations of $\boldsymbol{\alpha} \mathbin{\widehat{\bullet}} \boldsymbol{\gamma}$, instead examine continuations of $x \smile y$ where x and y are Poincaré duals of $\boldsymbol{\alpha}$ and $\boldsymbol{\gamma}$, respectively. As just formulated this does not quite work, although a modification obtained by composing the cup product at the isolating block level with a Lefschetz duality isomorphism on cohomology (assuming field coefficients) can be made to work. Essentially, that is the viewpoint taken by Montgomery in §§6–7 of [M]. In particular, the reader should compare Example 3.9 above with [M, Example 7.8]. However, in many examples it seems easier to directly compute intersection numbers and intersection pairings rather than cup products.

Duality of Conley indices, its relation to $\mathfrak{L}$, and by implication the continuability of cup products are examined in Chapter 9.

CHAPTER 4

CONSTRUCTION OF BILINEAR PAIRINGS ON CONLEY INDICES

In this section S denotes an isolated invariant set of a flow in M. The main goal is to give proofs of Theorems 2.4 and 2.11. The latter yields existence of the intersection class pairing $\mathfrak{L}$ on $\widetilde{H}_*\mathcal{C}(S) \otimes \widetilde{H}_*\mathcal{C}^*(S)$ into $\check{H}_*(S)$; the former yields existence of the homology intersection number pairing L on $\left[\widetilde{H}_*\mathcal{C}(S) \otimes \widetilde{H}_*\mathcal{C}^*(S)\right]_m$ in the special case $M := \mathbf{R}^m$. It is convenient in proving these results to obtain Theorem 2.4 as a simple corollary of a generalization, namely Theorem 4.7, showing that whenever there exists a functor on $\mathcal{AIP}(S)$ that assigns to each admissible pair of nested index pairs a bilinear pairing on the product of R-modules functorially produced from the index pairs in the admissible pair, then there is an associated induced pairing on the tensor product of the R-modules functorially produced in the manner of Definition 1.5(B) from the forward and reverse time Conley indices. In Theorem 4.7 all pairings take values in the same R-module, but the result can be extended to allow the pairings assigned to admissible pairs to take values in an inverse (direct) system of R-modules over the set of neighborhoods of the isolated invariant set directed by containment; however, the induced pairing then has its values in the inverse (direct) limit of the inverse (direct) system. This generalization is Corollary 4.8, and Theorem 2.11 follows straightforwardly from it.

A. The Existence of Admissible Pairs of Index Pairs. The aim of the next several definitions and propositions is to describe the class of regular index pairs and the class of index pairs with thick exit set. The former plays an important role in the constructions that make up the proofs of Theorems 2.4 and 2.11, the latter in the proof of Theorem 3.2. In particular, we show that an index pair in either class has the property that the quotient map from the pair to its index space induces an isomorphism on homology for any homology theory. It is not known to the author as of this writing if there exists an index pair relative to some flow that does not have this property. However, as noted in Remark 4.1.1 below, a slight enlargement of the exit set of any index pair yields a regular index pair that necessarily has the property. The definitions of these classes and of some auxiliary constructions are made relative to the forward time flow, but of course can be given relative to any continuous flow in M; in particular for the reverse time flow.

4.1 DEFINITION. (A) A *nested index pair* (L_1, L_0) for the forward time flow *is regular* if, and only if, the map $\tau_{(L_1,L_0)}\colon L_1 \to [0,\infty]$ is continuous where

$$\tau_{(L_1,L_0)}(\mathbf{u}) := \begin{cases} \sup\{t \geq 0 : \mathbf{u}\cdot[0,t] \cap L_0 = \emptyset\} & \text{if } \mathbf{u} \in L_1 \setminus L_0; \\ 0 & \text{otherwise.} \end{cases}$$

The notion of regular index pair is introduced by Salamon in [Sl] and is the immediate generalization of the pair formed by an isolating block and its exit set.

(B) A nested index pair (L_1, L_0) for S has *thick exit set* if, and only if, there exists $K_0 \subset \mathrm{int}_{L_1}(L_0)$ so that (L_1, K_0) is a nested index pair for S. The notion of an index pair with thick exit set was first defined in [K4] and is needed, for example, in the proof of invariance of the intersection number pairing on Conley indices under continuation of the isolated invariant set when the ambient manifold is at best C^0.

(C) To prove that the quotient map from an index pair with thick exit set to its index space induces isomorphisms on homology and for other important uses below, let us introduce auxiliary index pairs that can be constructed from a given pair. Accordingly, for $t \geq 0$, for $\langle N_1, N_0 \rangle$ an index pair for S (not necessarily nested) in the forward time direction, define

$$N_1^t := \{\mathbf{u} \in N_1 : \mathbf{u} \cdot [-t, 0] \subset N_1\},$$

and define

$$N_0^{-t} := N_0 \cup \{\mathbf{u} \in N_1 : \exists s \in [0,t],\ \mathbf{u} \cdot [0,s] \subset N_1 \text{ and } \mathbf{u} \cdot s \in N_0\}.$$

Then it follows, but not immediately, from [K1, Proposition 4.9] that given any index pair $\langle N_1, N_0 \rangle$ for S, for any $r, s \geq 0$, $(N_1^r, \cap N_0^{-s})$ is a nested index pair for S. The appeal to the result of [K1] just cited is not immediate because the notion of index pair used in the current work is much broader than that used in [K1]. Specifically, all index pairs in [K1] are taken relative to some isolating neighborhood; i.e., an index pair $\langle K_1, K_0 \rangle$ is an *index pair relative to an isolating neighborhood* N if, and only if, each member of the index pair is positively invariant relative to N, i.e., $\mathbf{u} \in K_i$ and $\mathbf{u} \cdot [0,t] \subset N$ imply $\mathbf{u} \cdot [0,t] \subset K_i$ for $i = 0, 1$. Unfortunately there are a number of instances below where an appeal to some result of [K1] is made where a similar problem arises. Fortunately, it is always possible to reduce the situation to one in which the results of [K1] apply directly and always in the same manner. This reduction will now be carried out to prove that $(N_1^r, \cap N_0^{-s})$ is indeed a nested index pair for S for any $r, s \geq 0$; however, all other such reductions will be left to the reader without further comment.

To proceed with the reduction, set $\bar{N}_1 := \mathrm{cl}\,(N_1 \setminus N_0)$ and $\bar{N}_0 := N_0 \cap \bar{N}_1$. It is easily verified that $(\bar{N}_1, \bar{N}_0)$ is an index pair relative to the isolating neighborhood $\bar{N}_1$, whence the results of [K1] apply. In particular, Proposition 4.9 of [K1] yields that $(\bar{N}_1^r, \cap \bar{N}_0^{-s})$ is a nested index pair for S for each $r, s \geq 0$. Then, because $N_1^r = \bar{N}_1^r \cup (N_1^r \cap N_0)$ and because $N_0^{-s} = \bar{N}_0^{-s} \cup N_0$, it is easy to verify that $(N_1^r, \cap N_0^{-s})$ is a nested index pair for S.

Further, for $s > 0$ sufficiently large, $(N_1^r, \cap N_0^{-s})$ has thick exit set. For it follows from [K1, Proposition 4.9] that for each $r \geq 0$ and for $s > 0$ sufficiently large,

$$N_0 \cap N_1^r \subset \mathrm{int}_{N_1^r}(N_0^{-s} \cap N_1^r). \tag{4.1}$$

4.1.1 Remark. A key observation in the development of the intersection pairing is that given any nested index pair (N_1, N_0) for S and a neighborhood W of N_0, then there exists a compact N_0' so that $W \supset N_0' \supset N_0$ and (N_1, N_0') is a regular index pair for S. For it is proved in [Sl, Lemma 5.3 and Remark 5.4] that given any nested index pair (N_1, N_0) for S, there exists a continuous Lyapunov function $g\colon N_1 \to [0,1]$ so that where $N_\varepsilon := g^{-1}([0,\varepsilon])$, for $0 < \varepsilon < 1$, (N_1, N_ε) is a regular index pair for S and $N_0 = g^{-1}(0) \subset N_\varepsilon$. Then, with $V := \text{int}(W)$, the intersection of the nested family of compact sets $\{N_\varepsilon \setminus V : 0 < \varepsilon < 1\}$ equals $N_0 \setminus V = \emptyset$, whence $N_\varepsilon \subset V$ for all sufficiently small $\varepsilon > 0$.

4.1.2 Remark. A second important observation in the development of the intersection pairing on Conley indices is that both the class of regular index pairs and the class of index pairs with thick exit set are closed under the formation of the index pair $(N_1^r, \cap N_0^{-s})$ from the nested pair (N_1, N_0). Regularity propagates because $\tau_{(N_1^r, \cap N_0^{-s})} = \max\{0, -s + \tau_{(N_1,N_0)} | N_1^r\}$ and thick exit sets because $K \subset \text{int}_{N_1}(N_0)$ implies $K \cap N_1^r \subset \text{int}_{N_1^r}(N_1^r \cap N_0^{-s})$.

4.2 Proposition. *If an index pair is either regular or has thick exit set, then the quotient map from the index pair to its index space induces isomorphisms on homology for any homology theory.*

Proof. Let us work for the sake of argument with a nested index pair (L_1, L_0) with respect to the forward time flow. Note that the quotient map $l^+\colon (L_1, L_0) \to (L_1/L_0, [L_0])$ has factorization

$$(L_1, L_0) \overset{e}{\hookrightarrow} (L_1 \cup CL_0, CL_0) \overset{k}{\to} (L_1 \cup CL_0/CL_0, [CL_0]) \equiv (L_1/L_0, [L_0])$$

where CL_0 is the unreduced cone on L_0 and $L_1 \cup CL_0$ is the mapping cone of the inclusion $L_0 \subset L_1$. Because the excision axiom implies e induces isomorphisms on homology, the proposition follows if k too induces isomorphisms, and as CL_0 is contractible, the Five Lemma applied to k and the sequences of the pairs $(L_1 \cup CL_0, CL_0)$ and $(L_1 \cup CL_0/CL_0, [CL_0])$ shows it sufficient to prove that the collapse map

$$k'\colon L_1 \cup CL_0 \to L_1 \cup CL_0/CL_0$$

is a homotopy equivalence as a map of spaces without basepoint.

According to [K2, Lemma 3.1], k' will be a homotopy equivalence if we can exhibit (i) a deformation D of L_1 whose restriction to $U \times [0,1]$ for some closed neighborhood U of L_0 is a weak deformation retraction of U into L_0 and (ii) a continuous function $\varphi\colon L_1\colon \to [0,1]$ satisfying $L_0 \subset \varphi^{-1}(1)$ and $L_1 \setminus U \subset \varphi^{-1}(0)$. In the present context L_1 is a normal space so that a Urysohn function φ satisfying (ii) can always be found. If (L_1, L_0) is a regular index pair, then choose any $c > 0$, take $U := \tau^{-1}_{(L_1,L_0)}([0,c])$ and define $D(\mathbf{u}, s) := \mathbf{u} \cdot s \min\{c, \tau_{(L_1,L_0)}(\mathbf{u})\}$. If, however, (L_1, L_0) is not regular but has thick exit set, then by Remark 4.1.1 choose $K_0 \subset \text{int}_{L_1}(L_0)$ with (L_1, K_0) an index pair which without loss of generality can be assumed regular; else choose K_0' with $K_0 \subset K_0' \subset \text{int}_{L_1}(L_0)$ and (L_1, K_0') regular. As was done to get (4.1) above, choose $t > 0$ so that $L_0 \subset \text{int}_{L_1}(L_0^{-t} \cap L_1)$, take $U := L_0^{-t} \cap L_1$, set $c := \max\{\tau_{(L_1,K_0)}(\mathbf{u}) : \mathbf{u} \in U\}$, and define $D(\mathbf{u}, s) := \mathbf{u} \cdot s \min\{c, \tau_{(L_1,K_0)}(\mathbf{u})\}$.

Note that when (L_1, L_0) is regular, D is actually a strong deformation retraction of U onto L_0 and we could instead invoke [Sp, Theorem 4.8.9] to get l^+ inducing isomorphisms on homology. □

4.3 DEFINITION. When M is C^r $(r \geq 1)$ and the flow in M is generated by a continuous vectorfield X, call an isolating block B for S a C^r *isolating block with corners* if B is a topological manifold with boundary and there exist C^r submanifolds Σ^+ and Σ^- of M which together with B have the following additional properties:

(1) $\Sigma^\pm$ if non-void is of codimension one in M;
(2) X is transverse to $\Sigma^\pm$;
(3) $B^\pm$ is a codimension zero C^r submanifold with boundary of $\Sigma^\pm$;
(4) $\partial B = B^+ \cup B^-$ and $B^+ \cap B^- = \partial B^\pm$.

The existence of isolating blocks with corners is proved in [WY]. The submanifold $B^+ \cap B^-$ is the set of "corners" of B and is sometimes referred to as the tangency set of B since the orbit through a point $\mathbf{u} \in B^+ \cap B^-$ locally lies exterior to B intersecting it only in $\mathbf{u}$. As just defined a block with corners need not be a C^r manifold with corners since it is not assumed that Σ^+ and Σ^- can be simultaneously linearized at points of $B^+ \cap B^-$. Such simultaneous linearization can be carried out, however, if Σ^+ and Σ^- are transverse in which case B will be a C^r manifold with corners.

It is convenient in what follows to use the notation AIP(regular) to denote the class of pairs of index pairs that are AIP for S wherein both pairs are regular index pairs. Similarly, the notation AIP(thick) is used below to denote the class of AIP pairs for S wherein both pairs have thick exit set.

4.4 PROPOSITION. *There exist AIP(regular) and AIP(thick) pairs of index pairs for S. If U is any neighborhood of S, there exists N AIP(regular) for S so that $N_1 \cap N_1^* \subset U$. If M is C^r, $r \geq 1$, and the flow is generated by a vectorfield of class C^s, $0 \leq s \leq r-1$, then N_1, N_1^*, and $N_1 \cap N_1^*$ can be chosen to be C^r isolating blocks with corners.*

Proof. Let U be a neighborhood of S and choose $K \subset U$ so that K isolates S. The construction of isolating blocks in [Ch] although carried out for a complete flow on a compact metric space works equally well in the present setting and yields a block B for S with $B \subset \mathrm{int}_M(K)$. For any $c, d \geq 0$ define

$$B_{c\,d} := \tau_B^{-1}([c,\infty]) \cap \tau_B^{*\,-1}([d,\infty]) .$$

Then $B_{c\,d}$ is an isolating block for S and is called a squeeze of B. Fix $d > 0$ and for $0 \leq c < d$ set

$$(P_{1\,c}, P_{0\,c}) := \left(B_{0\,d}, \tau_{B_{0\,d}}^{-1}([0,c])\right), \quad (P_{1\,c}^*, P_{0\,c}^*) := \left(B_{d\,0}, \tau_{B_{d\,0}}^{*\,-1}([0,c])\right)$$

and define $P_c := \left((P_{1\,c}, P_{0\,c}), (P_{1\,c}^*, P_{0\,c}^*)\right)$. Note that $B_{0\,d}$ (respectively, $B_{d\,0}$) is a squeeze of B along its entrance (respectively, exit) set. It is easily verified that if $0 \leq c < d$, then P_c is an AIP(regular) pair of index pairs for S and is also AIP(thick) when $0 < c < d$ because

$$P_{0\,0} \subset \mathrm{int}_{B_{0\,d}}(P_{0\,c}) \quad \text{and} \quad P_{0\,0}^* \subset \mathrm{int}_{B_{d\,0}}(P_{0\,c}^*) .$$

Set $N := P_0$. Then N is AIP(regular) for S and $N_1 \cap N_1^* = B_{d,d} \subset U$. The isolating block B as constructed in [Ch] has B^+ and B^-disjoint, whence if d is chosen sufficiently small, $N_0 = B^+$ and $N_0^* = B^-$. Also as N_1, N_1^* and $N_1 \cap N_1^*$ are squeezes of a block, they are themselves blocks.

When M is C^r, $r \geq 1$, and the flow is generated by a continuous vectorfield X, then a result of W. Wilson and and J. Yorke (see Theorem 2.1 and the Remark at the top of p. 114 of [WY]) shows that for some open neighborhood Ω of S with $\Omega \subset K$, there exist continuous Lyapunov functions $g_+, g_- : \Omega \to \mathbf{R}^+$ (i.e., the Lie derivative $Xg_\pm$ exists and is continuous on Ω) with the properties

(1) $A^\pm(\Omega) = g_\pm^{-1}(0)$;
(2) $Xg_+ > 0$ on $\Omega \setminus A^+(\Omega)$ and $Xg_- < 0$ on $\Omega \setminus A^-(\Omega)$;
(3) $g_\pm$ is C^r on $\Omega \setminus A^\pm(\Omega)$.

The prototypical examples of g_+ and g_- arise from the system $\dot{x} = x$, $\dot{y} = -y$ for $x \in \mathbf{R}^n$, $y \in \mathbf{R}^p$ which has $S = \{0\}$ as isolated invariant set: Take $g_+(x,y) = \|x\|^2$ and $g_-(x,y) = \|y\|^2$ where the norms are the Euclidean norms. The reader will probably find it helpful to keep this example in mind while reading the following construction of blocks.

The necessary C^r isolating blocks with corners can be constructed from g_+ and g_- via a minor variation on the method of [WY, Theorem 2.4]. In particular, the blocks finally produced will be selected from the family of neighborhoods of S whose members have the form

$$C_{pq} := g_+^{-1}([0,q]) \cap g_-^{-1}([0,p]) \qquad \text{for } p, q > 0.$$

To eventually obtain restrictions on p and q ensuring that C_{pq} is compact and isolates S, choose a relatively compact, open neighborhood W of S with $\mathrm{cl}_M(W) \subset \Omega$. Then, since $g_\pm \geq 0$, condition (1) implies that some member of the nested family of compact sets

$$\{(g_+ + g_-)^{-1}([0,(i+1)^{-1}]) \cap K \setminus W\}_{i \in \mathbf{N}}$$

must be void. Therefore, for some $d > 0$, $(g_+ + g_-)^{-1}([0,2d]) \subset \mathrm{cl}_M(W)$, whence $(g_+ + g_-)^{-1}([0,2d])$ is compact and isolates S. It follows that C_{pq} is compact and isolates S whenever $p, q \in \,]0,d[$. Further, for such p and q, note that the topological boundary of C_{pq} (i.e., its frontier relative to M) satisfies

$$\partial_M(C_{pq}) = C_{pq} \cap (g_+^{-1}(q) \cup g_-^{-1}(p)).$$

Observe too that condition (2) implies that

$$C_{pq}^+ := (\tau_{C_{pq}})^{-1}(0) = g_+^{-1}(q) \cap C_{pq} \quad \text{and} \quad C_{pq}^- := (\tau_{C_{pq}}^*)^{-1}(0) = g_-^{-1}(p) \cap C_{pq},$$

whence the exit and entrance sets of C_{pq} are closed relative to M. As observed in Definition 1.4(C) this implies that C_{pq} is a block for S.

Set $\Omega^\pm := \Omega \setminus A^\pm(\Omega)$, set $\Omega^{\mathbf{R}} := \Omega^+ \cap \Omega^-$, set $\Omega_0 := \Omega^+ \cup \Omega^- = \Omega \setminus S$ and observe that conditions (2) and (3) on $g_\pm$ imply that each positive number is a regular value of $g_\pm|\Omega^\pm$. Define

$$\Sigma_\varepsilon^\pm := (g_\pm|\Omega^\pm)^{-1}(\varepsilon) = g_\pm^{-1}(\varepsilon) \qquad \text{for } \varepsilon > 0.$$

For $p, q \in \,]0, d[$, it follows that C_{pq} will be a C^r isolating block with corners for S if $C^+_{pq} \cap C^-_{pq}$ has the structure of a C^r submanifold in the topology inherited from M.

To obtain such structure, first, observe that conditions (2) and (3) on $g_\pm$ imply that each $c \in \mathbf{R}$ is a regular value of $(g_+ - g_-)|\Omega^{\mathbf{R}}$ since $X(g_+ - g_-) > 0$ on $\Omega_0 \supset \Omega^{\mathbf{R}}$. Hence

$$h_c := \left((g_+ - g_-)|\Omega^{\mathbf{R}}\right)^{-1}(c)$$

is either void or a codimension one submanifold of $\Omega^{\mathbf{R}}$; hence of M.

The blocks N_1 and N_1^* can now be defined as follows. Choose $c \in \,]0, d[$; choose $\varepsilon_1 \in \,]0, d - c[$ so that ε_1 is a regular value of $g_-|h_c$, $g_-|h_0$, and $g_+|h_{-c}$; set $\varepsilon_2 := c + \varepsilon_1$. Set $N_1 := C_{\varepsilon_1 \varepsilon_2}$, set $N_1^* := C_{\varepsilon_2 \varepsilon_1}$, note that $N_1 \cap N_1^* = C_{\varepsilon_1 \varepsilon_1}$, and observe that

$$\begin{aligned} N_1^+ \cap N_1^- &= (g_-|h_c)^{-1}(\varepsilon_1)\,, \\ N_1^{*+} \cap N_1^{*-} &= (g_+|h_{-c})^{-1}(\varepsilon_1)\,, \\ (N_1 \cap N_1^*)^+ \cap (N_1 \cap N_1^*)^- &= (g_-|h_0)^{-1}(\varepsilon_1)\,. \end{aligned}$$

Hence, by the choice of ε_1 as regular value, each of these three intersections of exit and entrance set is a C^r manifold in M that is either void or of codimension two in M. Finally, observe that the pair of regular index pairs $\left((N_1, N_1^+), (N_1^*, N_1^{*-})\right)$ satisfies the off-diagonal condition since $\Sigma^\pm_{\varepsilon_2}$ is disjoint from $g_\pm^{-1}([0, \varepsilon_1])$ by virtue of the inequality $\varepsilon_1 < \varepsilon_2$. □

B. Functorially Produced Pairings on the Conley Indices. The next lemma proves the existence of several AIP(regular) pairs of index pairs that mediate between two such pairs via inclusion maps and is the basis for showing that a functor on $\mathcal{AIP}(S)$ with values that are bilinear pairings on the product of (co)homology modules of the index pairs in an admissible pair induces a pairing on the tensor product of (co)homology modules of the forward and reverse time Conley indices. Prior to the statement of the lemma we need the following.

4.5 Definition. (*A*) $F \subset \operatorname{ob}(\mathcal{AIP}(S))$ is a *pre-cofinal family for* S if, and only if, for each $U \in \mathcal{N}(S)$, there exists $P \in F$ with $P_1 \cap P_1^* \subset U$. Proposition 4.4 implies that there exist pre-cofinal families for S, and in particular, $\operatorname{ob}(\mathcal{AIP}(S))$ is one such family.

(*B*) A pre-cofinal family F for S is a *saturated pre-cofinal family* if, and only if, it satisfies the following condition:

(†) $Q \in \operatorname{ob}(\mathcal{AIP}(S))$ and $Q_1 \cap Q_1^* \subset P_1 \cap P_1^*$ for some $P \in F$ implies $Q \in F$.

Evidently, if F is pre-cofinal for S, then the set of $Q \in \operatorname{ob}(\mathcal{AIP}(S))$ satisfying the premise of condition (†), viz., $Q_1 \cap Q_1^* \subset P_1 \cap P_1^*$ for some $P \in F$, is a saturated pre-cofinal family for S and is the smallest one containing F.

4.6 LEMMA. *Let F be a saturated pre-cofinal family for S. If, for $j = 1, 2$, $N_j \in F$ is AIP(regular) for S and has the additional property that both N_{1j} and N^*_{1j} are isolating neighborhoods for S, then there exists $L, P_1, P_2, Q_1, Q_2 \in F$ all AIP(regular) for S and inclusion morphisms*

$$L \hookrightarrow P_j \hookleftarrow Q_j \hookrightarrow N_j \qquad (j = 1, 2). \tag{4.2}$$

Also, if AIP(thick) is substituted for AIP(regular) in the previous sentence, the resulting statement is true.

In the proof of Lemma 4.6, local stable and unstable sets of S are always defined relative to the forward time flow, never the reverse time flow.

Proof of Lemma 4.6. For $j = 1, 2$, let N_j be an AIP(regular) pair of index pairs for S and assume $N_j \in F$. As noted in Remark 4.2.1, for any $r, s \geq 0$, each index pair in the pair of auxiliary index pairs

$$((N^r_{1j}, \cap N^{-s}_{0j}), (N^{*r}_{1j}, \cap N^{*-s}_{0j})) \tag{4.3}$$

is regular with respect to the forward or reverse time flow as appropriate.

Next, let us show that for each $s > 0$ there exists $T(s) > 0$ so that if $r \geq T(s)$, then the pair of index pairs given in (4.3) satisfies the off-diagonal condition; hence is AIP(regular) for S. Also this pair of index pairs will be in F because F is saturated and because for all $r > 0$, $N^r_{1j} \cap N^{*r}_{1j} \subset N_{1j} \cap N^*_{1j}$. Fix $s > 0$ and note that

$$W^u(S; N_{1j}) \cap N^{*-s}_{0j} \cap N^*_{1j} = \emptyset \qquad (j = 1, 2). \tag{4.4}$$

For if $\mathbf{u}$ is in the intersection, then for some t, $0 < t \leq s$ and $\mathbf{u} \cdot [-t, 0] \subset N^*_{1j}$ and $\mathbf{u} \cdot -t \in N^*_{0j}$, but also $\mathbf{u} \cdot -t \in N_{1j}$, a contradiction as N_{1j} and N^*_{0j} are disjoint. The analogous argument with respect to the reverse time flow shows that

$$W^s(S; N^*_{1j}) \cap N^{-s}_{0j} \cap N_{1j} = \emptyset \qquad (j = 1, 2). \tag{4.5}$$

Because N_{1j} and N^*_{1j} isolate S, the sets $W^u(S; N_{1j})$ and $W^s(S; N^*_{1j})$ are closed. (This need not be true if N_{1j} or N^*_{1j} does not isolate S; e.g., if S is the repellor of a repellor-attractor pair and N_{1j} is also a neighborhood of the attractor, then $W^u(S; N_{1j})$ will not be closed if their exist heteroclinic orbits in N_{1j} from S to its complementary attractor—see [K2, Lemma 4.1].) By normality, (4.4) and (4.5) yield sets U_j and U^*_j, open in M, satisfying

$$W^u(S; N_{1j}) \subset U_j, \quad W^s(S; N^*_{1j}) \subset U^*_j \quad (j = 1, 2) \tag{4.6}$$

and

$$U_j \cap N^{*-s}_{0j} \cap N^*_{1j} = \emptyset, \quad U^*_j \cap N^{-s}_{0j} \cap N_{1j} = \emptyset \quad (j = 1, 2). \tag{4.7}$$

Then by (4.6) and [K1, Lemma 3.3(1)], there exists $T(s) > 0$ so that for $r \geq T(s)$,

$$N^r_{1j} \subset U_j \quad \text{and} \quad N^{*r}_{1j} \subset U^*_j \quad (j = 1, 2).$$

It then follows from (4.7) and Remark 4.1.2 on the regularity of auxiliary index pairs that the pair of index pairs in (4.3) is AIP(regular) for S.

The index pairs (L_1, L_0) and (L_1^*, L_0^*) will be constructed from index pairs (N_{13}, N_{03}) and (N_{13}^*, N_{03}^*) defined as follows:

$$\begin{aligned} N_{13} &:= N_{11} \cap N_{12}, \quad N_{03} := N_{13} \cap (N_{01} \cup N_{02}), \\ N_{13}^* &:= N_{11}^* \cap N_{12}^*, \quad N_{03}^* := N_{13}^* \cap (N_{01}^* \cup N_{02}^*). \end{aligned}$$

It follows from the proof of Proposition A1 in the appendix to [K3] that (N_{13}, N_{03}) and (N_{13}^*, N_{03}^*) are index pairs for S, the first with respect to the forward time flow, the second with respect to the reverse time flow. Further, both of these index pairs are regular. For example, $\tau_{(N_{13}, N_{03})}$ equals $\min\{\tau_{(N_{1j}, N_{0j})}|N_{13} : j = 1, 2\}$, hence is continuous; similarly $\tau_{(N_{13}^*, N_{03}^*)}$ is continuous.

We proceed to show that we can find $s > 0$ and $r \geq T(s)$ so that if P_j is the pair of index pairs given in (4.3) above ($j = 1, 2$) and if the pair of index pairs L is defined by the equalities

$$\begin{aligned} L_1 &:= N_{13}^r, \quad L_0 := N_{13}^r \cap N_{03}, \\ L_1^* &:= N_{13}^{*r}, \quad L_0^* := N_{13}^{*r} \cap N_{03}^*, \end{aligned} \tag{4.8}$$

then L and P_j are AIP(regular) for S and there exists an inclusion morphism $L \hookrightarrow P_j$. Because $N_{13} \subset N_{1j}$ ($j = 1, 2$) and because both isolate S, by [K1, Lemma 3.3(4)] there exist $s' > r' > 0$ so that

$$N_{13}^{r'} \cap N_{03}^{-r'} \cap N_{1j} \setminus N_{0j}^{-s'} = \emptyset \quad (j = 1, 2),$$

and since $N_{03} \subset N_{03}^{-r'}$ and because for each $r \geq r'$ and $s \geq s'$

$$N_{13}^r \subset N_{13}^{r'} \subset N_{1j} \quad \text{and} \quad N_{0j}^{-s'} \subset N_{0j}^{-s} \quad (j = 1, 2),$$

it follows that for each $r \geq r'$ and $s \geq s'$,

$$N_{13}^r \cap N_{03} \subset N_{0j}^{-s} \quad (j = 1, 2). \tag{4.9}$$

Similarly there exist $s'' > r'' > 0$ so that for $r \geq r''$ and $s \geq s''$,

$$N_{13}^{*r} \cap N_{03}^* \subset N_{0j}^{*-s} \quad (j = 1, 2). \tag{4.10}$$

Choose $s \geq \max\{s', s''\}$ and choose $r \geq \max\{r', r'', T(s)\}$. Use these values of s and r to define (L_1, L_0) and (L_1^*, L_0^*) according to the prescription of (4.8), and define P_j ($j = 1, 2$) to be the pair of index pairs in (4.3) for these values of s and r. Then P_j is AIP(regular) for S by choice of $r > T(s)$ and the index pairs in L satisfy the off-diagonal condition as follows straightforwardly from the definition of these index pairs because the pair of index pairs N_j satisfies this condition for $j = 1, 2$. Hence L is AIP(regular) for S, and for $j = 1, 2$, there exists an inclusion morphism $L \hookrightarrow P_j$ as a consequence of (4.8), (4.9), and (4.10). Note that the existence of the inclusion morphism implies that $L \in F$ because F is saturated and $P_j \in F$.

To finish, use the chosen value of r to define

$$Q_j := ((N^r_{1j}, \cap N_{0j}), (N^{*r}_{1j}, \cap N^*_{0j})) \quad (j = 1, 2).$$

It follows easily that Q_j so defined is AIP(regular) for S because N_j is and that the inclusion morphisms indicated in (4.2) exist. As with L, the existence of the inclusion morphisms implies that $Q_j \in F$, $j = 1, 2$.

It is straightforward to check that if the initial pairs are AIP(thick) rather than AIP(regular), the construction given above yields pairs that are AIP(thick). For example, if $K_j \subset \operatorname{int}_{N_{1j}}(N_{0j})$, then $K_j \subset V_j \cap N_{1j} \subset N_{0j}$ for some V_j open in M $(j = 1, 2)$. Hence, $N_{13} \cap (K_1 \cup K_2) \subset N_{13} \cap (V_1 \cup V_2) \subset N_{03}$ from which it follows that (N_{13}, N_{03}) has thick exit set if (N_{1j}, N_{0j}) does for $j = 1, 2$. □

For the main result of this section, assume that F and G are covariant functors on the homotopy category of compact Hausdorff pairs to the category of left R-modules and that K is a left R-module.

4.7 THEOREM. *Let F be a saturated pre-cofinal family for S. For each $P \in F$ assume given a pairing of R-modules*

$$B_P \colon \mathsf{F}(P_1, P_0) \otimes \mathsf{G}(P^*_1, P^*_0) \to K$$

so that for any $Q \in F$, if there exist inclusions

$$i \colon (Q_1, Q_0) \subset (P_1, P_0) \quad \text{and} \quad j \colon (Q^*_1, Q^*_0) \subset (P^*_1, P^*_0) \tag{4.11}$$

then the diagram

$$\begin{array}{ccc} \mathsf{F}(Q_1, Q_0) \otimes \mathsf{G}(Q^*_1, Q^*_0) & \xrightarrow{B_Q} & K \\ {\scriptstyle \mathsf{F}(i) \otimes \mathsf{G}(j)} \big\downarrow & & \big\downarrow {\scriptstyle 1_K} \\ \mathsf{F}(P_1, P_0) \otimes \mathsf{G}(P^*_1, P^*_0) & \xrightarrow{B_P} & K \end{array} \tag{4.12}$$

commutes. Also assume that for each $P \in F$, the quotient maps

$$p^+ \colon (P_1, P_0) \to (P_1/P_0, [P_0]) \quad \text{and} \quad p^- \colon (P^*_1, P^*_0) \to (P^*_1/P^*_0, [P^*_0])$$

transform to R-module isomorphisms under F and G respectively. Then $\overline{\mathsf{F}(p^+)} \otimes \overline{\mathsf{G}(p^-)}$ is an isomorphism, and there exists a unique pairing of R-modules

$$B \colon \mathsf{F}\mathcal{C}(S) \otimes \mathsf{G}\mathcal{C}^*(S) \to K$$

satisfying

$$B = B_P \circ \left(\overline{\mathsf{F}(p^+)} \otimes \overline{\mathsf{G}(p^-)}\right)^{-1} \quad \text{for each } P \in F. \tag{4.13}$$

4.7.1 Remark. If F and G were assumed to be contravariant rather than covariant, then the statement dual to that of Theorem 4.7 obtained by reversing the direction of the vertical arrows in diagram (4.12) and replacing statement (4.13) with the statement

$$B = B_P \circ \overline{\mathsf{F}(p^+)} \otimes \overline{\mathsf{G}(p^-)} \qquad \text{for each } P \in F \tag{4.13*}$$

is true and has a proof analogous to the one given below for Theorem 4.7; details are left to the reader.

4.7.2 Remark. If $F = \text{ob}(\mathcal{AIP}(S))$, the hypotheses on the family of pairings B_P, $P \in \text{ob}(\mathcal{AIP}(S))$, imply that this family defines a covariant functor on $\mathcal{AIP}(S)$ with values in the category of R-bilinear maps into K.

The proof of the theorem is delayed until after the proof of the following corollary whose proof is the reason for introducing the notion of saturated pre-cofinal families into the previous theorem rather than just stating and proving it for $F = \text{ob}(\mathcal{AIP}(S))$.

For the corollary, let R, F and G be as for the theorem. Also, direct $\mathcal{N}(S)$ by containment and let $\{\{K_U\}_{U \in \mathcal{N}(S)}, \{i_U^V\}_{U \preceq V \in \mathcal{N}(S)}\}$ be an inverse system of R-modules and homomorphisms.

4.8 Corollary. *Let F be a saturated pre-cofinal family for S. For each $P \in F$ assume given a pairing of R-modules*

$$B_P \colon \mathsf{F}(P_1, P_0) \otimes \mathsf{G}(P_1^*, P_0^*) \to K_{P_1 \cap P_1^*}$$

so that for any $Q \in F$, if there exist inclusions

$$i \colon (Q_1, Q_0) \subset (P_1, P_0) \quad \text{and} \quad j \colon (Q_1^*, Q_0^*) \subset (P_1^*, P_0^*) \tag{4.14}$$

then the diagram

$$\begin{array}{ccc} \mathsf{F}(Q_1, Q_0) \otimes \mathsf{G}(Q_1^*, Q_0^*) & \xrightarrow{B_Q} & K_{Q_1 \cap Q_1^*} \\ {\scriptstyle \mathsf{F}(i) \otimes \mathsf{G}(j)} \big\downarrow & & \big\downarrow {\scriptstyle i_{P_1 \cap P_1^*}^{Q_1 \cap Q_1^*}} \\ \mathsf{F}(P_1, P_0) \otimes \mathsf{G}(P_1^*, P_0^*) & \xrightarrow{B_P} & K_{P_1 \cap P_1^*} \end{array} \tag{4.15}$$

commutes. Also assume that for each $P \in F$, the quotient maps

$$p^+ \colon (P_1, P_0) \to (P_1/P_0, [P_0]) \quad \text{and} \quad p^- \colon (P_1^*, P_0^*) \to (P_1^*/P_0^*, [P_0^*])$$

transform to R-module isomorphisms under F and G, respectively. For each $V \in \mathcal{N}(S)$, choose $P \in F$ with $P_1 \cap P_1^* \subset V$ and define $B_V^S \colon \widetilde{\mathsf{F}\mathcal{C}}(S) \otimes \widetilde{\mathsf{G}\mathcal{C}}^*(S) \to K_V$ by

$$B_V^S := i_V^{P_1 \cap P_1^*} \circ B_P \circ \left(\overline{\mathsf{F}(p^+)} \otimes \overline{\mathsf{G}(p^-)}\right)^{-1}. \tag{4.16}$$

Then the definition of B_V^S is independent of the choice of $P \in F$ satisfying $P_1 \cap P_1^* \subset V$, and

$$i_U^V \circ B_V^S = B_U^S \qquad \text{for } U \supset V \in \mathcal{N}(S), \tag{4.17}$$

whence passage to the inverse limit yields a pairing

$$B \colon \mathsf{F}\mathcal{C}(S) \otimes \mathsf{G}\mathcal{C}^*(S) \to \varprojlim K_U\,.$$

4.8.1 Remark. As for the theorem, there is a version of the corollary for F and G contravariant where the inverse system of R-modules and its limit is replaced by a direct system and its limit, where the vertical arrows of (4.15) are reversed, where B_V^S in (4.16) is replaced by

$$B_S^V := i_V^{P_1 \cap P_1^*} \circ B_P \circ \left(\overline{\mathsf{F}(p^+)} \otimes \overline{\mathsf{G}(p^-)}\right), \tag{4.16*}$$

and where (4.17) is replaced by

$$i_U^V \circ B_S^U = B_S^V \quad \text{for } U \supset V \in \mathcal{N}(S). \tag{4.17*}$$

Proof of Corollary 4.8. For each $V \in \mathcal{N}(S)$, set $F_V := \{P \in F : P_1 \cap P_1^* \subset V\}$ and note that F_V is a saturated pre-cofinal family for S because F is. For each $P \in F_V$, define $B_V^P := i_V^{P_1 \cap P_1^*} \circ B_P$, a pairing on $\mathsf{F}(P_1, P_0) \otimes \mathsf{G}(P_1^*, P_0^*)$ to K_V. Application of Theorem 4.7 to the family of pairings B_V^P, $P \in F_V$, yields a pairing B_V^S for which the equality of (4.16) holds for each $P \in F_V$; i.e., the definition of B_V^S as given by (4.16) is independent of the choice of $P \in F_V$ as desired. Thus, if $U \supset V \in \mathcal{N}(S)$, by choosing $P \in F$ satisfying $P_1 \cap P_1^* \subset V$ to define B_U^S, we see immediately from the properties of an inverse system that (4.17) holds, whence the family of transformations B_V^S, $V \in \mathcal{N}(S)$, pass to the inverse limit yielding $B := \varprojlim B_V^S$. □

Proof of Theorem 4.7. Let $P \in F$ and let p^+ and p^- be the quotient maps as given in the statement of the theorem. Because $\mathsf{F}(p^+)$ and $\mathsf{G}(p^-)$ are by hypothesis isomorphisms, so too (e.g., see [Sp, Lemmas 4.1.4–5]) is $\overline{\mathsf{F}(p^+)} \otimes \overline{\mathsf{G}(p^-)}$ where $\overline{\mathsf{F}(p^+)}$ and $\overline{\mathsf{G}(p^-)}$ are defined as in Definition 4.1(A). Define

$$\overline{B}_P := B_P \circ \left(\overline{\mathsf{F}(p^+)} \otimes \overline{\mathsf{G}(p^-)}\right)^{-1} \qquad \text{for} \quad P \in F$$

and define an equivalence relation on F by $P \overset{B}{\sim} Q$ if, and only if, $\overline{B}_P = \overline{B}_Q$.

To complete the proof it suffices to show that there is exactly one equivalence class for this equivalence relation, for we can then define $B := \overline{B}_P$ for any $P \in F$ yet have (4.13) hold. Obviously, (4.13) implies the uniqueness of B.

Basic to our proof is the observation that if $P, Q \in F$ and there exists an inclusion morphism $Q \hookrightarrow P$, i.e., P and Q satisfy the inclusions (4.11), then $P \overset{B}{\sim} Q$. We call this observation *"invariance of B under inclusions."* To prove it, first observe that the statement $P \overset{B}{\sim} Q$ is equivalent to the commutativity of the diagram

$$\begin{array}{ccccc}
\widetilde{\mathsf{F}C}(S) \otimes \widetilde{\mathsf{G}C}^*(S) & \xleftarrow{\overline{\mathsf{F}(p^+)} \otimes \overline{\mathsf{G}(p^-)}} & \mathsf{F}(P_1, P_0) \otimes \mathsf{G}(P_1^*, P_0^*) & \xrightarrow{B_P} & K \\
{\scriptstyle 1_{\widetilde{\mathsf{F}C}(S) \otimes \widetilde{\mathsf{G}C}^*(S)}} \uparrow & & {\scriptstyle \mathsf{F}(i) \otimes \mathsf{G}(j)} \uparrow & & \uparrow {\scriptstyle 1_K} \\
\widetilde{\mathsf{F}C}(S) \otimes \widetilde{\mathsf{G}C}^*(S) & \xleftarrow{\overline{\mathsf{F}(q^+)} \otimes \overline{\mathsf{G}(q^-)}} & \mathsf{F}(Q_1, Q_0) \otimes \mathsf{G}(Q_1^*, Q_0^*) & \xrightarrow{B_Q} & K\,.
\end{array} \tag{4.18}$$

The right-hand rectangle is just diagram (4.12) which is commutative by hypothesis if $P, Q \in F$ and satisfy the inclusions (4.11). The left-hand rectangle commutes when the inclusions (4.11) are satisfied because it expands (after rotating it counterclockwise 90°) to the diagram

$$
\text{(4.19)}\qquad
\begin{array}{ccc}
\mathsf{F}(P_1,P_0)\otimes\mathsf{G}(P_1^*,P_0^*) & \xleftarrow{\mathsf{F}(i)\otimes\mathsf{G}(j)} & \mathsf{F}(Q_1,Q_0)\otimes\mathsf{G}(Q_1^*,Q_0^*)\\
{\scriptstyle \mathsf{F}(p^+)\otimes\mathsf{G}(p^-)}\downarrow & & \downarrow{\scriptstyle \mathsf{F}(q^+)\otimes\mathsf{G}(q^-)}\\
\widetilde{\mathsf{F}}(P_1/P_0)\otimes\widetilde{\mathsf{G}}(P_1^*/P_0^*) & \xleftarrow{\widetilde{\mathsf{F}}(\bar\imath)\otimes\widetilde{\mathsf{G}}(\bar\jmath)} & \widetilde{\mathsf{F}}(Q_1/Q_0)\otimes\widetilde{\mathsf{G}}(Q_1^*/Q_0^*)\\
\downarrow & & \downarrow\\
\widetilde{\mathsf{F}}\mathcal{C}\,(S)\otimes\widetilde{\mathsf{G}}\mathcal{C}^*(S) & \xleftarrow{1_{\widetilde{\mathsf{F}}\mathcal{C}(S)\otimes\widetilde{\mathsf{G}}\mathcal{C}^*(S)}} & \widetilde{\mathsf{F}}\mathcal{C}\,(S)\otimes\widetilde{\mathsf{G}}\mathcal{C}^*(S)
\end{array}
$$

where $\bar\imath$ and $\bar\jmath$ are the maps on quotients induced by the inclusions i and j and where the unlabeled vertical arrows are the tensor product of canonical maps sending an element to its equivalence class: the bottom rectangle commutes by definition of the equivalence relations defining $\widetilde{\mathsf{F}}\mathcal{C}\,(S)$ and $\widetilde{\mathsf{G}}\mathcal{C}^*(S)$; the top rectangle commutes because it is the transform by the functor $\mathsf{F}\otimes\mathsf{G}$ of a commutative diagram of pairs of compact Hausdorff pairs and pairs of continuous maps between such pairs of pairs.

CLAIM I. *For each $Q\in F$, there exists $P\in F$ an AIP(regular) pair of index pairs for S and an inclusion morphism $Q\hookrightarrow P$; hence each pair of index pairs in F that is AIP for S is equivalent to an AIP(regular) pair of index pairs for S in F.*

Proof of Claim I. Let $Q\in F$; write $Q=((Q_1,Q_0),(Q_1^*,Q_0^*))$. Choose an open neighborhood U of Q_0 that is disjoint from Q_1^* and choose an open neighborhood U^* of Q_0^* that is disjoint from Q_1. This choice of neighborhoods is possible because $Q\in M^{(2)}_{\Delta c}$ and M is normal. As follows from Remark 4.1.1, we can find compact sets Q_0' and $Q_0'^*$ satisfying the inclusions

$$Q_0\subset Q_0'\subset U\quad\text{and}\quad Q_0^*\subset Q_0'^*\subset U^*$$

so that (Q_1,Q_0') is a regular index pair for S in forward time and $(Q_1^*,Q_0'^*)$ is a regular index pair for S in reverse time. Set $P:=((Q_1,Q_0'),(Q_1^*,Q_0'^*))$. By the choice of U and U^* it follows that P is AIP(regular) for S, and by construction, $P_1\cap P_1^*=Q_1\cap Q_1^*$, whence $P\in F$ because F is saturated. Also by construction, there exists an inclusion morphism $Q\hookrightarrow P$. Hence by the invariance of B under inclusions $Q\overset{B}{\sim}P$. □

CLAIM II. *Any two AIP(regular) pairs of index pairs for S, both elements of F, are $\overset{B}{\sim}$-equivalent.*

Proof of Claim II. In proving the claim we need only consider AIP(regular) pairs of index pairs for S wherein the first compact set of each index pair in the pair of index pairs is an isolating neighborhood for S. For suppose N is AIP(regular) for S. If say N_1 does not isolate S, then define

$$N_1':=\operatorname{cl}(N_1\setminus N_0)\quad\text{and}\quad N_0':=N_0\cap N_1'.$$

Then (N_1', N_0') is a regular index pair since $\tau_{(N_1',N_0')}$ is the restriction to N_1' of $\tau_{(N_1,N_0)}$ and also N_1' isolates S. Thus $((N_1', N_0'), (N_1^*, N_0^*))$ is AIP(regular) for S, and by construction $(N_1', N_0') \subset (N_1, N_0)$. Because F is saturated and $N \in F$, the inclusion implies that $((N_1', N_0'), (N_1^*, N_0^*)) \in F$, whence $((N_1, N_0), (N_1^*, N_0^*))$ is $\overset{B}{\sim}$-equivalent to $((N_1', N_0'), (N_1^*, N_0^*))$ by invariance of B under inclusions. Similarly we can assume that N_1^* isolates S.

Thus assume given $N_1, N_2 \in F$ both AIP(regular) for S and with the property that $N_{1\,j}$ and $N_{1\,j}^*$ are isolating neighborhoods of S for $j = 1, 2$. From the transitivity and symmetry of the equivalence relation $\overset{B}{\sim}$, it is then a simple consequence of Lemma 4.6 and invariance of B under inclusions that $N_1 \overset{B}{\sim} N_2$. □

By Claims I and II there is exactly one $\overset{B}{\sim}$-equivalence class. □

C. The Proofs of Theorems 2.4 and 2.11. For the proof of Theorem 2.4 let S be an isolated invariant set of a flow in $\mathbf{R}^m$. All homology modules occurring in the proof are taken with coefficients in R.

4.9 Proof of Theorem 2.4. Let $P, Q \in \operatorname{ob}(\mathcal{AIP}(S))$ and assume there exists an inclusion morphism $(i, j)\colon Q \hookrightarrow P$. It follows that there is a commutative diagram of inclusion maps

$$\begin{array}{ccc} (Q_1, Q_0) \times (Q_1^*, Q_0^*) & \xhookrightarrow{\qquad\qquad} & (P_1, P_0) \times (P_1^*, P_0^*) \\ & \searrow \qquad \swarrow & \\ & (\mathbf{R}^m \times \mathbf{R}^m, \mathbf{R}^m \times \mathbf{R}^m \setminus \Delta) & \end{array}$$

whence naturality of homology cross-products and Definition 2.2 imply that

$$\tag{4.20} \begin{array}{ccc} \big[H_*(Q_1, Q_0) \otimes H_*(Q_1^*, Q_0^*)\big]_m & \xrightarrow{\mathsf{L}_Q} & H_m(\mathbf{R}^m, \mathbf{R}^m \setminus \{0\}) \\ {\scriptstyle i_* \otimes j_*}\big\downarrow & & \big\downarrow{\scriptstyle 1_*} \\ \big[H_*(P_1, P_0) \otimes H_*(P_1^*, P_0^*)\big]_m & \xrightarrow{\mathsf{L}_P} & H_m(\mathbf{R}^m, \mathbf{R}^m \setminus \{0\}) \end{array}$$

is a commutative diagram; i.e., intersection numbers as defined in Definition 2.2 are invariant under inclusion maps. As diagram (4.20) corresponds in the present context to diagram (4.12) of Theorem 4.7, the desired result is immediate from that theorem by applying it to each family of summands $\{\mathsf{L}_{N\,k} : N \in \operatorname{ob}(\mathcal{AIP}(S))\}$ for each integer $k \in [0, m]$ thereby obtaining a pairing $\mathsf{L}_k\colon \widetilde{H}_{m-k}\mathcal{C}(S) \otimes \widetilde{H}_k\mathcal{C}^*(S) \to H_m(\mathbf{R}^m, \mathbf{R}^m \setminus \{0\})$ and then setting $\mathsf{L} := \bigoplus_{k=0}^m \mathsf{L}_k$. □

For the proofs of Lemma 2.10 and Theorem 2.11 assume G and G' are R-modules; for the proof of Theorem 2.11, S is an isolated invariant set of a flow in M.

4.10 Proof of Lemma 2.10. The naturality of the pairings $\mathfrak{L}_X$, $X \in M^{(2)}_{\Delta}$ will be proved; the naturality of the pairings ${}^{\#}\mathfrak{L}_X$, $X \in M^{(2)}_{\Delta}$ follows trivially. The proof of naturality of the pairings $\mathfrak{L}_X$, $X \in M^{(2)}$ is entirely analogous to that for $X \in M^{(2)}_{\not\Delta}$ and is omitted. Accordingly, let $X, Y \in M^{(2)}_{\not\Delta}$ with $X \hookrightarrow Y$. It must be shown that

$$
(4.21)\qquad
\begin{array}{ccc}
H_*(Y_1, Y_0; G) \otimes H_*(Y_1^*, Y_0^*; G') & \xrightarrow{\mathfrak{L}_Y} & \check{H}_*(Y_1 \cap Y_1^*; G \otimes G') \\
\uparrow & & \uparrow \\
H_*(X_1, X_0; G) \otimes H_*(X_1^*, X_0^*; G') & \xrightarrow{\mathfrak{L}_X} & \check{H}_*(X_1 \cap X_1^*; G \otimes G')
\end{array}
$$

is a commutative diagram where the vertical arrows are inclusion induced. It will suffice to show that (4.21) commutes in the special case $X \in M^{(2)}_{\not\Delta c}(Y)$. For if so, it follows for the general case that if $K \in M^{(2)}_{\not\Delta c}(X)$ is chosen, then also $K \in M^{(2)}_{\not\Delta c}(Y)$ as a consequence of the inclusion $j_Y^X = \left(j_{(Y_1,Y_0)}^{(X_1,X_0)}, j_{(Y_1^*,Y_0^*)}^{(X_1^*,X_0^*)}\right): X \hookrightarrow Y$, whence the diagram

$$
(4.22)\qquad
\begin{array}{ccccc}
H_*(Y_1, Y_0) \otimes H_*(Y_1^*, Y_0^*) & & \xrightarrow{\quad\mathfrak{L}_Y\quad} & & \check{H}_*(Y_1 \cap Y_1^*) \\
\uparrow & \nwarrow & & \nearrow & \uparrow \\
 & H_*(K_1, K_0) \otimes H_*(K_1^*, K_0^*) & \xrightarrow{\mathfrak{L}_K} & \check{H}_*(K_1 \cap K_1^*) & \\
 & \swarrow & & \searrow & \\
H_*(X_1, X_0) \otimes H_*(X_1^*, X_0^*) & & \xrightarrow{\quad\mathfrak{L}_X\quad} & & \check{H}_*(X_1 \cap X_1^*)
\end{array}
$$

is commutative where all the non-horizontal arrows are inclusion induced and where coefficients have been suppressed to facilitate the printing. The commutativity of (4.21) in the general case then follows element by element. That is, if $\alpha \otimes \gamma \in H_*(X_1, X_0; G) \otimes H_*(X_1^*, X_0^*; G')$, then by taking (K_1, K_0) to equal $(|c|, |\partial c|)$ where c is a chain representing α and similarly defining (K_1^*, K_0^*) using a chain representing γ, it follows that $\alpha \otimes \gamma$ is the image under the inclusion induced map of an element of $H_*(K_1, K_0; G) \otimes H_*(K_1^*, K_0^*; G')$. Thus, the commutativity of diagram (4.22) yields that diagram (4.21) commutes on $\alpha \otimes \gamma$, therefore, on each generating element, and is therefore commutative.

To establish commutativity of (4.21) in the special case $X \in M^{(2)}_{\not\Delta c}(Y)$, it will suffice to show that for each neighborhood $W \in \mathcal{N}(Y_1 \cap Y_1^*)$,

$$
(4.23)\qquad \pi_W^Y \circ \check{\imath}_Y^X \circ \mathfrak{L}_X = \pi_W^Y \circ \mathfrak{L}_Y \circ \left(j_{(Y_1,Y_0)*}^{(X_1,X_0)} \otimes j_{(Y_1^*,Y_0^*)*}^{(X_1^*,X_0^*)}\right)
$$

where $\pi_W^Y : \check{H}_*(Y_1 \cap Y_1^*) \to H_*(W)$ is the canonical projection. To see that (4.23) holds, note that the left-hand side of (4.23) equals $\mathfrak{L}_W^X$ by definition of $\mathfrak{L}_X^X$ as the inverse limit over $U \in \mathcal{N}(X_1 \cap X_1^*)$ of the pairings $\mathfrak{L}_U^X$ because $\pi_W^Y \circ \check{\imath}_Y^X = \pi_W^X$ and because $\mathfrak{L}_X = \mathfrak{L}_X^X$ for $X \in M^{(2)}_{\not\Delta c}$ as observed in Definition 2.8(C) where the right-hand side of the last equality corresponds to the map in (2.8). On the other hand, by definition of $\mathfrak{L}_Y$ as the direct limit over $K \in M^{(2)}_{\not\Delta c}(Y)$ of the pairings $\mathfrak{L}_Y^K$, it follows that

$$
\mathfrak{L}_Y \circ \left(j_{(Y_1,Y_0)*}^{(X_1,X_0)} \otimes j_{(Y_1^*,Y_0^*)*}^{(X_1^*,X_0^*)}\right) = \mathfrak{L}_Y^X .
$$

Hence, by definition of $\mathfrak{L}^X_Y$ as the inverse limit over $U \in \mathcal{N}(Y_1 \cap Y_1^*)$ of the pairings $\mathfrak{L}^X_U$, it follows that the right-hand side of (4.23) also equals $\mathfrak{L}^X_W$. $\square$

4.11 The Proof of Theorem 2.11. That $\mathcal{N}(S; \mathcal{AIP})$ is cofinal in $\mathcal{N}(S)$ is a simple consequence of Proposition 4.4.

Set $K_U := \check{H}_{i+j-m}(U; G \otimes G')$ for each $U \in \mathcal{N}(S)$ and let $\check{\iota}^V_U : K_V \to K_U$ be the inclusion induced map for $U \supset V \in \mathcal{N}(S)$. Then

$$\{\{K_U\}_{U \in \mathcal{N}(S)}, \{\check{\iota}^V_U\}_{U \preceq V \in \mathcal{N}(S)}\}$$

is an inverse system of R-modules and homomorphisms. Also set $B_P := \mathfrak{L}_{P\,(i,j)}$ for each $P \in \text{ob}(\mathcal{AIP}(S))$. By Lemma 2.10, the pairings B_P are natural relative to the inclusion morphisms of $\mathcal{AIP}(S)$. Thus, Corollary 4.8 applied to the family of maps B_P yields a pairing

$$B : \widetilde{H}_i\mathcal{C}(S; G) \otimes \widetilde{H}_j\mathcal{C}^*(S; G') \to \varprojlim \check{H}_{i+j-m}(U; G \otimes G')$$

where $\varprojlim \check{H}_{i+j-m}(U; G \otimes G')$ is taken over $U \in \mathcal{N}(S)$. Further, statement (4.17) of Corollary 4.8 applied in the current context to neighborhoods $P_1 \cap P_1^* \supset Q_1 \cap Q_1^* \in \mathcal{N}(S; \mathcal{AIP})$ yields that statement (2.13) of Theorem 2.11 holds. Because $\mathcal{N}(S; \mathcal{AIP})$ is cofinal in $\mathcal{N}(S)$, it is immediate that passage to the limit in (2.13) over $\mathcal{N}(S; \mathcal{AIP})$ yields the same pairing B as above.

To finish, note that the family of natural transformations $\psi_U : H_*(U; G \otimes G') \to \check{H}_*(U; G \otimes G')$, $U \in \mathcal{N}(S)$, pass to the limit and yield a natural transformation $\psi : \check{H}_*(S) := \varprojlim H_*(U) \to \varprojlim \check{H}_*(U)$ which is easily seen to be an isomorphism using the fact derived from the normality of M that if $U \in \mathcal{N}(S)$ then there exists $W \in \mathcal{N}(S)$ with $U \in \mathcal{N}(W)$. Thus, we can define $\mathfrak{L}_{(i,j)} := \psi^{-1} \circ B$. $\square$

CHAPTER 5

PROOFS OF THE CONTINUATION RESULTS

This chapter provides proofs of Theorems 3.1–3.3, yielding the invariance of intersection numbers under continuation, and of Theorem 3.7, showing that a path in a space of isolated invariant sets can be lifted to the corresponding space of Čech homology classes of the isolated invariant sets if the initial point of the lift is an intersection class of homology classes in the Conley indices of the invariant set below it. These proofs are inextricably bound up with the definition of the map between Conley indices defined by a path in a space of isolated invariant sets as given in [K3]. Thus, in preparation for the proofs we (*i*) review the notion of a morphism between connected simple systems (maps between Conley indices are a special case) which leads to the notion of a category of connected simple systems that are subcategories of a category $\mathcal{H}$, denoted by $\mathcal{CSS}(\mathcal{H})$, (*ii*) review how a path β in a space of isolated invariant sets $\mathcal{S}(\varphi)$ defines maps between Conley indices

$$\mathcal{C}\left(S_\beta(0)\right) \xrightarrow{\mathcal{C}(\beta)} \mathcal{C}\left(S_\beta(1)\right) \quad \text{and} \quad \mathcal{C}^*\left(S_\beta(0)\right) \xrightarrow{\mathcal{C}^*(\beta)} \mathcal{C}^*\left(S_\beta(1)\right)$$

both of which are equivalences in $\mathcal{CSS}(\mathcal{T}^{*\prime})$, (*iii*) show that the singular homology functor can be extended to a functor on $\mathcal{CSS}(\mathcal{T}^{*\prime})$ to $\mathcal{G}_R\mathcal{M}$ whose application to $\mathcal{C}\,(\beta)$ and $\mathcal{C}^*(\beta)$ yields the graded R-module isomorphisms

$$\widetilde{H}_*\mathcal{C}\left(S_\beta(0)\right) \xrightarrow{\mathcal{C}(\beta)_*} \widetilde{H}_*\mathcal{C}\left(S_\beta(1)\right) \quad \text{and} \quad \widetilde{H}_*\mathcal{C}^*\left(S_\beta(0)\right) \xrightarrow{\mathcal{C}^*(\beta)_*} \widetilde{H}_*\mathcal{C}^*\left(S_\beta(1)\right)$$

used to state the continuation results of Chapter 3. To facilitate the exposition, the preparatory material will be given in the order (*i*), (*iii*), (*ii*).

A. Maps between Conley Indices from Paths of Invariant Sets. As shown in [K3], the collection of connected simple systems that are subcategories of some category $\mathcal{H}$ is itself the class of objects of a category, denoted $\mathcal{CSS}(\mathcal{H})$ and called *the category of connected simple systems that are subcategories of* $\mathcal{H}$. A morphism in this new category is called *a map of connected simple systems.* Formally, as defined in [K3], a map of connected simple systems $F\colon \mathcal{A} \to \mathcal{B}$, with $\mathcal{A}$ and $\mathcal{B}$ both subcategories of $\mathcal{H}$, is a covariant functor on the product category $\mathcal{A} \times \mathcal{B}$ to the morphism category of $\mathcal{H}$ satisfying certain additional properties. This formal definition and the definition of the composition of two maps of connected simple systems as given in [K3] together are rather long and are not really needed for the proofs of the continuation results; hence, only an informal characterization of the complete definitions will be given.

5.1 DEFINITION. (A) Informally, if $\mathcal{A}$ and $\mathcal{B}$ are connected simple systems in $\mathcal{CSS}(\mathcal{H})$, *a map of connected simple systems* $F: \mathcal{A} \to \mathcal{B}$ is a collection of commutative diagrams of morphisms in $\mathcal{H}$ of the form

$$\begin{array}{ccc} A & \xrightarrow{F_A^B} & B \\ {\scriptstyle h_{\mathcal{A}}^{A\,A'}}\downarrow & & \downarrow{\scriptstyle h_{\mathcal{B}}^{B\,B'}} \\ A' & \xrightarrow[F_{A'}^{B'}]{} & B' \end{array} \tag{5.1}$$

where A and A' are objects of $\mathcal{A}$ and B and B' are objects of $\mathcal{B}$. As the notation in (5.1) implies, for any two such pairs of objects (A, B) and (A', B'), there must be one, and only one, such diagram in the collection. Note that this uniqueness implies that the collection is completely specified if F_A^B is given for a single pair of objects (A, B). For necessarily,

$$F_{A'}^{B'} = h_{\mathcal{B}}^{B\,B'} \circ F_A^B \circ h_{\mathcal{A}}^{A'\,A}. \tag{5.2}$$

(B) In particular, for any connected simple system $\mathcal{A}$ in $\mathcal{CSS}(\mathcal{H})$, *the identity morphism of* $\mathcal{A}$ in $\mathcal{CSS}(\mathcal{H})$, denoted $1_{\mathcal{A}}$, is specified by $(1_{\mathcal{A}})_A^A = 1_A$ for any $A \in \operatorname{ob}(\mathcal{A})$. It follows from (5.2) and the connected simple system identities (1.1) that if A' and A'' are any objects of $\mathcal{A}$, then $(1_{\mathcal{A}})_{A'}^{A''} = h_{\mathcal{A}}^{A'\,A''}$.

(C) If $G: \mathcal{B} \to \mathcal{C}$ is also a map of connected simple systems in $\mathcal{CSS}(\mathcal{H})$, then informally *the composition* $G \circ F$ is obtained by juxtaposing a diagram in G to the right of a diagram in F whenever the right vertical arrow of the diagram in F coincides with the left vertical arrow of the diagram in G and then deleting this common arrow from the juxtaposed diagrams. Some work must be done to show the resulting collection of diagrams has the necessary uniqueness property. See [K3] for details. Clearly this composition is then associative.

It is easily verified that with this definition of composition, the self-map of each connected simple system defined in (B) indeed functions as an identity morphism. Hence, (A), (B), and (C) together yield that $\mathcal{CSS}(\mathcal{H})$ is a well-defined category.

5.1.1 Remark. From (5.2), (B), and (C) it follows that a map of connected simple systems $F: \mathcal{A} \to \mathcal{B}$ in $\mathcal{CSS}(\mathcal{H})$ is an equivalence, i.e., there exists $G: \mathcal{B} \to \mathcal{A}$ in $\mathcal{CSS}(\mathcal{H})$ so that $G \circ F = 1_{\mathcal{A}}$ and $F \circ G = 1_{\mathcal{B}}$, if, and only if, F_A^B is an equivalence in $\mathcal{H}$ for one, hence every, pair $(A, B) \in \operatorname{ob}(\mathcal{A}) \times \operatorname{ob}(\mathcal{B})$.

Recall that in Definition 1.5 we observed that every connected simple system is a directed set and used this to show that to every functor on $\mathcal{T}^{*\prime}$, the homotopy category of pointed topological spaces, into ${}_R\mathcal{M}$ (or $\mathcal{G}_R\mathcal{M}$, or $\partial_R\mathcal{M}'$) one could assign limit modules to a connected simple system that is a subcategory of $\mathcal{T}^{*\prime}$. The following result completes that construction in that it shows that each such functor induces a functor on $\mathcal{CSS}(\mathcal{T}^{*\prime})$ to ${}_R\mathcal{M}$ (or $\mathcal{G}_R\mathcal{M}$, or $\partial_R\mathcal{M}'$) and generalizes it by replacing $\mathcal{T}^{*\prime}$ with any category $\mathcal{H}$. In particular, for $\mathcal{C}$ a connected simple system and a subcategory of a category $\mathcal{H}$ and F a functor on $\mathcal{H}$ to a category $\mathcal{X}$,

$$\{\{\mathsf{F}(X)\}_{X \in \operatorname{ob}(\mathcal{C})}, \{\mathsf{F}(h_{\mathcal{C}}^{X\,X'}\}_{X \preceq X' \in \operatorname{ob}(\mathcal{C})}\} \tag{5.3}$$

is as F is covariant or contravariant a direct or inverse system of objects and isomorphisms in $\mathcal{X}$ directed over $\operatorname{ob}(\mathcal{C})$. The limits taken in the following theorem are those of the system (5.3).

5.2 THEOREM. *Associated to each functor* F *on a category* $\mathcal{H}$ *to* $\mathcal{X}$ *where* $\mathcal{X}$ *is one of the categories* ${}_R\mathcal{M}$, $\mathcal{G}_R\mathcal{M}$, *or* $\partial_R\mathcal{M}'$ *is an induced functor of the same variance also denoted by* F *on* $\mathcal{CSS}(\mathcal{H})$ *to* $\mathcal{X}$ *with the following properties:*

(1) *If* F *is covariant,* $\mathsf{F}(\mathcal{C}) := \varinjlim \mathsf{F}(X)$ *and for each* $X \in \mathrm{ob}(\mathcal{C})$, *the canonical map* $\iota^X : \mathsf{F}(X) \to \mathsf{F}(\mathcal{C})$ *is an* $\mathcal{X}$*-isomorphism;*

(2) *if* F *is contravariant,* $\mathsf{F}(\mathcal{C}) := \varprojlim \mathsf{F}(X)$ *and for each* $X \in \mathrm{ob}(\mathcal{C})$, *the canonical map* $\pi_X : \mathsf{F}(\mathcal{C}) \to \mathsf{F}(X)$ *is an* $\mathcal{X}$*-isomorphism;*

(3) *if* $f : \mathcal{A} \to \mathcal{B}$ *is a map of connected simple systems in* $\mathcal{CSS}(\mathcal{H})$, *then* $\mathsf{F}(f)$ *is a morphism in* $\mathcal{X}$ *with the property that*

(*i*) *if* F *is covariant there is a commutative diagram*

$$\begin{array}{ccc} \mathsf{F}(\mathcal{A}) & \xrightarrow{\mathsf{F}(f)} & \mathsf{F}(\mathcal{B}) \\ {\scriptstyle i^A}\uparrow & & \uparrow{\scriptstyle i^B} \\ \mathsf{F}(A) & \xrightarrow{\mathsf{F}(f_A^B)} & \mathsf{F}(B) \end{array} \qquad \textit{for } A \in \mathrm{ob}(\mathcal{A}),\ B \in \mathrm{ob}(\mathcal{B});$$

(*ii*) *if* F *is contravariant there is a commutative diagram*

$$\begin{array}{ccc} \mathsf{F}(\mathcal{A}) & \xleftarrow{\mathsf{F}(f)} & \mathsf{F}(\mathcal{B}) \\ {\scriptstyle \pi_A}\downarrow & & \downarrow{\scriptstyle \pi_B} \\ \mathsf{F}(A) & \xleftarrow{\mathsf{F}(f_A^B)} & \mathsf{F}(B) \end{array} \qquad \textit{for } A \in \mathrm{ob}(\mathcal{A}),\ B \in \mathrm{ob}(\mathcal{B}).$$

Proof. Parts (1) and (2) of the theorem follow from the constructions given in Definition 1.5(A) by substituting the objects and morphisms of $\mathcal{H}$ for those of $\mathcal{T}^{*\prime}$ as no categorical properties of $\mathcal{T}^{*\prime}$ specific to it were used in the construction. Note, however, that F here corresponds to $\widetilde{\mathsf{F}}$ in the construction of Definition 1.5(A) and that there is no correspondent to the functor F in that definition. It also follows from the observations made in Definition 1.5(A) that, for any $\mathcal{C} \in \mathrm{ob}(\mathcal{CSS}(\mathcal{H}))$, for any $X \in \mathrm{ob}(\mathcal{C})$, when F is covariant the canonical map ι^X is an $\mathcal{X}$-isomorphism, but when F is contravariant the canonical map π_X is an $\mathcal{X}$-isomorphism.

To avoid confusion in giving the definition of how the induced functor acts on maps of connected simple systems, let us temporarily use $\bar{\mathsf{F}}$ to denote the induced functor.

For each $f : \mathcal{A} \to \mathcal{B}$ a map of connected simple systems in $\mathcal{CSS}(\mathcal{H})$, the morphism $\bar{\mathsf{F}}(f)$ is defined as follows: Choose $A \in \mathrm{ob}(\mathcal{A})$ and $B \in \mathrm{ob}(\mathcal{B})$ and if F is covariant set

$$\bar{\mathsf{F}}(f) := i^B \circ \mathsf{F}(f_A^B) \circ (i^A)^{-1} : \bar{\mathsf{F}}(\mathcal{A}) \to \bar{\mathsf{F}}(\mathcal{B}),$$

however, if F is contravariant, set

$$\bar{\mathsf{F}}(f) := (\pi_A)^{-1} \circ \mathsf{F}(f_A^B) \circ \pi_B : \bar{\mathsf{F}}(\mathcal{B}) \to \bar{\mathsf{F}}(\mathcal{A}).$$

It is immediate from the definition of f as a collection of commutative diagrams of the same type as in (5.1) that the definition of $\bar{\mathsf{F}}(f)$ is independent of the choice of

A and B. It is then easily verified that so defined $\bar{\mathsf{F}}$ is a functor on $\mathcal{CSS}(\mathcal{H})$ to $\mathcal{X}$ of the same variance as F. □

From now until the proof of Theorem 3.1 is completed, in the notation of Chapter 1, let $\varphi: G_\Lambda \to M$ denote a continuous family of C^s flows in M $(0 \leq s \leq \infty)$ and let $\pi_\varphi: \mathcal{S}(\varphi) \to \Lambda$ be the natural projection on the associated space of isolated invariant sets.

Before we can define the map of connected simple systems induced by a path in $\mathcal{S}(\varphi)$ between the Conley indices at the endpoints of the path, we must first define a class of maps between Conley indices of isolated invariant sets where one invariant set is a slice of the other and so require the following preliminary definition and lemma.

5.3 Definition. Let $\tau: M \times \mathbf{R} \times \Lambda \to M \times \Lambda \times \mathbf{R}$ be the homeomorphism transposing the second and third coordinates, and for K a compact subset of Λ, set G_K equal to $G_\Lambda \cap (M \times \mathbf{R} \times K)$ and let τG_K be the image of G_K under τ. Then define $\varphi_K: \tau G_K \to M \times K$ by $\varphi_K((\mathbf{u}, \lambda), t) := (\mathbf{u} \overset{\lambda}{\cdot} t, \lambda)$. The definition of a C^0 flow on a manifold made in Chapter 1 works equally well if "manifold" is replaced by "locally compact Hausdorff space", and so defined, it follows that φ_K is a continuous flow in the locally compact Hausdorff space $M \times K$ because φ is a continuous family of flows in M.

Let us also note that the definitions of "isolated invariant set" and "Conley index" given in Chapter 1 also carry over to the setting of a continuous flow in a locally compact Hausdorff space, which in fact is the setting used in [C]. However, in that setting, see [K5] as regards the definition of the morphisms in a Conley index given in Definition 1.3(C).

The following result is due to Conley; for a proof see [C, Theorem IV.2.1.B] or [Sl, Lemma 6.4(i)].

5.4 Lemma. *For K a compact subset of Λ, S_K is an isolated invariant set of φ_K if, and only if, there exists a continuous section $\sigma: K \to \mathcal{S}(\varphi)$ of π_φ with the property*

$$S_K = \bigcup_{\lambda \in K} S_\sigma(\lambda) \times \{\lambda\}.$$

As a consequence of this lemma we can make the following definition.

5.5 Definition. For compact $K \subset M$ and S_K an isolated invariant set of the flow φ_K, let σ be the section of π_φ guaranteed by Lemma 5.4 and for each $\lambda \in K$ define a map of connected simple systems $F(\lambda, K): \mathcal{C}(S_\sigma(\lambda)) \to \mathcal{C}(S_K)$ as follows. Choose an index pair $\langle N_1, N_0 \rangle$ for S_K and for each $\lambda \in K$, for $i = 0, 1$, define $N_i(\lambda)$ by the equality

$$N_i(\lambda) \times \{\lambda\} = N_i \cap (M \times \{\lambda\}).$$

So defined, it is easy to verify that $\langle N_1(\lambda), N_0(\lambda) \rangle$ is an index pair for $S_\sigma(\lambda)$ for each $\lambda \in K$. Hence, for each $\lambda \in K$ the embedding $\iota_{\lambda K}: M \to M \times K$ given by $\iota_{\lambda K}(\mathbf{u}) = (\mathbf{u}, \lambda)$ induces a continuous map of quotients

$$\bar{\iota}_{\lambda K}: N_1(\lambda)/N_0(\lambda) \to N_1/N_0$$

and via formula (5.2) applied to the situation at hand thereby defines a map of connected simple systems from $\mathcal{C}(S_\sigma(\lambda))$ to $\mathcal{C}(S_K)$ which we take to be $F(\lambda, K)$.

N.B. *It is quite important to the proof of the invariance of intersection numbers under continuation given below that for each $\lambda \in K$, the map of connected simple simple systems $F(\lambda, K)$ as just defined is independent of the index pair $\langle N_1, N_0\rangle$ chosen and used to make the definition.* See [K3, Proposition 2.5 and Corollary 2.22] or [Sl, Proposition 6.5] for a proof of this fact.

The definition of $\mathcal{C}(\beta)$ rests heavily on the fact that "small" compact K can be found for which $F(\lambda, K)$ is an equivalence. Note that by Remark 5.1.1, $F(\lambda, K)$ will be an equivalence in $\mathcal{CSS}(\mathcal{T}^{*\prime})$ if, and only if, $\bar{\iota}_{\lambda K}$ is a homotopy equivalence, and for some compact $K \subset \Lambda$, this is indeed the case. Specifically, the following existence lemma is proved in [C,Theorem IV.2.2C] given W open in Λ and $\sigma: W \to \mathcal{S}(\varphi)$ a local section of π_φ.

5.6 LEMMA. *For each $\nu \in W$ given N_ν that ν-isolates $S_\sigma(\nu)$, there exists V_ν open, $\nu \in V_\nu \subset W \cap \Lambda(N_\nu)$, with the property that if $\nu \in K \subset V_\nu$ and K is compact and contractible, then $F(\lambda, K): \mathcal{C}(S_\sigma(\lambda)) \to \mathcal{C}(S_K)$ is an equivalence in $\mathcal{CSS}(\mathcal{T}^{*\prime})$.*

5.7 DEFINITION. Suppose that β is a path in $\mathcal{S}(\varphi)$. Let us first define $\mathcal{C}(\beta)$ in the following special case.

(*A*) Assume $\Lambda = [0,1]$ and β is a section of π_φ. For each $\nu \in \Lambda$ choose N_ν that ν-isolates $S_\beta(\nu)$ and apply Lemma 5.6 to get an open cover $\{V_\nu : \nu \in \Lambda\}$ of $[0,1]$ so that each element V_λ of this cover satisfies $\lambda \in V_\lambda \subset \Lambda(N_\lambda)$ and has the property that if $\lambda \in K \subset V_\lambda$ and K is compact and contractible, then $F(\lambda, K)$ is an equivalence.

From the compactness of $[0,1]$ it follows that there exists a partition of $[0,1]$, $\mathcal{P}: 0 = \lambda_0 < \ldots < \lambda_n = 1$, subordinate to the cover $\{V_\nu : \nu \in \Lambda\}$. Set $K_j := [\lambda_{j-1}, \lambda_j]$ and define

$$S_j := S_{K_j} := \bigcup_{\lambda \in K_j} S_\beta(\lambda) \times \{\lambda\} \qquad (j = 1, \ldots, n)$$

so that S_j is an isolated invariant set of the flow φ_{K_j} as a consequence of Lemma 5.4. Then the map of Conley indices

$$F_{ij} := F(\lambda_i, K_j): \mathcal{C}(S_\sigma(\lambda_i)) \to \mathcal{C}(S_j) \qquad (i = j-1, j;\ j = 1, \ldots, n)$$

is an equivalence in $\mathcal{CSS}(\mathcal{T}^{*\prime})$ by virtue of which there is a well-defined equivalence $F_{jj}^{-1} \circ F_{j-1\,j}: \mathcal{C}(S(\lambda_{j-1})) \to \mathcal{C}(S(\lambda_j))$ for $j = 1, \ldots, n$. Thus, there is a well-defined equivalence in $\mathcal{CSS}(\mathcal{T}^{*\prime})$

$$F(\beta, \mathcal{P}): \mathcal{C}(S_\beta(0)) \to \mathcal{C}(S_\beta(1))$$

given by the composite

$$F(\beta, \mathcal{P}) := \left(F_{nn}^{-1} \circ F_{n-1\,n}\right) \circ \cdots \circ \left(F_{11}^{-1} \circ F_{01}\right), \tag{5.4}$$

and we define

$$\mathcal{C}(\beta) := F(\beta, \mathcal{P}).$$

This definition of $\mathcal{C}(\beta)$ is in fact independent of (i) for $\nu \in \Lambda$, the choice of isolating neighborhood N_ν that ν-isolates $S_\beta(\nu)$, (ii) the choice of cover $\{V_\nu : \nu \in \Lambda\}$, and (iii) the choice of partition $\mathcal{P}$ subordinate to that cover so long as F_{ij} $(i = j-1, j;$ $j = 1, \dots, n)$ is an equivalence; see [C, §IV.2.3] and [K3,Proposition 2.11].

(B) Let us now treat the general case: Λ is some locally compact Hausdorff space and β is an arbitrary path in $\mathcal{S}(\varphi)$. In essence, we reduce this case to the previous case by pulling back π_φ by λ_β. Specifically, set $\Lambda' = [0,1]$ and without loss of generality we can assume Λ' and Λ are disjoint; otherwise identify Λ' with a copy of itself that is disjoint. For each $\lambda' \in \Lambda'$, set $G_{\lambda'} := G_{\lambda_\beta(\lambda')}$ and define a C^s flow $\psi_{\lambda'}$ in M with domain $G_{\lambda'}$ by

$$\psi_{\lambda'}(\mathbf{u}, t) := \varphi_{\lambda_\beta(\lambda')}(\mathbf{u}, t)\,.$$

Then $\{\psi_{\lambda'}\}_{\lambda' \in \Lambda'}$ is a continuous family of C^s flows in M because $\{\varphi_\lambda\}_{\lambda \in \Lambda}$ is. By definition, $\psi = \lambda_\beta^* \varphi$; i.e., $\psi(\mathbf{u}, t, \lambda') = \varphi(\mathbf{u}, t, \lambda_\beta(\lambda')\}$. Note that S is isolated with respect to the flow $\psi_{\lambda'}$ if, and only if, S is isolated with respect to the flow $\varphi_{\lambda_\beta(\lambda')}$. Define a path β' in $\mathcal{S}(\psi)$ by $\beta'(\lambda') = (S_\beta(\lambda'), \lambda')$ and note that β' is a section of $\mathcal{S}(\lambda_\beta^*\varphi)$, whence $\mathcal{C}(\beta')$ has already been defined by part (A) of this definition. Then define $\mathcal{C}(\beta) := \mathcal{C}(\beta')$. Since by definition $S_{\beta'}(\lambda') = S_\beta(\lambda')$ we get that $\mathcal{C}(\beta)$ is indeed a map between $\mathcal{C}(S_\beta(0))$ and $\mathcal{C}(S_\beta(1))$.

The definition of $\mathcal{C}(\beta)$ just given is equivalent to the definition given in [K3][1] although the formalism used to describe continuous families of flows in [K3] follows that of [C] and is based on the idea of embedding each flow in M into the space of regular curves in $M \times \Lambda$ making the notation used there much more complicated. It is also shown in [K3] that $\mathcal{C}(\beta)$ is actually an invariant of the homotopy class of a path β (endpoints fixed) and in fact is a contravariant functor on $\Pi(\mathcal{S}(\varphi))$, the fundamental groupoid of $\mathcal{S}(\varphi)$, to $\mathcal{CSS}(\mathcal{T}^{*\prime})$; see [K3, Theorem 2.19 and Corollary 2.22]. (The fundamental groupoid of a space X is the small category $\Pi(X)$ with X the set of objects and for points $x, x' \in X$, the set of morphisms $\Pi(x, x')$ is the set of homotopy classes of paths (endpoints fixed) in X from x' to x with composition the usual multiplication of paths.)

(C) Similarly there is a contravariant functor $\mathcal{C}^*\colon \Pi(\mathcal{S}(\varphi)) \to \mathcal{CSS}(\mathcal{T}^{*\prime})$ that assigns the reverse time Conley index $\mathcal{C}^*(S)$ to a point $(S, \lambda) \in \mathcal{S}(\varphi)$ and assigns to a path class β in $\mathcal{S}(\varphi)$ a map of connected simple systems $\mathcal{C}^*(\beta)\colon \mathcal{C}^*(S_\beta(0)) \to \mathcal{C}^*(S_\beta(1))$ where $\mathcal{C}^*(\beta)$ is defined just as $\mathcal{C}(\beta)$ is, but using the family of time-reversed flows $\{\varphi_\lambda^*\}_{\lambda \in \Lambda}$; i.e., index pairs relative to the family of time-reversed flows are used in the definition rather than index pairs relative to the given flows.

(D) As noted at the beginning of the chapter, for β a path or the homotopy class of a path in $\mathcal{S}(\varphi)$, the graded R-module homomorphisms (coefficients taken in an R-module G)

$$\widetilde{H}_*\mathcal{C}(S_\beta(0)) \xrightarrow{\mathcal{C}(\beta)_*} \widetilde{H}_*\mathcal{C}(S_\beta(1)) \text{ and } \widetilde{H}_*\mathcal{C}^*(S_\beta(0)) \xrightarrow{\mathcal{C}^*(\beta)_*} \widetilde{H}_*\mathcal{C}^*(S_\beta(1))$$

[1] However, there is an error in [K3] in the definition of the pullback of a product parameterization. Part (2) of that definition beginning on line 6 of page 305 should be replaced with the following:

(2) Define $f^*\Psi$ to be the subset of $\Psi_{f(\Lambda'')} \times \Lambda''$ consisting of those points (γ', λ'') satisfying $\gamma' \in \Psi_{f(\lambda'')}$.

are obtained by applying Theorem 5.2 to the reduced singular homology functor (coefficients in G) and the maps of connected simple systems $\mathcal{C}(\beta)$ and $\mathcal{C}^*(\beta)$. Because the path class β regarded as a morphism of $\Pi(\mathcal{S}(\varphi))$ is invertible with the inverse being given by running the path backwards, the fact that $\mathcal{C}$ and $\mathcal{C}^*$ are functors on $\Pi(\mathcal{S}(\varphi))$ implies that $\mathcal{C}(\beta)$ and $\mathcal{C}^*(\beta)$ are invertible, hence so too are $\mathcal{C}(\beta)_*$ and $\mathcal{C}^*(\beta)_*$.

B. The Proof of Theorems 3.1, 3.2, 3.3, and 3.7.

5.8 Proof of Theorem 3.3. For each $\lambda \in [0,1]$, choose C^r isolating blocks with corners N_λ and N^*_λ for $S_{\sigma(\lambda)}$ as guaranteed by Proposition 4.4 so that the pair of index pairs $\big((N_\lambda, N^+_\lambda), (N^*_\lambda, N^{*-}_\lambda)\big)$ is AIP(regular) for $S_{\sigma(\lambda)}$. Because of the continuity of the curve of isolated invariant sets, the transversality of the vectorfield X_λ to the codimension one submanifolds containing the exit and entrance sets of N_λ and N^*_λ, and the continuity of $\lambda \to X_\lambda$, about each $\lambda \in [0,1]$ is centered an interval J_λ so that $\lambda' \in J_\lambda$ implies $\big((N_\lambda, N^+_\lambda), (N^*_\lambda, N^{*-}_\lambda)\big)$ is AIP(regular) for $S_\sigma(\lambda')$.

Choose a partition of $[0,1]$, $\mathcal{P} : 0 = \lambda_0 < \dots, \lambda_n = 1$, subordinate to the cover of $[0,1]$ by the intervals J_λ ($\lambda \in [0,1]$). Let K_j and S_j ($j = 1,\dots,n$) be defined relative to the partition $\mathcal{P}$ as in part (A) of Definition 5.7. Because K_j is contractible, it follows that for each $\lambda \in K_j$, the embedding of pairs

$$
\begin{aligned}
\iota_{\lambda j} \colon (N_{\lambda_j}, N^+_{\lambda_j}) &\to (N_{\lambda_j} \times K_j, N^+_{\lambda_j} \times K_j) \\
\mathbf{u} &\mapsto (\mathbf{u}, \lambda)
\end{aligned}
\tag{5.5}
$$

induces a homotopy equivalence $\bar\iota_{\lambda j} \colon N_{\lambda_j}/N^+_{\lambda_j} \to (N_{\lambda_j} \times K_j)/(N^+_{\lambda_j} \times K_j)$ between index spaces and thereby determines the equivalence of connected simple systems $F(\lambda, K_j) \colon \mathcal{C}(S_\sigma(\lambda)) \to \mathcal{C}(S_j)$. Thus with $F_{ij} := F(\lambda_i, K_j)$ for $i = j-1, j$ and $j = 1,\dots,n$,

$$\mathcal{C}(\beta) = (F^{-1}_{nn} \circ F_{n-1\,n}) \circ \dots \circ (F^{-1}_{11} \circ F_{01}).$$

Analogously, the embedding of pairs

$$
\begin{aligned}
\jmath_{\lambda i} \colon (N^*_{\lambda_i}, N^{*-}_{\lambda_i}) &\to (N^*_{\lambda_i} \times K_i, N^{*-}_{\lambda_i} \times K_i) \\
\mathbf{u} &\mapsto (\mathbf{u}, \lambda)
\end{aligned}
\tag{5.6}
$$

induces a homotopy equivalence $\bar\jmath_{\lambda i} \colon N^*_{\lambda_i}/N^{*-}_{\lambda_i} \to (N^*_{\lambda_i} \times K_i)/(N^{*-}_{\lambda_i} \times K_i)$ of index spaces for the reverse time flow that in turn determines an equivalence of connected simple systems $G(\lambda, K_i) \colon \mathcal{C}^*(S_\sigma(\lambda)) \to \mathcal{C}^*(S_i)$. Thus with $G_{\ell i} := G(\lambda_\ell, K_i)$ for $\ell = i-1, i$ and $i = 1,\dots,n$,

$$\mathcal{C}^*(\beta) = (G^{-1}_{nn} \circ G_{n-1\,n}) \circ \dots \circ (G^{-1}_{11} \circ G_{01}).$$

Let us note that the rule $r(\mathbf{u}, \lambda) = \mathbf{u}$ defines a left inverse to the embedding (5.5); hence induces a map on quotients $\bar r$ that is a homotopy inverse to $\bar\iota_{\lambda j}$; hence determines the map of connected simple systems $F(\lambda, K_j)^{-1}$. In the same way the rule r induces a map on quotients that is a homotopy inverse to $\bar\jmath_{\lambda i}$ and determines $G(\lambda, K_i)^{-1}$.

Set $\boldsymbol{\alpha}_0 := \boldsymbol{\alpha}$ and $\boldsymbol{\gamma}_0 := \boldsymbol{\gamma}$ and inductively define $\boldsymbol{\alpha}_j := F_{jj*}^{-1} F_{j-1\,j*} \boldsymbol{\alpha}_{j-1}$ and $\boldsymbol{\gamma}_j := G_{jj*}^{-1} G_{j-1\,j*} \boldsymbol{\gamma}_{j-1}$ for $j = 1, \ldots, n$. To finish, it suffices to show that

$$(5.7) \qquad {}^{\#}(\boldsymbol{\alpha}_{j-1} \mathbin{\hat{\bullet}} \boldsymbol{\gamma}_{j-1}) = {}^{\#}(\boldsymbol{\alpha}_j \mathbin{\hat{\bullet}} \boldsymbol{\gamma}_j) \qquad (j = 1, \ldots, n).$$

To show that (5.7) holds, let us first make the simple observation that if $V \subset M$ and $\xi \in H_0(V)$ and $i_M^V \colon H_*(V) \to H_*(M)$ is the inclusion induced homomorphism, then

$$\langle 1, i_M^V \xi \rangle = \langle 1, \xi \rangle$$

where on the left $1 \in H^0(M)$, but on the right $1 \in H^0(V)$ and is in both cases the cohomology class of the augmentation on 0-chains into R. Hence for each compact $K \subset M$, for each $\check{\xi} \in \check{H}_0(K)$, and for each $V \in \mathcal{N}(K)$,

$$(5.8) \qquad \langle 1, \check{\xi}_M \rangle = \langle 1, \check{\xi}_V \rangle .$$

Suppose then α is a relative cycle in N_{λ_j} modulo $N_{\lambda_j}^+$ representing $\boldsymbol{\alpha}_{j-1}$. Because the homotopy class $\bar{\iota}_{\lambda_j j}^{-1} \circ \bar{\iota}_{\lambda_{j-1} j}$ is represented by $\bar{r} \circ \bar{\iota}_{\lambda_{j-1} j} = 1_{N_{\lambda_j}/N_{\lambda_j}^+}$ it follows that the map of connected simple systems $F_{jj}^{-1} \circ F_{j-1\,j}$ is determined by the identity map $1_{N_{\lambda_j}/N_{\lambda_j}^+}$. It is then immediate that α also represents $\boldsymbol{\alpha}_j$. Similarly, if γ is a relative cycle in $N_{\lambda_j}^*$ modulo $N_{\lambda_j}^{*-}$ representing $\boldsymbol{\gamma}_{j-1}$, then also γ represents $\boldsymbol{\gamma}_j$. Choose $V \in M_{\text{\AA} o}^{(2)}(N_{\lambda_j})$. Then from Definition 2.12, from Theorem 2.11 (in particular Remark 2.11.1 with $P := N_{\lambda_j}$), from the cofinality of $\{N_{\lambda_j}\}$ in $M_{\text{\AA} c}^{(2)}(N_{\lambda_j})$, from the definition of the pairing $\mathfrak{L}_{N_{\lambda_j}}^{N_{\lambda_j}}$ (a particular case of (2.8)), and from the cofinality of $\{V_1 \cap V_1^*\}$ in $\mathcal{N}(V_1 \cap V_1^*)$, it follows that

$$(5.9) \qquad (\boldsymbol{\alpha}_k \mathbin{\hat{\bullet}} \boldsymbol{\gamma}_k)_{V_1 \cap V_1^*} = \mathfrak{L}_{V_1 \cap V_1^*}^{N_{\lambda_j}} (\alpha \otimes \gamma) \qquad \text{for } k = j-1, j.$$

It is immediate from (5.9) and the observation (5.8) that (5.7) holds as desired. □

To prove the analogous result without the smoothness assumption on M and the flows φ_λ $(\lambda \in \Lambda)$ requires a technical result that describes how certain pairs of index pairs that are AIP(thick) for $S_\sigma(\lambda)$ can be perturbed to index pairs that are AIP(thick) for $S_\sigma(\lambda')$ for λ' sufficiently close to λ. This will allow us to replace the use of $1_{N_{\lambda_j}/N_{\lambda_j}^+}$ in the preceding demonstration by a suitable inclusion induced map of index spaces, but otherwise the proof is essentially unchanged. The proof of the following lemma is deferred until after the proof of Theorem 3.1.

5.9 LEMMA. *Given a local section σ of $\mathcal{S}(\varphi)$ defined on W open in Λ, for each integer $k \geq 2$, for each $\nu \in W$, there exist ν-isolating neighborhoods N_ν and N_ν^* of $S_\sigma(\nu)$, and an open neighborhood V_ν of ν, $V_\nu \subset W \cap \Lambda(N_\nu) \cap \Lambda(N_\nu^*)$, so that for each $\lambda \in V_\nu$, for $j = 1, \ldots, k$, there are index pairs $\langle P_{1j}(\lambda), P_{0j}(\lambda) \rangle$ relative to N_ν and $\langle P_{1j}^*(\lambda), P_{0j}^*(\lambda) \rangle$ relative to N_ν^* for $S_\sigma(\lambda)$, the former with respect to the forward time flow, the latter with respect to the reverse time flow, satisfying the following statements:*

(1) $((P_{1j}(\lambda), {\cap} P_{0j}(\lambda)), (P_{1j}^*(\lambda), {\cap} P_{0j}^*(\lambda)))$ *is AIP(thick) for* $S_\sigma(\lambda)$;

(2) $P_{ij-1}(\lambda) \subset \text{int}_{N_\nu}(P_{ij}(\nu))$ and $P^*_{ij-1}(\lambda) \subset \text{int}_{N^*_\nu}(P^*_{ij}(\nu))$ $(j = 2, \ldots, k)$;
(3) $P_{ij-1}(\nu) \subset \text{int}_{N_\nu}(P_{ij}(\lambda))$ and $P^*_{ij-1}(\nu) \subset \text{int}_{N^*_\nu}(P^*_{ij}(\lambda))$ $(j = 2, \ldots, k)$;
(4) $K \subset V_\nu$ and K is compact and contractible implies
  (i) S_K has as index pairs $\langle \bar{P}_{1j}(K), \bar{P}_{0j}(K) \rangle$ and $\langle \bar{P}^*_{1j}(K), \bar{P}^*_{0j}(K) \rangle$, the former with respect to the forward time flow, the latter with respect to the reverse time flow, where $\bar{P}_{ij}(K) := \bigcup_{\lambda \in K} P_{ij}(\lambda) \times \{\lambda\}$ and $\bar{P}^*_{ij}(K) := \bigcup_{\lambda \in K} P^*_{ij}(\lambda) \times \{\lambda\}$ $(i = 0, 1; j = 1, \ldots, k)$;
  (ii) $\lambda \in K$ implies that for $j = 1, \ldots, k$, the map $\mathbf{u} \mapsto (\mathbf{u}, \lambda)$ induces homotopy equivalences

$$\bar{\iota}_{\lambda K j} : P_{1j}(\lambda)/P_{0j}(\lambda) \to \bar{P}_{1j}(K)/\bar{P}_{0j}(K),$$

and

$$\bar{\jmath}_{\lambda K j} : P^*_{1j}(\lambda)/P^*_{0j}(\lambda) \to \bar{P}^*_{1j}(K)/\bar{P}^*_{0j}(K).$$

5.10 Proof of Theorem 3.2. By Lemma 5.9, there is a cover of $[0,1]$ by open sets $V_\nu \subset W$, $\nu \in W$, and index pairs $\langle P_{1j}(\lambda), P_{0j}(\lambda) \rangle$ and $\langle P^*_{1j}(\lambda), P^*_{0j}(\lambda) \rangle$ $(j = 1, 2)$ satisfying the properties listed in that result. Choose a partition of $[0, 1]$, $\mathcal{P} : 0 = \lambda_0 < \ldots < \lambda_n = 1$, subordinate to the cover $\{V_\nu : \nu \in W\}$, and define K_j and S_j, $j = 1, \ldots, n$, relative to this partition as in part (A) of Definition 5.7.

For $i = j - 1, j$, for $j = 1, \ldots, n$, let

$$F_{ij} : \mathcal{C}(S_\sigma(\lambda_i)) \to \mathcal{C}(S_j)$$

and

$$G_{ij} : \mathcal{C}^*(S_\sigma(\lambda_i)) \to \mathcal{C}^*(S_j)$$

be the equivalences of Conley indices induced respectively by the homotopy equivalences

$$\bar{\iota}_{\lambda_i K_j 1} : P_{11}(\lambda_i)/P_{01}(\lambda_i) \to \bar{P}_{11}(K_j)/\bar{P}_{01}(K_j)$$

and

$$\bar{\jmath}_{\lambda_i K_j 1} : P^*_{11}(\lambda_i)/P^*_{01}(\lambda_i) \to \bar{P}^*_{11}(K_j)/\bar{P}^*_{01}(K_j).$$

It follows that

$$\mathcal{C}(\beta) = (F^{-1}_{nn} \circ F_{n-1\,n}) \circ \cdots \circ (F^{-1}_{11} \circ F_{01})$$

and

$$\mathcal{C}^*(\beta) = (G^{-1}_{nn} \circ G_{n-1\,n}) \circ \cdots \circ (G^{-1}_{11} \circ G_{01}).$$

For $j = 0, \ldots, n$, let $\boldsymbol{\alpha}_j$ and $\boldsymbol{\gamma}_j$ be defined as in the proof of Theorem 3.3. Again it suffices to show that (5.7) holds.

Claim $F^{-1}_{jj} \circ F_{j-1\,j}$ and $G^{-1}_{jj} \circ G_{j-1\,j}$ are respectively determined by the inclusion induced maps on quotients

$$\bar{f}_j : \frac{P_{11}(\lambda_{j-1})}{P_{01}(\lambda_{j-1})} \to \frac{P_{12}(\lambda_j)}{P_{02}(\lambda_j)} \quad \text{and} \quad \bar{g}_j : \frac{P^*_{11}(\lambda_{j-1})}{P^*_{01}(\lambda_{j-1})} \to \frac{P^*_{12}(\lambda_j)}{P^*_{02}(\lambda_j)}$$

where the inducing inclusions are respectively

$$f_j\colon (P_{1\,1}(\lambda_{j-1}), {}_\cap P_{0\,1}(\lambda_{j-1})) \hookrightarrow (P_{1\,2}(\lambda_j), {}_\cap P_{0\,2}(\lambda_j))$$

and

$$g_j\colon (P^*_{1\,1}(\lambda_{j-1}), {}_\cap P^*_{0\,1}(\lambda_{j-1})) \hookrightarrow (P^*_{1\,2}(\lambda_j), {}_\cap P^*_{0\,2}(\lambda_j))\,.$$

Furthermore, for $i = 0, 1$, $j = 1, 2$, and $k = j-1, j$, setting $P_{i\,j\,k} := P_{i\,j}(\lambda_k)$ and $P^*_{i\,j\,k} := P^*_{i\,j}(\lambda_k)$ the diagram
(5.10)

$$\begin{array}{ccc} H(P_{1\,2\,j}, {}_\cap P_{0\,2\,j}) \otimes H(P^*_{1\,2\,j}, {}_\cap P^*_{0\,2\,j}) & \xrightarrow{\mathfrak{L}_{P_2}} & \check{H}(P_{1\,2\,j} \cap P^*_{1\,2\,j}) \\ {\scriptstyle f_{j\,*} \otimes g_{j\,*}} \big\uparrow & & \big\uparrow {\scriptstyle \check{\imath}^{P_1}_{P_2}} \\ H(P_{1\,1\,j-1}, {}_\cap P_{0\,1\,j-1}) \otimes H(P^*_{1\,1\,j-1}, {}_\cap P^*_{0\,1\,j-1}) & \xrightarrow{\mathfrak{L}_{P_1}} & \check{H}(P_{1\,1\,j-1} \cap P^*_{1\,1\,j-1}) \end{array}$$

commutes where $\check{\imath}^{P_1}_{P_2}$ is the inclusion induced homomorphism as follows immediately from Lemma 2.10.

Assuming the claim that $\bar{f}_j$ and $\bar{g}_j$ respectively induce $F^{-1}_{j\,j} \circ F_{j-1\,j}$ and $G^{-1}_{j\,j} \circ G_{j-1\,j}$, choose

$$\alpha \in H_{m-k}(P_{1\,1}(\lambda_{j-1}), {}_\cap P_{0\,1}(\lambda_{j-1})) \quad \text{representing } \boldsymbol{\alpha}_{j-1}\,,$$

and

$$\gamma \in H_k(P^*_{1\,1}(\lambda_{j-1}), {}_\cap P^*_{0\,1}(\lambda_{j-1})) \qquad \text{representing } \boldsymbol{\gamma}_{j-1}\,.$$

Also choose $V \in M^{(2)}_{\not{\Delta}\,o}(P_2)$ and note that also $V \in M^{(2)}_{\not{\Delta}\,o}(P_1)$. Then

$$f_{j\,*}\alpha \in H_{m-k}(P_{1\,2}(\lambda_j), {}_\cap P_{0\,2}(\lambda_j)) \quad \text{and represents } \boldsymbol{\alpha}_j\,,$$

and

$$g_{j\,*}\gamma \in H_k(P^*_{1\,2}(\lambda_j), {}_\cap P^*_{0\,2}(\lambda_j)) \qquad \text{and represents } \boldsymbol{\gamma}_j\,.$$

Also, by definition of $\mathfrak{L}^{P_k}_V$ $(k = 1, 2)$ and the commutativity of diagram (5.10)

$$(5.11)\quad (\boldsymbol{\alpha}_{j-1} \,\widehat{\bullet}\, \boldsymbol{\gamma}_{j-1})_{V_1 \cap V_1^*} = \mathfrak{L}^{P_1}_V(\alpha \otimes \gamma) = \mathfrak{L}^{P_2}_V(f_{j\,*}\alpha \otimes g_{j\,*}\gamma) = (\boldsymbol{\alpha}_j \,\widehat{\bullet}\, \boldsymbol{\gamma}_j)_{V_1 \cap V_1^*}$$

where the two leftmost (rightmost) intersection pairings are for the isolated invariant set S_{j-1} (respectively, S_j) of the flow $\varphi_{\lambda_{j-1}}$ (respectively, φ_{λ_j}). Combined with our observation (5.8), the string of equalities (5.11) yields

$$\langle 1, (\boldsymbol{\alpha}_{j-1} \,\widehat{\bullet}\, \boldsymbol{\gamma}_{j-1})_M \rangle = \langle 1, (\boldsymbol{\alpha}_j \,\widehat{\bullet}\, \boldsymbol{\gamma}_j)_M \rangle\,,$$

whence (5.7) follows immediately.

It remains to prove the claim. Accordingly, consider the diagram of maps

$$\begin{array}{ccc} \dfrac{P_{1\,1}(\lambda_{j-1})}{P_{0\,1}(\lambda_{j-1})} & \xrightarrow{\bar{f}_j} & \dfrac{P_{1\,2}(\lambda_j)}{P_{0\,2}(\lambda_j)} \\ {\scriptstyle \bar{\iota}_{\lambda_{j-1}\,K_j\,1}} \big\downarrow & & \big\downarrow {\scriptstyle \bar{\iota}_{\lambda_j\,K_j\,2}} \\ \dfrac{\bar{P}_{1\,1}(K_j)}{\bar{P}_{0\,1}(K_j)} & \xrightarrow{\bar{\varphi}_j} & \dfrac{\bar{P}_{1\,2}(K_j)}{\bar{P}_{0\,2}(K_j)} \end{array}$$

where $\bar{\varphi}_j$ is induced by the functorial inclusion

$$\varphi_j: (\bar{P}_{1\,1}(K_j), \cap\bar{P}_{0\,1}(K_j)) \hookrightarrow (\bar{P}_{1\,2}(K_j), \cap\bar{P}_{0\,2}(K_j))$$

and therefore via [K5, Proposition 3.1] has homotopy class a morphism in $\mathcal{C}\,(S_j)$.

This diagram is homotopy commutative. To see this, define a homotopy H from $\bar{\varphi}_j \circ \bar{\iota}_{\lambda_{j-1}\,K_j\,1}$ to $\bar{\iota}_{\lambda_j\,K_j\,2} \circ \bar{f}_j$ via the formula

$$H([\mathbf{u}], s) := H_s([\mathbf{u}]) := [(\mathbf{u}, (1-s)\lambda_{j-1} + s\lambda_j)] \in \frac{\bar{P}_{1\,2}(K_j)}{\bar{P}_{0\,2}(K_j)}$$

for $[\mathbf{u}] \in P_{1\,1}(\lambda_{j-1})/P_{0\,1}(\lambda_{j-1})$ and $s \in [0,1]$. Note that H_s is well-defined for each $s \in [0,1]$ since it factors through the quotient space $(P_{1\,2}(\nu_j) \times K_j)/(P_{0\,2}(\nu_j) \times K_j)$ via the maps on quotients induced by the functorial inclusions

$$\begin{aligned}(\bar{P}_{1\,1}(K_j), \cap\bar{P}_{0\,1}(K_j)) &\subset (P_{1\,2}(\nu_j) \times K_j, P_{0\,2}(\nu_j) \cap P_{1\,2}(\nu_j) \times K_j)\\ &\subset (\bar{P}_{1\,2}(K_j), \cap\bar{P}_{0\,2}(K_j))\end{aligned}$$

where $\nu_j \in W$ is chosen so that $K_j \subset V_{\nu_j}$.

It follows that

$$\bar{f}_j \sim \bar{\iota}^{-1}_{\lambda_j\,K_j\,2} \circ \bar{\varphi}_j \circ \bar{\iota}_{\lambda_{j-1}\,K_j\,1}$$

where $\bar{\iota}^{-1}_{\lambda_j\,K_j\,2}$ denotes any homotopy inverse of $\bar{\iota}_{\lambda_j\,K_j\,2}$. Then as the homotopy class of $\bar{\varphi}_j$ is a morphism in $\mathcal{C}\,(S_j)$, by definition of $F_{j-1\,j}$ and the definition of a map of connected simple systems, it follows that the homotopy class of $\bar{\varphi}_j \circ \bar{\iota}_{\lambda_{j-1}\,K_j\,1}$ determines $F_{j-1\,j}: \mathcal{C}\,(S_\sigma(\lambda_{j-1})) \to \mathcal{C}\,(S_j)$. Thus, as $\bar{\iota}_{\lambda_j\,K_j\,2}$ determines $F_{j\,j}: \mathcal{C}\,(S_\sigma(\lambda_j)) \to \mathcal{C}\,(S_j)$, necessarily the homotopy class of $\bar{\iota}^{-1}_{\lambda_j\,K_j\,2}$ determines $F^{-1}_{j\,j}$, whence the homotopy class of $\bar{\iota}^{-1}_{\lambda_j\,K_j\,2} \circ \bar{\varphi}_j \circ \bar{\iota}_{\lambda_{j-1}\,K_j\,1}$, hence of $\bar{f}_j$, determines $F^{-1}_{j\,j} \circ F_{j-1\,j}$ as claimed. Similarly, the homotopy class of $\bar{g}_j$ determines $G^{-1}_{j\,j} \circ G_{j-1\,j}$. □

5.11 *Proof of Theorem 3.1.* Let $\beta: [0,1] \to \mathcal{S}(\varphi)$ be an arbitrary continuous path, let $\Lambda' = [0,1]$, and let the family of flows $\lambda^*_\beta\varphi$ and the path β' that is a section of $\pi_{\lambda^*_\beta\varphi}: \mathcal{S}(\lambda^*_\beta\varphi) \to \Lambda'$ be defined using β as in part (B) of Definition 5.7. By that definition, $\mathcal{C}\,(\beta) := \mathcal{C}\,(\beta')$; thus application of Theorem 3.2 to the section β' immediately yields the desired result. □

It remains to prove Lemma 5.9. The following result is needed first.

5.12 Lemma. *Suppose S is an isolated invariant set of a continuous flow on a locally compact Hausdorff space X, (N_1^*, N_0^*) is a nested index pair for S with respect to the reverse time flow with N_1^* isolating S, $\mathcal{O}$ is open in X, and $N_1^* \subset \mathcal{O}$. Then there exists an isolating neighborhood N of S so that $N \subset \mathcal{O}$ and (N_1^*, N_0^*) is an index pair relative to N, and there exist index pairs $\langle N_1, N_0\rangle$ and $\langle K_1, K_0\rangle$ for S relative to N with respect to the forward time flow satisfying*

(1) $N_1^* \cap N_0 = \emptyset = N_1 \cap N_0^*$,
(2) $K_i \subset \operatorname{int}_N(N_i)$, $i = 0, 1$,
(3) $\operatorname{cl}_X(N_1 \setminus K_0) \subset \operatorname{int}_X(N)$.

5.12.1 Definition. Call $\langle K_1, K_0\rangle$ a *perturbable index pair* if it is an index pair relative to some isolating neighborhood N and if there exists an index pair $\langle N_1, N_0\rangle$ relative to N so that conditions (2) and (3) above are satisfied. Note that it is generally impossible for a nested index pair to be perturbable because of conditions (2) and (3). Since it is the index pair $\langle K_1, K_0\rangle$ that is being perturbed, this definition of perturbable makes more sense than the definition of perturbable index pair in [K4] where $\langle N_1, N_0\rangle$ was called perturbable if (2) and (3) were satisfied for some index pair $\langle K_1, K_0\rangle$.

5.12.2 Remark. Lemma 5.12 yields another proof that there exists a pair of index pairs that is AIP(thick) for S. Choose any nested index pair (M_1^*, M_0^*) for S with respect to the reverse time flow. Then choose $s > 0$ so that $M_0^* \subset \operatorname{int}_{M_1^*}(M_1^* \cap M_0^{*-s})$. Set $N_1^* := M_1^*$ and $N_0^* := M_1^* \cap M_0^{*-s}$ and apply Lemma 5.12. Then the pair $((N_1, \cap N_0), (N_1^*, N_0^*))$ is AIP for S because both index pairs have thick exit set.

Proof of Lemma 5.12. For M any isolating neighborhood, the notations $A^+(M)$ and $A^-(M)$ denote respectively the forward and backward asymptotic sets of M with respect to the forward time flow; i.e., $\mathbf{u} \in A^\pm(M)$, if, and only if, $\mathbf{u}\cdot\mathbf{R}^\pm \subset M$. Similarly, $A^{+*}(M)$ and $A^{-*}(M)$ denote respectively the forward and backward asymptotic sets of M with respect to the reverse time flow. Thus, $A^{\pm *}(M) = A^\mp(M)$. With this last equality in mind, interpreting the defining properties of the reverse time index pair (N_1^*, N_0^*) in terms of the forward time flow as will be done throughout the proof, it follows that N_0^* and $A^-(N_1^*)$ are disjoint closed sets. Thus, choose disjoint open sets $V_1 \supset A^-(N_1^*)$ and $V_0 \supset N_0^*$ with $V_i \subset \mathcal{O}$ $(i = 0, 1)$, and set $F_0 := \partial_X(N_1^*) \cap A^-(N_1^*)$.

For each $\mathbf{u} \in F_0$, there exists $t > 0$ so that $\mathbf{u}\cdot t \in X \setminus N_1^*$. Then compactness of F_0 and continuity of the flow, imply that there exist strictly positive times $t_1, \dots, t_p$ and a finite cover of F_0 by compact neighborhoods $B_1, \dots, B_p$, each B_i a subset of V_1, so that $B_i\cdot t_i$ lies in $X\setminus N_1^*$. Let F_0' be the union of the $B_i\cdot t_i$, $i = 1, \dots, p$. Then F_0' is compact and disjoint from F_0. Then we can choose a compact neighborhood L_0 of F_0 that is disjoint from F_0', but is a subset of the union of the B_i, $i = 1, \dots, p$. Thus if $\mathbf{u} \in L_0$, for some i, $\mathbf{u}\cdot t_i$ lies in F_0', hence does not lie in $N_1^* \cup L_0$. It follows that $N_1^* \cup L_0$ is an isolating neighborhood of S contained in $\mathcal{O}$. Also N_1^* is negatively invariant relative to $N_1^* \cup L_0$ because any negative orbit segment originating in N_1^* and contained in $N_1^* \cup L_0$, but not in N_1^*, must exit N_1^* through N_0^*; hence there must be points on the orbit segment in $V_0 \cap L_0$ which is impossible. It follows that (N_1^*, N_0^*) is an index pair relative to $N_1^* \cup L_0$.

Set $F_1 := \partial_X(N_1^* \cup L_0) \cap A^-(N_1^* \cup L_0)$, and note F_1 is disjoint from N_1^*. For as L_0 is a neighborhood of F_0, certainly $F_1 \cap F_0 = \emptyset$, whence if $\mathbf{u} \in N_1^* \cap F_1$, then $\mathbf{u} \notin A^-(N_1^*)$. Thus, where $\tau_{N_1^*}^*$ is the entrance time map of N_1^* relative to the forward time flow, i.e., the exit time map relative to the reverse time flow,

$$\mathbf{u}' := \mathbf{u}\cdot -\tau_{N_1^*}^*(\mathbf{u}) \in N_0^* \subset V_0$$

and there exists $\varepsilon > 0$ so that $\mathbf{u}'\cdot[-\varepsilon, 0] \subset V_0$. Also, by definition of $\mathbf{u}'$, $\mathbf{u}'\cdot[-\varepsilon, 0]$ intersects $V_0 \setminus N_1^*$. However, as $\mathbf{u}' \in A^-(N_1^* \cup L_0)$, it follows that V_0 intersects L_0, an impossibility.

Thus by an argument similar to that used to choose L_0, we can choose a compact neighborhood L_1 of F_1 with the properties: *(i)* $L_1 \subset \mathcal{O}$ and is disjoint from $N_1^* \cup V_0 \cup F_0'$, *(ii)* there exists a compact neighborhood F_1', disjoint from $N_1^* \cup L_0 \cup L_1$, and strictly positive times $s_1, \ldots, s_q$ so that for each $\mathbf{u} \in L_1$, for some i, $\mathbf{u} \cdot s_i \in F_1'$.

Define N by

$$N := N_1^* \cup L_0 \cup L_1 .$$

Then from the properties of L_1, it follows, as before, that N is an isolating neighborhood for S and is a subset of $\mathcal{O}$, and also (N_1^*, N_0^*) is an index pair relative to N for the reverse time flow.

We note that N_1^*, hence also $N \setminus L_1$, is an N-neighborhood of $A^+(N)$. For if $\mathbf{u} \in L_0 \cup L_1$, then for some $t > 0$, $\mathbf{u} \cdot t \in F_0' \cup F_1'$, which is disjoint from N. Thus

$$A^+(N) \subset N \setminus (L_0 \cup L_1) \subset N_1^* ,$$

which gives the desired result. Thus L_1 is disjoint from $A^+(N)$. Thus, recalling that for general $K \subset N$, $P'(K,N)$ (respectively, $P(K,N)$) denotes the smallest (respectively, smallest compact) positively invariant relative to N subset of N containing K, defining $K_0 := P(L_1, N)$, we get via [K4, Proposition 4.1] that

$$L_1 \subset K_0 = P'(L_1, N) .$$

Also, K_0 is disjoint from N_1^*. For suppose $\mathbf{u} \in K_0 \cap N_1^*$. Then for some $\mathbf{u}' \in L_1$ and $t' > 0$,

$$\mathbf{u} = \mathbf{u}' \cdot t' \quad \text{and} \quad \mathbf{u}' \cdot [0, t'] \subset N .$$

Set $t'' := \sup\{t > 0 : \mathbf{u}' \cdot [0,t] \subset N \setminus N_1^*\}$. Because L_1 is disjoint from N_1^*, necessarily $0 < t'' \leq t'$. Set $\mathbf{u}'' = \mathbf{u}' \cdot t''$, and note $\mathbf{u}'' \in N_1^*$. By definition of t'' it is immediate that $\mathbf{u}'' \notin A^-(N_1^*)$, whence by the exit property of index pairs, necessarily $\mathbf{u}'' \in N_0^*$. However, as $\mathrm{cl}_X \left(N \setminus N_1^*\right) \subset L_0 \cup L_1$, also $\mathbf{u}'' \in L_0 \cup L_1$. This is impossible since N_0^* and $L_0 \cup L_1$ are disjoint.

As K_0 is disjoint from N_1^*, choose Z_0, a closed N-neighborhood of K_0, disjoint from N_1^*. Set $N_0 := P(Z_0, N)$. Then N_0 is an N-neighborhood of K_0, and by an argument analogous to the one just given, N_0 is disjoint from N_1^*.

As $A^-(N)$ is disjoint from N_0^* and by the construction of L_0 and L_1 as neighborhoods of F_0 and F_1 respectively, we can choose an open neighborhood W of $A^-(N)$ disjoint from a neighborhood of N_0^* and which has the additional properties that $\mathrm{cl}_X\,(W \cap N)$ is compact and

$$\partial_X(N) \cap \mathrm{cl}_X\,(W \cap N) \subset N \setminus (N_1^* \cup L_0) = L_1 \setminus L_0 .$$

Now by [K4, Proposition 4.1(5)], there exists Z' a compact N-neighborhood of $A^-(N)$ so that $A^-(N) \subset Z \subset Z'$ implies $P(Z,N) \subset W$. Thus choose Z_0 to be a compact N-neighborhood of $A^-(N)$ and an X-neighborhood of S satisfying $Z_0 \subset Z'$. Set $N_1 := P(Z_0, N)$. Then N_1 isolates S, is positively invariant relative to N, and is a subset of $W \cap N$. Also, N_1 is an N-neighborhood of $A^-(N)$ since it contains Z_0; hence, for some V an open subset of W,

$$A^-(N) \subset V \cap N \subset \mathrm{int}_N(N_1) .$$

Applying an analogous argument with V replacing W, allows us to find $K_1 \subset V \cap N$ such that K_1 isolates S and is positively invariant relative to N . By definition of V, it is immediate that K_1 lies in the interior of N_1 relative to N.

It follows that $\langle N_1, N_0 \rangle$ and $\langle K_1, K_0 \rangle$ are index pairs for S with the desired properties. The only property which is not obviously satisfied is the exit property. However, if $\mathbf{u} \in N_1 \setminus A^+(N)$, and also $\mathbf{u} \notin K_0$, set

$$\mathbf{u}' := \mathbf{u} \cdot \tau_N(\mathbf{u}).$$

Note $\mathbf{u}' \in \partial_X(N) \cap \mathrm{cl}_X(W \cap N)$, whence by our choice of W,

$$\mathbf{u}' \in L_1 \setminus L_0 \subset K_0.$$

Since N_1 is positively invariant relative to N and since $K_0 \subset N_0$ and $K_1 \subset N_1$, it follows immediately that the exit property is satisfied for both pairs. □

5.13 Proof of Lemma 5.9. Choose an isolating neighborhood N_ν^* of $S_\sigma(\nu)$. By the construction of index pairs given in [C] and by [K4, Lemma 4.9] (see Remark 3 following Theorem 4.2 of [K4]) we can find, for $j = 0, \dots, k+1$, index pairs $\langle N_{1j}^*, N_{0j}^* \rangle$ for $S_\sigma(\nu)$ relative to N_ν^* and the reverse time flow satisfying:

(1) $N_{ij}^* \subset \mathrm{int}_{N_\nu^*}(N_{ij+1}^*)$ for $i = 0, 1$, for $j = 0, \dots, k+1$;
(2) $\mathrm{cl}\left(N_{1j+1}^* \setminus N_{0j}^*\right) \subset \mathrm{int}_X(N_\nu^*)$ for $j = 0, \dots, k$.

Then by Lemma 5.12 we can find an isolating neighborhood N_ν of $S_\sigma(\nu)$ and index pairs $\langle N_1, N_0 \rangle$ and $\langle K_1, K_0 \rangle$ for $S_\sigma(\nu)$ relative to N_ν and the forward time flow satisfying properties (2) and (3) in the statement of that lemma and also

$$N_1 \cap N_{0\,k+1}^* \cap N_{1\,k+1}^* = \emptyset = N_0 \cap N_{1\,k+1}^*. \tag{5.12}$$

Then setting $N_{i\,k+1} := N_i$ and $N_{i\,0} := K_i$ $(i = 0, 1)$, by [K4, Lemma 4.9] we can find additional sets $N_{i\,j}$ $(i = 0, 1,\ j = 1, \dots, k)$ so that

(3) $N_{i\,j} \subset \mathrm{int}_{N_\nu}(N_{i\,j+1})$ for $i = 0, 1$, for $j = 0, \dots, k$;
(4) $\mathrm{cl}\,(N_{1\,j+1} \setminus N_{0\,j}) \subset \mathrm{int}_X(N_\nu)$ for $j = 0, \dots, k$.
(5) $\langle N_{1\,j}, N_{0\,j} \rangle$ is an index pair for $S_\sigma(\nu)$ relative to N_ν and the forward time flow for $j = 0, \dots, k+1$.

Note that as a consequence of (5.12) and the inclusion relations in items (1) and (3), the pair of index pairs

$$((N_{1\,j}, \cap N_{0\,j}), (N_{1j}^*, \cap N_{0j}^*)) \qquad j = 1, \dots, k+1$$

is AIP(thick) for $S_\sigma(\nu)$.

For every compact pair (N, K) in X and $\lambda \in W$ define $P(K, N : \lambda)$ to be the intersection of all the subsets of N that contain K and are compact and positively invariant relative to N relative to the flow φ_λ. The set $P^*(K, N : \lambda)$ is defined analogously with respect to the reverse time flow associated to φ_λ. Then for $\lambda \in W$, for $i = 0, 1$, for $j = 0, \dots, k$, define

$$P_{i\,j}(\lambda) := P(N_{i\,j}, N_\nu : \lambda) \quad \text{and} \quad P_{ij}^*(\lambda) := P^*(N_{ij}^*, N_\nu^* : \lambda).$$

Note that $P_{ij}(\nu) = N_{ij}$ and $P^*_{ij}(\nu) = N^*_{ij}$.

By the Perturbation Theorem for Index Pairs, [K4,Theorem 4.2] also see [C, §IV1.2A], we get that there exists V_ν open, $\nu \in V_\nu \subset W \cap \Lambda(N_\nu) \cap \Lambda(N^*_\nu)$ so that for $\lambda \in V_\nu$,

(6) for $j = 0, \dots, k$, $\langle P_{1j}(\lambda), P_{0j}(\lambda)\rangle$ and $\langle P^*_{1j}(\lambda), P^*_{0j}(\lambda)\rangle$ are index pairs for $S_\sigma(\lambda)$, the former relative to N_ν and the flow φ_λ, the latter relative to N^*_ν and the reverse time flow associated to φ_λ;

(7) $N_{ij} \subset P_{ij}(\lambda) \subset \operatorname{int}_{N_\nu}(N_{i\,j+1})$ for $i = 0, 1$, for $j = 0, \dots, k$;

(8) $N^*_{ij} \subset P^*_{ij}(\lambda) \subset \operatorname{int}_{N^*_\nu}(N^*_{i\,j+1})$ for $i = 0, 1$, for $j = 0, \dots, k$.

By (5.12) and the inclusion relations in (1), (3), (7) and (8), it follows that for $\lambda \in V_\nu$, the pair of index pairs

$$((P_{1j}(\lambda), \cap P_{0j}(\lambda)), (P^*_{1j}(\lambda), \cap P^*_{0j}(\lambda))) \qquad (j = 1, \dots, k)$$

is AIP(thick) for $S_\sigma(\lambda)$.

For i, j fixed, the sets $P_{ij}(\lambda)$ and $P^*_{ij}(\lambda)$ vary upper semi-continuously with λ, so that for $K \subset V_\nu$ compact, the sets

$$\bar{P}_{ij}(K) := \bigcup_{\lambda \in K} P_{ij}(\lambda) \times \{\lambda\} \quad \text{and} \quad \bar{P}^*_{ij}(K) := \bigcup_{\lambda \in K} P_{ij}(\lambda) \times \{\lambda\}$$

are compact (see [K4, Lemma 4.8]). It follows easily that

$$\langle \bar{P}_{1j}(K), \bar{P}_{0j}(K)\rangle \quad \text{and} \quad \langle \bar{P}^*_{1j}(K), \bar{P}^*_{0j}(K)\rangle \quad (j = 1, \dots, k)$$

are index pairs for S_K, the former with respect to the forward time flow on $X \times K$, the latter with respect to the reverse time flow on $X \times K$.

Further, by Lemma 5.6 we can shrink V_ν if necessary so that for $\lambda \in V_\nu$, for $K \subset V_\nu$ compact and contractible, the maps on quotients

$$\bar{\iota}_{\lambda K j}: P_{1j}(\lambda)/P_{0j}(\lambda) \to \bar{P}_{1j}(K)/\bar{P}_{0j}(K)$$

and

$$\bar{\jmath}_{\lambda K j}: P^*_{1j}(\lambda)/P^*_{0j}(\lambda) \to \bar{P}^*_{1j}(K)/\bar{P}^*_{0j}(K)$$

are homotopy equivalences since $\bar{\iota}_{\lambda K j}$ and $\bar{\jmath}_{\lambda K j}$ respectively define the maps of connected simple systems $F(\lambda, K): \mathcal{C}(S_\sigma(\lambda)) \to \mathcal{C}(S_K)$ and $G(\lambda, K): \mathcal{C}^*(S_\sigma(\lambda)) \to \mathcal{C}^*(S_K)$ for $j = 1, \dots, k$. $\square$

5.14 Proof of Theorem 3.7. Without loss of generality let us assume that β is the restriction of a section σ as in the statement of Theorem 3.2. Also, let us assume that $0 \le s < 1$ and prove $\ell(\beta, \boldsymbol{\alpha}, \boldsymbol{\gamma})$ right continuous at s; for $0 < s \le 1$, left continuity at $s \in]0, 1]$ is proved similarly. The proof of right continuity at s is essentially already contained in the proof of Theorem 3.2 and in particular follows from the string of equalities (5.11) after some minor restrictions are placed on a few of the choices made in the course of the derivation of (5.11). Throughout the proof we use the same notation as in the proof of Theorem 3.2.

To begin, suppose $\ell(\beta, \alpha, \gamma)(s) \in \mathcal{N}(\check{\zeta}, S, \lambda, N, U) \in \mathbf{B}_k$, whence

(1) $s = \lambda_\beta(s) \in U$ and $S_\beta(s)$ has N as $\lambda_\beta(s)$-isolating neighborhood;

(2) $\pi_N^{S_\beta(s)}(\mathcal{C}(\beta^s)_* \boldsymbol{\alpha} \mathbin{\widehat{\bullet}} \mathcal{C}^*(\beta^s)_* \boldsymbol{\gamma}) = \pi_N^S \check{\zeta}$.

To prove right continuity at s_0 it suffices to exhibit $\delta > 0$ so that if $0 < s' - s \leq \delta$, then the analogues of (1) and (2) obtained by replacing s with s' hold.

By continuity of σ, choose δ_0 so that $s' \in [0,1]$ and $|s - s'| \leq \delta_0$ implies $\lambda_\sigma(s') \in U$ and $S_\sigma(s')$ has N as $\lambda_\beta(s')$-isolating neighborhood, whence $\sigma(\lambda) = \sigma_N(\lambda)$ if $|s - \lambda| \leq \delta_0$. Apply Lemma 5.9 and choose a partition $\mathcal{P}$ as in §5.10, the proof of Theorem 3.2, where without loss of generality we can additionally assume that $\operatorname{mesh}(\mathcal{P}) \leq \delta_0$ and that s is one of the partition points of $\mathcal{P}$, i.e, $s = \lambda_{i-1}$ where $i \in \{1, \ldots, n\}$. Observe that the proof of Lemma 5.9 allows us to assume without loss of generality that for each $\nu \in J_i := [\lambda_{i-1}, \lambda_i]$, the isolating neighborhoods N_ν and N_ν^* guaranteed by that lemma lie interior to N. It then follows that for each $\lambda \in J_i$, for $j = 1, 2$, the pair of admissible index pairs $P_j(\lambda)$ has the property $P_{1\,j}(\lambda) \cap P_{1\,j}^*(\lambda) \subset \operatorname{int}(N)$. It follows that when choosing $V \in M^{(2)}_{\mathbb{A}\,o}(P_2)$ as in the proof of Theorem 3.2, it can be assumed that $V_1 \cap V_1^* \subset N$.

Set $\delta := \operatorname{mesh}(\mathcal{P})$ and suppose $0 < s' - \lambda_{i-1} \leq \delta$. Then without loss of generality it can be assumed that $s' = \lambda_i$. For if not refine the partition $\mathcal{P}$ by adding s' as a partition point. Then as in the proof of Theorem 3.2 we arrive at the string of equalities (5.11) from which it is immediate that

$$(5.13) \qquad (\boldsymbol{\alpha}_{i-1} \mathbin{\widehat{\bullet}} \gamma_{i-1})_N = (\boldsymbol{\alpha}_i \mathbin{\widehat{\bullet}} \gamma_i)_N$$

because $V_1 \cap V_1^* \subset N$. The equality (5.13) and (2) together yield that

$$\pi_N^{S_\sigma(s')}\left(\mathcal{C}\left(\beta^{s'}\right)_* \boldsymbol{\alpha} \mathbin{\widehat{\bullet}} \mathcal{C}^*\left(\beta^{s'}\right)_* \gamma\right) = \pi_N^S(\check{\zeta}).$$

The choice of δ implies that $\lambda_\beta(s') \in U$ and $S_\beta(s')$ has N as $\lambda_\beta(s')$-isolating neighborhood. $\square$

CHAPTER 6

SOME BASIC COMPUTATIONAL TOOLS

For simplicity, all homology in this chapter is taken with coefficients in R, and coefficient modules are, therefore, suppressed from the notation.

A. Conditions on Singular Cycles for Computing $\mathfrak{L}$ and $^{\#}\mathfrak{L}$. The first result, Theorem 6.1, describes to what extent it is possible to express an intersection class of homology classes in the Conley indices of an isolated invariant set in terms of relative singular cycles supported in an isolating block for the invariant set. Corollary to it is the result mentioned at the end of §2.B; viz., it is possible to express an intersection number of homology classes in the Conley indices of an isolated invariant set directly in terms of a pair of relative singular cycles supported in a block modulo some minor restrictions on how the chains are situated within the block. The following hypotheses on the homology classes and representing chains in the block will be assumed:

HYPOTHESES CB.

(1) *B is an isolating block for S;*
(2) *$\boldsymbol{\alpha} \in \widetilde{H}_i\mathcal{C}(S)$ and $\boldsymbol{\gamma} \in \widetilde{H}_j\mathcal{C}^*(S)$;*
(3) *c^+ is a singular cycle in B modulo B^+ representing $\boldsymbol{\alpha}$ and c^- is a singular cycle in B modulo B^- representing $\boldsymbol{\gamma}$;*
(4) *$\alpha_0 := \langle c^+\rangle \in H_i(|c^+|, |\partial c^+|)$ and $\gamma_0 := \langle c^-\rangle \in H_j(|c^-|, |\partial c^-|)$.*

Note that if $|\partial\, c^{\pm}| \cap |c^{\mp}| = \emptyset$, then as follows from Definition 2.8(C), $\alpha_0 \,\widehat{\bullet}\, \gamma_0$ is a well-defined element of $\check{H}_{i+j-m}(|c^+| \cap |c^-|)$ where $m = \dim(M)$.

6.1 THEOREM. *Under hypotheses CB, if (i) $|c^+| \cap |c^-| \cap (B^- \cup B^+) = \emptyset$, then there exists $N \in \operatorname{ob}(\mathcal{AIP}(S))$ satisfying*

$$(ii)\ |c^+| \cap |c^-| \subset N_1 \cap N_1^*, \quad (iii)\ N \hookrightarrow ((B, B^+), (B, B^-))$$

and

$$(iv)\ V \in \mathcal{N}(N_1 \cap N_1^*) \qquad \text{implies} \qquad \pi_V^S(\boldsymbol{\alpha} \,\widehat{\bullet}\, \boldsymbol{\gamma}) = \pi_V^{|c^+|\cap|c^-|}(\alpha_0 \,\widehat{\bullet}\, \gamma_0)\,;$$

hence, for such V, $\pi_V^S(\boldsymbol{\alpha} \,\widehat{\bullet}\, \boldsymbol{\gamma}) \neq 0$ implies $\emptyset \neq |c^+| \cap |c^-| \subset B \setminus (B^+ \cup B^-)$.

Taking $V := M$ in (iv) we immediately get the following.

6.2 COROLLARY. *Assume $i + j = m$. Then (i) implies*

$$^{\#}(\boldsymbol{\alpha} \,\widehat{\bullet}\, \boldsymbol{\gamma}) = {}^{\#}(\alpha_0 \,\widehat{\bullet}\, \gamma_0)\,.$$

Hence, if $^{\#}(\boldsymbol{\alpha} \,\widehat{\bullet}\, \boldsymbol{\gamma}) \neq 0$, then $|c^+| \cap |c^-| \neq \emptyset$ whether or not (i) holds, but if it does, then $|c^+| \cap |c^-| \subset B \setminus (B^+ \cup B^-)$.

Corollary 6.2 together with Proposition 2.13 yields the corresponding result for homology intersection numbers.

6.3 COROLLARY. *Assume $M := \mathbf{R}^m$ and $i + j = m$. Then (i) implies*

$$\boldsymbol{\alpha} \# \boldsymbol{\gamma} = \alpha_0 \# \gamma_0 .$$

Hence, if $\boldsymbol{\alpha} \# \boldsymbol{\gamma} \neq 0$, then $|c^+| \cap |c^-| \neq \emptyset$ whether or not (i) holds, but if it does, then $|c^+| \cap |c^-| \subset B \setminus (B^+ \cup B^-)$.

Proof of Theorem 6.1. Observe that condition (i) in the statement of the theorem can be rewritten in the form

$$|c^+| \cap B^+ \cap |c^-| = \emptyset = |c^-| \cap B^- \cap |c^+| . \tag{6.1}$$

From (6.1) and the fact that $c^{\pm}$ is a relative cycle modulo $B^{\pm}$ it is immediate that

$$|\partial c^+| \cap |c^-| = \emptyset = |c^+| \cap |\partial c^-| .$$

Hence, by Definition 2.8(C), the intersection class $\alpha_0 \,\widehat{\bullet}\, \gamma_0$ is well-defined as an element of $\check{H}_{i+j-m}(|c^+| \cap |c^-|)$.

The basic idea of the proof is that when (6.1) holds it is possible to construct $N \in \mathrm{ob}(\mathcal{AIP}(S))$ satisfying

(1) $(N_1, N_0) \subset (B, B^+)$ and $(N_1^*, N_0^*) \subset (B, B^-)$;

(2) c^+ and c^- are relative cycles, the former in N_1 modulo N_0, the latter in N_1^* modulo N_0^*

with the caveat that to carry out this construction it may first be necessary to deform within B the chain $c^{\pm}$ to a chain $c_1^{\pm}$ inconsequentially in the sense that $|c^+| \cap |c^-|$ and the boundary chain $\partial c^{\pm}$ remain pointwise fixed by the deformation and where α_1 is the class of c_1^+ in $H_*(|c^+|, |\partial c_1^+|)$ and γ_1 is the class of c_1^- in $H_*(|c_1^-|, |\partial c_1^-|)$

$$\alpha_0 \,\widehat{\bullet}\, \gamma_0 = \alpha_1 \,\widehat{\bullet}\, \gamma_1 . \tag{6.2}$$

It follows from (1) and (2) that

$$S \cup (|c^+| \cap |c^-|) \subset N_1 \cap N_1^* \subset B \setminus (B^+ \cup B^-) .$$

From these inclusions and from (1), (2), Lemma 2.10, and Theorem 2.11, it follows that $V \in \mathcal{N}(N_1 \cap N_1^*)$ implies

$$\pi_V^{|c^+| \cap |c^-|}(\alpha_0 \,\widehat{\bullet}\, \gamma_0) = \pi_V^S(\boldsymbol{\alpha} \,\widehat{\bullet}\, \boldsymbol{\gamma}) . \tag{6.3}$$

Thus, if (6.1) holds and $\pi_V^S(\boldsymbol{\alpha} \,\widehat{\bullet}\, \boldsymbol{\gamma}) \neq 0$, also $0 \neq \alpha_0 \,\widehat{\bullet}\, \gamma_0 \in \check{H}_*(|c^+| \cap |c^-|)$, whence $|c^+| \cap |c^-| \neq \emptyset$ which of course is also true if (6.1) fails.

To prove the existence of $N \in \mathrm{ob}(\mathcal{AIP}(S))$ with the desired properties there are two cases to consider with the second having two subcases. These cases are determined by the emptiness or non-emptiness of the sets C^+ and C^- defined by $C^{\pm} := |c^{\pm}| \cap B^{\mp}$.

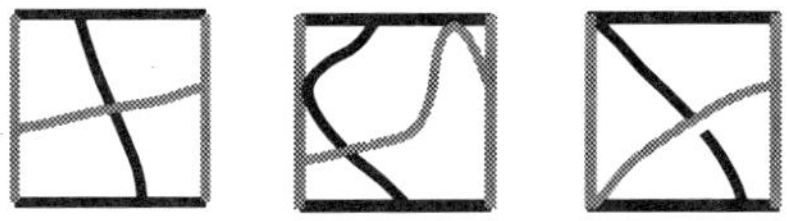

FIGURE 3. The three cases considered in Theorem 6.1. Each square represents an isolating block with exit set the black edges and entrance set the gray. The black arc corresponds to the relative cycle c^+, the gray to c^-.

Case I: $C^- = \emptyset = C^+$. Then

$$\emptyset = |\,c^+| \cap B^- = |\,c^+| \cap \tau_B^{*\,-1}(0)$$

and

$$\emptyset = |\,c^-| \cap B^+ = |\,c^-| \cap \tau_B^{-1}(0)\,.$$

Thus, τ_B^* restricted to $|\,c^+|$ and τ_B restricted to $|\,c^-|$ are positive continuous functions, each having compact domain, whence each has positive infimum. Therefore, $d > 0$ can be chosen small enough so that if N equals the AIP(regular) pair of index pairs constructed from B as in the proof of Proposition 4.4 for the non-smooth case,

$$(|\,c^+|, |\partial c^+|) \subset (N_1, N_0) \subset (B, B^+)$$

and

$$(|\,c^-|, |\partial c^-|) \subset (N_1^*, N_0^*) \subset (B, B^-)\,,$$

whence it is immediate that properties (1) and (2) are satisfied. As observed the desired conclusions then follow.

Case II: $C^+ \neq \emptyset$ or $C^- \neq \emptyset$. Essentially, this case can be reduced to the previous case by using the flow to pull $c^\pm$ off of $B^\mp$ as follows. Note that (6.1) is equivalent to $C^\mp \cap |\,c^\pm| = \emptyset$. Therefore we can find $V^\pm$ open, satisfying $C^\pm \subset V^\pm$ and $\mathrm{cl}\,(V^\pm)$ is disjoint from $|\,c^\mp| \cup \mathrm{cl}\,(V^\mp)$, and by compactness of $C^\pm$ and continuity of the flow, we can find $\varepsilon > 0$ and $W^\pm$ an open neighborhood of $C^\pm$ satisfying

$$W^+ \cdot [0, \varepsilon] \subset V^+ \quad \text{and} \quad W^- \cdot [-\varepsilon, 0] \subset V^-\,.$$

Also, choose $\rho^\pm \colon B \to [0,1]$, continuous, satisfying $\mathrm{supp}(\rho^\pm) \subset W^\pm$ and $\rho^\pm(C^\pm) = 1$. Then for each $\mathbf{u} \in B$ define

$$h^+(\mathbf{u}) := \rho^+(\mathbf{u}) \min\{\varepsilon, \tau_B(\mathbf{u})\}\,, \quad h^-(\mathbf{u}) := \rho^-(\mathbf{u}) \min\{\varepsilon, \tau_B^*(\mathbf{u})\}\,,$$

and for each $\mathbf{u} \in B$ and $s \in [0,1]$ define

$$D_s^\pm(\mathbf{u}) := D^\pm(\mathbf{u}, s) := \mathbf{u} \cdot \pm s h^\pm(\mathbf{u})\,.$$

Then $D^\pm$ is a continuous deformation of B into itself that leaves $B^\pm$ pointwise fixed throughout the deformation; in particular $D_1^\pm$ is a self-map of the pair $(B, B^\pm)$ inducing the identity isomorphism on homology.

Define

$$c_1^\pm := D_{1\sharp}^\pm c^\pm$$

where $D_{1\sharp}^\pm$ denotes the map on singular chains induced by $D_1^\pm$. It is immediate that $c^\pm$ and $c_1^\pm$ are homologous relative cycles in B modulo $B^\pm$, whence c_1^+ represents $\boldsymbol{\alpha}$ and c_1^- represents $\boldsymbol{\gamma}$. Further, from the given construction of $D^\pm$,

$$|c^+| \cap |c^-| = |c_1^+| \cap |c_1^-|. \tag{6.4}$$

For by the choice of ε, $W^\pm$, and $\rho^\pm$, the set theoretic symmetric difference of $|c^\pm|$ and $|c_1^\pm|$ lies within $V^\pm$, whence (6.4) follows because as chosen $\mathrm{cl}\,(V^+)$ and $\mathrm{cl}\,(V^-)$ are disjoint with $\mathrm{cl}\,(V^\pm)$ disjoint from $|c^\mp|$. Hence, with

$$(X^\pm, A^\pm) := (|c^\pm| \cup \mathrm{cl}\,(V^\pm)\,, |\partial c^\pm|)\,,$$

the construction of $D^\pm$ shows that its restriction to $X^\pm \times [0,1]$ defines a deformation of $X^\pm$ into itself leaving $A^\pm$ pointwise fixed. Thus, noting that $|\partial c^\pm| = |\partial c_1^\pm|$, we see that $c^\pm$ and $c_1^\pm$ are homologous relative cycles in $(X^\pm, A^\pm)$.

Let $\alpha_2 \in H_i(X^+, A^+)$ be the common homology class of c^+ and c_1^+ and let $\gamma_2 \in H_j(X^-, A^-)$ be the common homology class of c^- and c_1^-. Also, let c_1^+ have homology class $\alpha_1 \in H_i(|c_1^+|, |\partial c_1^+|)$ and let c_1^- have homology class $\gamma_1 \in H_j(|c_1^-|, |\partial c_1^-|)$. Too, note that $\big((X^+, A^+), (X^-, A^-)\big)$ satisfies the off-diagonal condition and $X^+ \cap X^- = |c^+| \cap |c^-|$. It follows from these definitions and observation as well as (6.4) and Lemma 2.10 that

$$\alpha_0 \,\hat{\bullet}\, \gamma_0 = \alpha_2 \,\hat{\bullet}\, \gamma_2 = \alpha_1 \,\hat{\bullet}\, \gamma_1 \tag{6.5}$$

as elements of $\check{H}_{i+j-m}(|c^+| \cap |c^-|)$. Thus (6.2) holds.

To complete the proof of Case II it will suffice to construct $N \in \mathrm{ob}(\mathcal{AIP}(S))$ satisfying the inclusions

$$(|c_1^+|, |\partial c_1^+|) \subset (N_1, N_0) \subset (B, B^+) \tag{6.6a}$$

and

$$(|c_1^-|, |\partial c_1^-|) \subset (N_1^*, N_0^*) \subset (B, B^-)\,. \tag{6.6b}$$

For given such an N, let c_1^+ have homology class $\alpha_3 \in H_i(N_1, N_0)$ and let c_1^- have homology class $\gamma_3 \in H_j(N_1^*, N_0^*)$. It is immediate from (6.6) and Theorem 2.11 that if V is any neighborhood of $N_1 \cap N_1^*$, then

$$\pi_V^{N_1 \cap N_1^*}(\alpha_3 \,\hat{\bullet}\, \gamma_3) = \pi_V^S(\boldsymbol{\alpha} \,\hat{\bullet}\, \boldsymbol{\gamma})\,, \tag{6.7}$$

and by (6.5) and Lemma 2.10,

$$\check{\imath}_*(\alpha_0 \,\hat{\bullet}\, \gamma_0) = \check{\imath}_*(\alpha_1 \,\hat{\bullet}\, \gamma_1) = \alpha_3 \,\hat{\bullet}\, \gamma_3 \tag{6.8}$$

where $\check{\imath}_* \colon \check{H}_*(|c^+| \cap |c^-|) \to \check{H}_*(N_1 \cap N_1^*)$ is inclusion induced. From (6.7) and (6.8) it is immediate that if $V \in \mathcal{N}(N_1 \cap N_1^*)$, then (6.3) holds.

It is convenient to divide the proof of existence of $N \in \mathrm{ob}(\mathcal{AIP}(S))$ satisfying the inclusions (6.6) into two subcases as follows. Set $C_1^{\pm} := |\,c_1^{\pm}| \cap B^{\mp}$.

Subcase (a): $C_1^+ = \emptyset = C_1^-$. Then their exists an AIP(regular) pair of index pairs satisfying the inclusions (6.6) as an immediate consequence of the argument used to prove Case I.

Subcase (b): $C_1^+ \neq \emptyset$ or $C_1^- \neq \emptyset$. The proof in this case is more difficult. First, it is straightforward to check that the construction of $D^{\pm}$ implies that $C_1^{\pm} \subset B^{\pm}$, and this inclusion implies that $|\,c_1^+| \cap B^- = |\,c_1^+| \cap B^- \cap B^+$ and $|\,c_1^-| \cap B^+ = |\,c_1^-| \cap B^+ \cap B^-$.

Using the notation of [C], for any compact $K \subset B$, let $P(K, B)$ denote the smallest compact subset of B containing K and positively invariant relative to B, i.e., $P(K, B)$ is the intersection of all the compact and positively invariant relative to B subsets of B containing K, and let $P'(K, B)$ denote the smallest subset of B containing K and positively invariant relative to B, i.e., $P'(K, B)$ is the union over $\mathbf{u} \in K$ of orbit segments $\mathbf{u} \cdot [0, t] \subset B$ where $t \geq 0$. Also let $P^*(K, B)$ and $P'^*(K, B)$ denote the analogues of $P(K, B)$ and $P'(K, B)$, respectively, when defined relative to the reverse time flow. Then $P(K, B) = P'(K, B)$ whenever $A^-(B) \subset \mathrm{int}_B(K)$ [C,III.4.1.C] where in the notation of [C], $A^-(B) = W^u(S; B)$; hence also $P^*(K, B) = P'^*(K, B)$ whenever $A^+(B) := W^s(S, B) \subset \mathrm{int}_B(K)$. As B is a block it is immediate that $A^{\pm}(B)$ and $B^{\mp}$ are disjoint closed sets. As a consequence of the lemmas in [C] used to demonstrate the existence of index pairs, in particular [C,III.4.1.D], we can find $\hat{Z}$ a compact B-neighborhood of $A^-(B)$ satisfying $P(\hat{Z}, B)$ is disjoint from B^-. Set $Z := \hat{Z} \cup |\,c_1^+|$ and define

$$N_1 := P(Z, B)\,, \qquad N_0 := B^+ \cap N_1\,.$$

It follows easily that (N_1, N_0) is a regular index pair for S satisfying the inclusions (6.6a).

Also,

$$N_1 \cap B^- \subset B^- \cap B^+\,. \tag{6.9}$$

For as Z is a compact B-neighborhood of $A^-(B)$, $P'(Z, B) = N_1$. Hence, if $\bar{\mathbf{u}} \in N_1 \cap B^-$, then $\bar{\mathbf{u}} = \mathbf{u} \cdot t$ where $\mathbf{u} \in Z$, $t \geq 0$, and $\mathbf{u} \cdot [0, t] \subset B$. By choice of $\hat{Z}$, $\mathbf{u} \notin \hat{Z}$, and the definition of B^- forces $t = 0$. Thus $\bar{\mathbf{u}} = \mathbf{u} \in |\,c_1^+| \cap B^- = C_1^+$, and as $C_1^+ \subset B^+$, the inclusion (6.9) follows.

An argument similar to that of the previous paragraph together with (6.9), the choice of $\hat{Z}$, and (6.1) yields that

$$|\,c_1^-| \cap B^- \cap N_1 = \emptyset\,. \tag{6.10}$$

From (6.9) and because $A^+(B)$ is disjoint from B^+, we can find U an open neighborhood of $A^+(B)$ satisfying $U \cap B$ is disjoint from B^+ and also from $N_1 \cap B^-$. Again using [C,III.4.1.D], but with respect to the reverse time flow, we can find $\hat{Z}^*$ a compact B-neighborhood of $A^+(B)$ so that $P^*(\hat{Z}^*, B) \subset U$. Set $Z^* := \hat{Z}^* \cup |\,c_1^-|$, and then define

$$N_1^* := P^*(Z^*, B)\,, \qquad N_0^* := B^- \cap N_1^*\,.$$

It follows easily that (N_1^*, N_0^*) is a regular index pair for S with respect to the reverse time flow satisfying the inclusions (6.6*b*).

To complete the proof of Subcase (b) and of the theorem, it only remains to check that $((N_1, N_0), (N_1^*, N_0^*))$ satisfies the off-diagonal condition, whence this pair of index pairs will be AIP(regular) for S.

To the contrary, suppose $\bar{\mathbf{u}} \in N_1 \cap N_0^*$. Then $\bar{\mathbf{u}} = \mathbf{u} \cdot -t$ where $t \geq 0$, $\mathbf{u} \in Z^*$, and $\mathbf{u} \cdot [-t, 0] \subset B$. Because $N_1 \cap B^- \subset B^+$, from the choice of $\hat{Z}^*$, it follows that $\mathbf{u} \in |c_1^-|$. Also, because $\bar{\mathbf{u}} \cdot [0, t] = \mathbf{u} \cdot [-t, 0] \subset B$ and because $\bar{\mathbf{u}} \in B^+$, the definition of B^+ forces $t = 0$. Thus $\bar{\mathbf{u}} = \mathbf{u} \in |c_1^-| \cap B^- \cap N_1 = \emptyset$ by (6.10). Thus $N_1 \cap N_0^* = \emptyset$.

Again to the contrary, suppose $\bar{\mathbf{u}} \in N_0 \cap N_1^*$. Again $\bar{\mathbf{u}} = \mathbf{u} \cdot -t$ where $t \geq 0$, $\mathbf{u} \in Z^*$ and $\mathbf{u} \cdot [-t, 0] \subset B$. Again $\bar{\mathbf{u}} \in B^+$, now because $N_0 \subset B^+$, whence as before $\mathbf{u} \in |c_1^-|$ and $t = 0$. Hence $\bar{\mathbf{u}} = \mathbf{u} \in |c_1^-| \cap B^+ = C_1^- \subset B^-$. Thus since $N_0 \subset N_1$, we get $\bar{\mathbf{u}} \in |c_1^-| \cap B^- \cap N_1 = \emptyset$. Hence also $N_0 \cap N_1^* = \emptyset$. □

Let us note that to actually have $\alpha_0 \frown \gamma_0 = \boldsymbol{\alpha} \frown \boldsymbol{\gamma}$ in the situation of Theorem 6.1 it would be necessary to have $|c^+| \cap |c^-| = S$, but apparently this by itself is not sufficient. The simple albeit quite restrictive hypotheses of the next proposition do ensure the above equality of intersection classes and that result suffices for the needs of the current work.

6.4 PROPOSITION. *Under the assumption of hypotheses* CB, *if* $|c^+| = W^u(S; B)$ *and* $|c^-| = W^s(S; B)$, *then in* $\check{H}_*(S)$,

$$\alpha_0 \frown \gamma_0 = \boldsymbol{\alpha} \frown \boldsymbol{\gamma}.$$

Remark. Note it is not assumed here that S has hyperbolic structure, nor is it even assumed that the local stable and unstable sets $W^s(S; B)$ and $W^u(S; B)$ are compact manifolds with boundary.

Proof. We must show that for each $V \in \mathcal{N}(S)$,

$$\pi_V^S(\alpha_0 \frown \gamma_0) = \pi_V^S(\boldsymbol{\alpha} \frown \boldsymbol{\gamma}). \tag{6.11}$$

However, from the nature of an inverse limit we need only consider arbitrarily small V. In particular, we may assume that V is open and $V \subset \operatorname{int}_M(B)$. Because $W^u(S; B)$ and $W^s(S; B)$ intersect in S, via normality it is simple to find B-open neighborhoods $V^u \supset W^u(S; B)$ and $V^s \supset W^s(S; B)$ satisfying $V^u \cap V^s = V$. From Lemma 3.3(2) of [K1] applied to both the forward and reverse time flows, it follows that there exists $T > 0$ so that if $t \geq T$, then $B_{0\,t} \subset V^u$ and $B_{t\,0} \subset V^s$ where we recall from the proof of Proposition 4.4 that $B_{0\,t}$ and $B_{t\,0}$ are squeezes of B for time t, the former along the entrance set, the latter along the exit set. Let (L_1, L_0) be the forward time regular index pair for S consisting of $B_{0\,T}$ and its exit set and let (L_1^*, L_0^*) be the reverse-time index pair for S consisting of $B_{T\,0}$ and its entrance set. Then $L \in \operatorname{ob}(\mathcal{AIP}(S))$ and by choice of T, $L_1 \cap L_1^* \subset V$. Further, because $W^u(S; B)$ is negatively invariant, c^+ is a relative cycle in L_1 modulo L_0 and let α_L be its homology class. Analogously, the positive invariance of $W^s(S; B)$ implies c^-

is a relative cycle in L_1^* modulo L_0^* and let γ_L be its homology class. It is immediate from Lemma 2.10 that

$$\check{\imath}^S_{L_1 \cap L_1^*}(\alpha_0 \mathbin{\widehat{\bullet}} \gamma_0) = \alpha_L \mathbin{\widehat{\bullet}} \gamma_L \tag{6.12}$$

Further, as $(L_1, L_0) \subset (B, B^+)$ and because (see [K5, Proposition 3.1]) an inclusion of nested index pairs induces a morphism in $\mathcal{C}(S)$ between the corresponding index spaces, it follows that α_L represents $\boldsymbol{\alpha}$ because c^+ as a cycle in B modulo B^+ does; similarly γ_L represents $\boldsymbol{\gamma}$. Finally, as $L_1 \cap L_1^* \subset V$ by construction, $V \in \mathcal{N}(L_1 \cap L_1^*)$ and it follows immediately from (6.12) and Theorem 2.11 that (6.11) holds. □

The next theorem provides a reasonable means of evaluating $\mathfrak{L}$ on a tensor product of homology classes each of which is represented by a compact submanifold. Its proof uses Theorem 6.1 and its corollaries and Proposition 6.4 to translate [D, Proposition VIII.13.23] to the Conley index setting and some preliminary discussion is necessary before the result from [D] can even be stated. In particular, we need the following general position hypotheses on the submanifolds:

HYPOTHESES GP (Part 1).

(1) *N^+ and N^- are submanifolds with boundary of M orientable over R and with $\dim(N^\pm) = n^\pm$.*
(2) *$N := N^+ \cap N^-$ is a connected, closed submanifold of M disjoint from $\partial N^\pm$ with $n := \dim(N) = n^+ + n^- - m$.*

Let $o_N^{N^\pm} \in H_{n^\pm}(N^\pm, N^\pm \setminus N)$ denote the fundamental class along N which exists because $N^\pm$ is R-orientable and N is compact, connected, and disjoint from $\partial N^\pm$. It follows that $o_N^{N^+} \mathbin{\widehat{\bullet}} o_N^{N^-} \in \check{H}_n(N)$. Moreover, as N is an ENR, the natural homomorphism $j_*^N \colon H_*(N) \to \check{H}_*(N)$ is an isomorphism (e.g., see [D, VIII.13.17]). Thus using this isomorphism to identify the Čech and singular homology modules of N, we get

$$o_N^{N^+} \mathbin{\widehat{\bullet}} o_N^{N^-} = \mu o_N$$

where o_N is the fundamental class of N and where $\mu \in R$ is called the intersection multiplicity (relative to R) of N^+ and N^-. Then [D, Proposition VIII.13.23] states, in part, that $\mu = \pm 1$ if N^+ and N^- have topologically transverse intersection at some $\mathbf{u} \in N$. Note that as stated in [D] the just cited proposition assumes $N^\pm$ to be boundaryless, but as the proof is local to a neighborhood of $\mathbf{u}$, it goes through unchanged since $N \cap \partial N^\pm = \emptyset$. To exploit the proposition in the context of the Conley index, requires the following additional hypotheses on N^+, N^-, and N:

HYPOTHESES GP (Part 2).

(3) *N^+ and N^- have topologically transverse intersection at $\mathbf{u} \in N$ and $N \subset S$.*
(4) *For some isolating block B for S, there are inclusions $j^\pm \colon (N^\pm, \partial N^\pm) \subset (B, B^\pm)$ and $N^\pm$ is compact.*
(5) *Either ∂N^+ is a weak deformation retraction of $N^+ \setminus S$ or $N^+ \subset S$, and either ∂N^- is a weak deformation retraction of $N^- \setminus S$ or $N^- \subset S$.*

The R-orientability and compactness of $N^\pm$ imply that there is a fundamental class $o_{N^\pm} \in H_{n^\pm}(N^\pm, \partial N^\pm)$ (note $\partial N^\pm = \emptyset$ if $N^\pm \subset S$); hence, by GP(4) $\mathbf{o}_{N^\pm} := \bar{b}_*^\pm j_*^\pm o_{N^\pm}$ is a class in the appropriate Conley index; viz., $\mathbf{o}_{N^+} \in \widetilde{H}_{n^+}\mathcal{C}(S)$ and $\mathbf{o}_{N^-} \in \widetilde{H}_{n^-}\mathcal{C}^*(S)$.

6.5 THEOREM. *Under the assumption of hypotheses* GP, *the following hold.*

(*i*) *If* $N^+ = W^u(S;B)$, $N^- = W^s(S;B)$, *and* $(N^\pm, \partial N^\pm)$ *is a triangulable pair, then*

$$\mathbf{o}_{N^+} \frown\!\!\bullet\ \mathbf{o}_{N^-} = \pm o_S\,.$$

(*ii*) *If* $N = \{\mathbf{u}\} \subset \operatorname{int}(B)$ *(so* $n = 0$*) and if either* $(N^\pm, \partial N^\pm)$ *is a triangulable pair or* $N^\pm \cap B^\mp = \emptyset$, *then*

$$^{\#}(\mathbf{o}_{N^+} \frown\!\!\bullet\ \mathbf{o}_{N^-}) = \pm 1\,.$$

Thus, in the case $M = \mathbf{R}^m$, $\mathbf{o}_{N^+} \# \mathbf{o}_{N^-} = \pm o_m$.

Proof. First, suppose $N^\pm \not\subset S$. Then, by hypothesis GP(5), the inclusion $\partial N^\pm \subset N^\pm \backslash S$ induces an isomorphism on homology. Hence, application of the Five Lemma to the sequences of the pairs in the inclusion $i\colon (N^\pm, \partial N^\pm) \subset (N^\pm, N^\pm \setminus S)$ shows that i induces an isomorphism on homology. On the other hand, if $N^\pm$ is a subset of S, then $\partial N^\pm = \emptyset$ as a consequence of hypothesis GP(4), whence it is trivial that i induces an isomorphism. In either case, the induced isomorphism i_* maps the fundamental class $o_{N^\pm}$ to $o^{N^\pm}_{S\cap N^\pm}$, the fundamental class of $N^\pm$ along $S \cap N^\pm$. Also, the homomorphism induced by the inclusion $(N^\pm, N^\pm \setminus S) \subset (N^\pm, N^\pm \setminus N)$ sends $o^{N^\pm}_{S\cap N^\pm}$ to $o^{N^\pm}_N$. Now as observed, the transversality hypothesis GP(3) implies $o^{N^+}_N \frown\!\!\bullet\ o^{N^-}_N = \pm o_N$. Hence from the naturality of intersection pairings with respect to inclusions expressed by Lemma 2.10,

$$o_{N^+} \frown\!\!\bullet\ o_{N^-} = o^{N^+}_{N^+\cap S} \frown\!\!\bullet\ o^{N^-}_{N^-\cap S} = o^{N^+}_N \frown\!\!\bullet\ o^{N^-}_N = \pm o_N\,. \tag{6.13}$$

First, assume that $N^+ = W^u(S;B)$, $N^- = W^s(S;B)$, and $(N^\pm, \partial N^\pm)$ is a polyhedral pair. Then there is a singular chain $c^\pm$ with support $N^\pm$ that is a relative cycle in $N^\pm$ modulo its boundary representing $o_{N^\pm}$; viz, $c^\pm$ is the sum of singular simplexes one for each geometric simplex of dimension $n^\pm$ in the triangulation of $N^\pm$ with each such singular simplex represented by a homeomorphism of the standard simplex $\Delta_{n^\pm}$ onto the geometric simplex. Thus $|c^+| \cap |c^-| = S \subset \operatorname{int}(B)$. It then follows from Proposition 6.4 and (6.13) that

$$\mathbf{o}_{N^+} \frown\!\!\bullet\ \mathbf{o}_{N^-} = \pm o_S\,.$$

Next, assume $N = \{\mathbf{u}\}$, whence $o_N \in H_0(N)$ and is represented by $\mathbf{u}$ regarded as a 0-simplex. If also $(N^\pm, \partial N^\pm)$ is a triangulable pair, then as above, $o_{N^\pm}$ is represented by a singular chain $c^\pm$ with support $N^\pm$ as a relative cycle in $N^\pm$ modulo $\partial N^\pm$. Then $|c^+| \cap |c^-| = \{\mathbf{u}\} \subset \operatorname{int}(B)$ so that condition (*i*) of Theorem 6.1 is satisfied. Because o_N is represented by the 0-simplex $\mathbf{u}$, it follows from Corollary 6.2 and (6.13) that

$$\pm 1 = \pm\langle 1, i^N_M o_N\rangle = {}^{\#}(\mathbf{o}_{N^+} \frown\!\!\bullet\ \mathbf{o}_{N^-})$$

as desired.

If instead of the triagulability assumption we assume that $N^\pm \cap B^\mp = \emptyset$, then via the constructions used in the proof of Proposition 4.4, we can find $P \in \operatorname{ob}(\mathcal{AIP}(S))$ satisfying

$$((N^+, \partial N^+), (N^-, \partial N^-)) \hookrightarrow P \hookrightarrow ((B, B^+), (B, B^-))\,.$$

In this case, Theorem 2.11, (6.13), and the fact that o_N is represented by a 0-simplex ensure that

$$\pm 1 = \pm\langle 1, i_M^N o_N\rangle = \pm\langle 1, \pi_M^{P_1\cap P_1^*} \check{\imath}^N_{P_1\cap P_1^*} o_N\rangle = {}^{\#}(\mathbf{o}_{N^+} \mathbin{\widehat{\bullet}} \mathbf{o}_{N^-})$$

as desired.

In either case, if $M = \mathbf{R}^m$, that $\mathbf{o}_{N^+} \# \mathbf{o}_{N^-} = \pm o_m$ then follows from Proposition 2.13. □

In most cases of interest where Theorem 6.5(i) applies, M and the flow are smooth (i.e., at least C^1) and S is a smooth invariant submanifold of the flow with hyperbolic structure, whence $N^- := W^s(S;B)$ and $N^+ := W^u(S;B)$ are smooth transverse submanifolds of M with boundary assuming B is a smooth isolating block with corners as defined in Definition 4.3. Note that with this choice of $N^\pm$, its boundary $\partial N^\pm$ is in fact a strong deformation retract of $N^\pm \setminus S$; for example, the map on $N^+ \setminus S \times [0,1]$ given by $(\mathbf{u}, s) \mapsto \mathbf{u}\cdot s\tau_B(\mathbf{u})$ is a strong deformation retraction of $N^+ \setminus S$ onto ∂N^+. The local invariant submanifolds $W^s(S;B)$ and $W^u(S;B)$ are geometric representatives for the classes $\mathbf{o}_{N^-}$ and $\mathbf{o}_{N^+}$, respectively, in the sense that in this smooth setting $(N^\pm, \partial N^\pm)$ is a simplicial pair and the fundamental class $o_{N^\pm}$ (coefficients in R) is represented by a singular $n^\pm$-chain $c^\pm = \sum \sigma_\nu$ where the summation is indexed by the simplexes in the triangulation of $N^\pm$ of dimension $n^\pm$ and σ_ν is a suitably oriented homeomorphism of the standard $n^\pm$-simplex onto ν.

Analogous remarks apply to the application of Theorem 6.5(ii) except that there typically at least one of N^+ or N^- is a fiber in some fibering of $W^u(S;B)$ or $W^s(S;B)$ respectively with N^+ and N^- of complementary dimension. Such a situation arises in computing the intersection pairing on the Conley indices of a hyperbolic periodic orbit; see Theorem 7.6 and Lemma 7.7 below.

Let us note that a priori the classes $\mathbf{o}_{N^\pm}$ in the statement of Theorem 6.5 might depend on the choice of block B satisfying hypothesis GP(4). However, in those situations where N^+ and N^- are contained in the local invariant sets within the block that is not the case as shown by the following.

6.6 PROPOSITION. *Let W be a compact submanifold with boundary of M of dimension k, assume W admits a fundamental class $o_W \in H_k(W, \partial W)$, and for $i = 1, 2$, where $(P_{1\,i}, P_{0\,i})$ is a forward time index pair for S, suppose that there is an inclusion $j_i\colon (W, \partial W) \subset (P_{1\,i}, P_{0\,i})$ where also $W \subset W^u(S; P_{1\,i})$. Then*

$$\bar{p}^+_{1\,*} j_{1\,*} o_W = \bar{p}^+_{2\,*} j_{2\,*} o_W \in \widetilde{H}_k \mathcal{C}(S)\ . \tag{6.14}$$

Proof. First, suppose that there is an inclusion

$$k\colon (P_{1\,1}, P_{0\,1}) \subset (P_{1\,2}, P_{0\,2})$$

and let $\bar{k}$ be the induced map on the index spaces. Then clearly

$$\bar{k}_* p^+_{1\,*} j_{1\,*} = p^+_{2\,*} j_{2\,*}\ .$$

This, the fact that $\bar{k}$ is the unique morphism in $\mathcal{C}(S)$ from $P_{1\,1}/P_{0\,1}$ to $P_{1\,2}/P_{0\,2}$ by [K5, Proposition 3.1], and the fact that $\widetilde{H}_* \mathcal{C}(S)$ is the direct limit of the homology modules of the index spaces in $\mathcal{C}(S)$, immediately imply that (6.14) holds.

The general case where there is no inclusion relation between the given index pairs can be reduced to the case where there is as follows from Proposition 1 in the Appendix of [K3] and the constructions given in its proof. In particular, it follows easily from the constructions in the proof just cited that there exist nested index pairs (L_1, L_0), $(N_{1\,1}, N_{0\,1})$, and $(N_{1\,2}, N_{0\,2})$ each containing the pair $(W, \partial W)$ and for $i = 1, 2$, inclusions

$$(P_{1\,i}, P_{0\,i}) \hookrightarrow (N_{1\,i}, N_{0\,i}) \hookleftarrow (L_1, L_0),$$

thereby accomplishing the desired reduction. Note it is crucial that W be a subset of $W^u(S; P_{1\,i})$, $i = 1, 2$, in order to obtain from the constructions in [K3] that each of the pairs (L_1, L_0), $(N_{1\,1}, N_{0\,1})$, and $(N_{1\,2}, N_{0\,2})$ contains the pair $(W, \partial W)$. □

More generally, Proposition 6.6 allows us to make the following:

6.7 DEFINITION. Let W be a compact, k-dimensional submanifold with boundary of M, assume W admits a fundamental class $o_W \in H_k(W, \partial W)$, and suppose there is an inclusion $k\colon (W, \partial W) \subset (P_1, P_0)$ for some nested index pair (P_1, P_0) for an isolated invariant set S of some flow in M with $W \subset W^u(S; P_1)$, then $\bar{p}_*^+ k_* o_W \in \widetilde{H}_k \mathcal{C}(S)$ is called *the canonical image of the fundamental class* o_W. By Proposition 6.6, the definition of the canonical image does not depend on the choice of nested index pair (P_1, P_0) satisfying the constraints of the definition. If $\boldsymbol{\alpha} \in \widetilde{H}_k \mathcal{C}(S)$ and $\boldsymbol{\alpha}$ is the canonical image of o_W, then let us say that W *geometrically represents* $\boldsymbol{\alpha}$.

B. The Behavior of $\mathfrak{L}$ under Orbit Preserving Maps. The following definition describes the basic conditions under which transformation of the intersection pairing shall be studied.

6.8 DEFINITION. (A) For $i = 1, 2$, let M_i be a manifold and let S_i be an isolated invariant set of a flow in M_i. If, for $i = 1, 2$, there exists an open neighborhood U_i of S_i and a continuous, proper surjection $f\colon U_1 \to U_2$ satisfying (i) $S_1 = f^{-1}(S_2)$ and (ii) each orbit of the flow in U_1 induced by that in M_1 is mapped by f onto an orbit of the flow in U_2 induced by that in M_2 in such manner that positive semi-orbits go to positive semi-orbits and negative to negative, then, and only then, call S_1 *the f-pre-image of* S_2. It is not assumed that f preserves time-parameterization.

(B) If in addition to the conditions imposed on f in part (A), also f is required to be a homeomorphism of U_1 onto U_2, then call f a *local orbital equivalence of* S_1 *and* S_2 and say that S_1 *and* S_2 *are locally orbitally equivalent.*

The following proposition makes possible the main results of this section. Its proof follows straightforwardly from the definitions and is left to the reader.

6.9 PROPOSITION. *Under the assumptions of Definition 6.8, the following statements hold:*

(1) *if* S_2 *has isolating neighborhood* N *relative to the flow in* U_2, *then* S_1 *has isolating neighborhood* $f^{-1}(N)$ *relative to the flow in* U_1;
(2) *if* S_2 *has index pair* $\langle N_1, N_0 \rangle$ *relative to the flow in* U_2, *then* S_1 *has index pair* $\langle f^{-1}(N_1), f^{-1}(N_0) \rangle$ *relative to the flow in* U_1;

(3) *if P is AIP(thick) for S_2, then $f^{-1}(P)$ is AIP(thick) for S_1 where $f^{-1}(P)$ is defined componentwise by taking the inverse image under f of the components of P;*

(4) *if f is a local orbital equivalence of S_1 and S_2 and $P \in \mathrm{ob}(\mathcal{AIP}(S_2))$ with respect to the flow in U_2, then $f^{-1}(P) \in \mathrm{ob}(\mathcal{AIP}(S_1))$ with respect to the flow in U_1.*

6.9.1 Remark. Thus, there exists a contravariant functor assigning to a manifold M' and flow φ' in M' the set $Iso(M', \varphi')$ of isolated invariant sets of φ' and assigning to a continuous, proper, orbit-preserving surjection $f: M'' \to M'$ the set map $f^{-1}: Iso(M', \varphi') \to Iso(M, \varphi)$. Further, for each $S' \in Iso(M', \varphi)$, there exists a map between Conley indices $\mathcal{C}\left(f^{-1}\right) : \mathcal{C}\left(S'\right) \to \mathcal{C}\left(f^{-1}(S')\right)$ so that the Conley index may be viewed as a functor on the sets of isolated invariant sets of flows.

To state the main result of this section, make the following observations. With f, S_1 and S_2 as in Definition 6.8(A), application of part (3) of Theorem 5.2 to the map between Conley indices mentioned in Remark 6.9.1 and the singular homology functor yields a functorial homomorphism $\mathcal{C}\left(f^{-1}\right)_* : \widetilde{H}_*\mathcal{C}\left(S_2\right) \to \widetilde{H}_*\mathcal{C}\left(S_1\right)$. When f is a local orbital equivalence, application of this construction to its inverse yields an isomorphism $\mathcal{C}\left(f\right)_* : \widetilde{H}_*\mathcal{C}\left(S_1\right) \to \widetilde{H}_*\mathcal{C}\left(S_2\right)$ whose inverse is $\mathcal{C}\left(f^{-1}\right)_*$. Analogous remarks apply to the reverse time flow and yield an isomorphism $\mathcal{C}^*\left(f\right)_* : \widetilde{H}_*\mathcal{C}^*\left(S_1\right) \to \widetilde{H}_*\mathcal{C}^*\left(S_2\right)$.

6.10 Proposition. *For $i = 1, 2$, let M_i be a manifold oriented over the ring R and let S_i be an isolated invariant set of a flow in M_i. If S_1 and S_2 are locally orbitally equivalent and if the homeomorphism $h: U_1 \to U_2$ describing the orbital equivalence is orientation preserving, then where $\mathfrak{L}^i$ is the intersection pairing associated with S_i,*

$$\check{h}^{-1} \circ \mathfrak{L}^2 = \mathfrak{L}^1 \circ \left(\mathcal{C}\left(h^{-1}\right)_* \otimes \mathcal{C}^*\left(h^{-1}\right)_*\right),$$

6.10.1 Remark. Note that the conclusion of the proposition could as well be written in the form

$$\mathfrak{L}^1 = \check{h}^{-1} \circ \mathfrak{L}^2 \circ \left(\mathcal{C}\left(h\right)_* \otimes \mathcal{C}^*\left(h\right)_*\right).$$

suggesting the abbreviation $\mathfrak{L}^1 = h^*\mathfrak{L}^2$; however, the conclusion of the proposition as written is preferable in the sense that it generalizes to continuous, proper, local surjections; see Proposition 6.13.

Proof. Let $P \in \mathrm{ob}(\mathcal{AIP}(S_1))$. Then for each $V \in M^{(2)}_{\Delta o}(P)$ as a consequence of [D, Corollary VIII.13.7], for each $\alpha \in H_*(h(V_1), h(V_0))$ and $\gamma \in H_*(h(V_1^*), h(V_0^*))$,

$$h_*((h_*)^{-1}\alpha \mathbin{\widehat{\bullet}}^V (h_*)^{-1}\gamma) = \alpha \mathbin{\widehat{\bullet}}^{h(V)} \gamma .$$

Thus, from the definitions—see diagram (2.6)—it follows that

$$h_*\mathfrak{L}^P_{V_1 \cap V_1^*} \circ (h_* \otimes h_*)^{-1} = \mathfrak{L}^{h(P)}_{h(V_1 \cap V_1^*)},$$

whence passage to the limit over the cofinal family of neighborhoods of S_1 having the form $V_1 \cap V_1^*$ for some $V \in M^{(2)}_{\mathcal{A}o}(P)$ yields

$$\check{h} \circ \mathfrak{L}^1_P \circ (h_* \otimes h_*)^{-1} = \mathfrak{L}^2_{h(P)} \tag{6.15}$$

because the sets in P are compact so that $\mathfrak{L}^P_P = \mathfrak{L}^1_P$ and $\mathfrak{L}^{h(P)}_{h(P)} = \mathfrak{L}^2_{h(P)}$. Set $Q = h(P)$ and observe that

$$\mathcal{C}\left(h^{-1}\right)_* = \bar{p}_*^+ h_*^{-1} (\bar{q}_*^+)^{-1} \qquad \text{and} \qquad \mathcal{C}^*\left(h^{-1}\right)_* = \bar{p}_*^- h_*^{-1} (\bar{q}_*^-)^{-1}.$$

From this and (6.15), it follows that

$$\check{h} \circ \mathfrak{L}^1_P \circ (\bar{p}_*^+ \otimes \bar{p}_*^-)^{-1} \circ (\mathcal{C}\left(h^{-1}\right)_* \otimes \mathcal{C}^*\left(h^{-1}\right)_*) = \mathfrak{L}^2_Q \circ (\bar{q}_*^+ \otimes \bar{q}_*^-)^{-1}. \tag{6.16}$$

As $\mathcal{N}(S_1, \mathcal{AIP})$ is cofinal in $\mathcal{N}(S_1)$ by Theorem 2.11, also the family of neighborhoods of S_2 of the form $h(P_1) \cap h(P_1^*)$ is cofinal in $\mathcal{N}(S_2)$. Thus, it follows from Theorem 2.11 that passage to the limit in (6.16) over $P \in \mathrm{ob}(\mathcal{AIP}(S))$ yields the desired result. □

To state and prove the generalization mentioned in Remark 6.10.1 requires a definition of intersection pairing on the tensor product of the Čech homology modules of the Conley indices of S to $\check{H}_*(S)$ and an extension of the notion of homology transfer map to obtain a homomorphism from the Čech homology of a compact pair to the Čech homology of its pre-image under a continuous, proper map between manifolds.

6.11 Definition. The intersection pairing on the Čech modules of the Conley indices can be obtained as follows. Using normality, one can show that if $P \in \mathrm{ob}(\mathcal{AIP}(S))$, then the family of open sets $\{V_1 \cap V_1^* : V \in M^{(2)}_{\mathcal{A}o}(P)\}$ is cofinal in $\mathcal{N}(P_1 \cap P_1^*)$. Hence, from the naturality expressed by diagram (2.4), taking inverse limits one can show that there is a pairing

$$\check{\mathfrak{L}}_P \colon \check{H}_*(P_1, P_0) \otimes \check{H}_*(P_1^*, P_0^*) \to \check{H}_*(P_1 \cap P_1^*).$$

It is not difficult to show that the family of pairings $\check{\mathfrak{L}}_P$, $P \in \mathrm{ob}(\mathcal{AIP}(S))$, is natural relative to inclusion morphisms. A proof similar to that of Theorem 2.11 then shows that the analogue of equation (2.13) holds for this family of pairings, whence as in Theorem 2.11 passage to the inverse limit yields a pairing

$$\check{\mathfrak{L}} \colon \check{\widetilde{H}}_*\mathcal{C}(S) \otimes \check{\widetilde{H}}_*\mathcal{C}^*(S) \to \check{H}_*(S),$$

called the *intersection class pairing on the Čech homology modules of the Conley indices of* S.

6.11.1 Remark. It is a simple exercise that there are natural maps

$$i_S^+ \colon \widetilde{H}_*\mathcal{C}(S) \to \check{\widetilde{H}}_*\mathcal{C}(S) \quad \text{and} \quad i_S^- \colon \widetilde{H}_*\mathcal{C}^*(S) \to \check{\widetilde{H}}_*\mathcal{C}^*(S)$$

and that because diagram (2.6) commutes $\mathfrak{L}$ and $\check{\mathfrak{L}}$ are related by

$$\mathfrak{L} = \check{\mathfrak{L}} \circ (i_S^+ \otimes i_S^-).$$

6.12 DEFINITION. The extension of the notion of homology transfer is obtained as follows. Let $f\colon M_1 \to M_2$ be a continuous, proper map between manifolds and (P_1, P_0) a compact pair in M_2. Then for any open pair $(V_1, V_0) \supset (P_1, P_0)$ there is a transfer map $f_!\colon H_*(V_1, V_0) \to H_*(f^{-1}(V_1), f^{-1}(V_0))$ and the family of these is natural relative to inclusions of the open pairs. Passage to the inverse limit over the open neighborhood pairs of (P_1, P_0) yields the desired extension, denote it $\check{f}_!$, once it is observed that the family of open sets $f^{-1}(V_i)$, V_i an open neighborhood of P_i, is cofinal in $\mathcal{N}(f^{-1}(P_i))$ for $i = 1, 2$. This is an exercise left to the reader.

6.13 PROPOSITION. *For $i = 1, 2$, let M_i be a manifold of dimension m_i oriented over the PID R, let S_i be an isolated invariant set of a flow in M_i, and suppose that S_1 is the f-pre-image of S_2. Then f induces graded module homomorphisms of degree $m_1 - m_2$,*

$$\mathcal{C}\left(\check{f}_!\right) : \check{\widetilde{H}}_*\mathcal{C}(S_2) \to \check{\widetilde{H}}_*\mathcal{C}(S_1) \qquad \text{and} \qquad \mathcal{C}^*\left(\check{f}_!\right) : \check{\widetilde{H}}_*\mathcal{C}^*(S_2) \to \check{\widetilde{H}}_*\mathcal{C}^*(S_1)$$

so that for $i = 1, 2$ where $\check{\mathfrak{L}}^i$ is the intersection pairing on the tensor product of the Čech homology modules of the forward and reverse time indices of S_i to the Čech homology of S_i

$$\check{f}_! \circ \check{\mathfrak{L}}^2 = \check{\mathfrak{L}}^1 \circ \left(\mathcal{C}\left(\check{f}_!\right) \otimes \mathcal{C}^*\left(\check{f}_!\right)\right). \tag{6.17}$$

Proof. Let P be AIP(thick) for S_2, and set $Q_i := f^{-1}(P_i)$ and $Q_i^* = f^{-1}(P_i^*)$ for $i = 0, 1$. By Proposition 6.9, Q is AIP(thick) for S_1.

Now define $\mathcal{C}\left(\check{f}_!\right) : \check{\widetilde{H}}_*\mathcal{C}(S_2) \to \check{\widetilde{H}}_*\mathcal{C}(S_1)$ by

$$\mathcal{C}\left(\check{f}_!\right) = \check{\bar{q}}^+_* \check{f}_! (\check{\bar{p}}^+_*)^{-1}.$$

Arguments similar to those in Proposition 6.6 show that this definition of $\mathcal{C}\left(\check{f}_!\right)$ is independent of the particular choice of P so long as P is AIP(thick). The homomorphism $\mathcal{C}^*\left(\check{f}_!\right) : \check{\widetilde{H}}_*\mathcal{C}^*(S_2) \to \check{\widetilde{H}}_*\mathcal{C}^*(S_1)$ is defined in an analogous manner.

The remainder of the proof being similar to the corresponding part of the proof of Proposition 6.10 is left to the reader, but now one invokes [D, Proposition VIII.13.6] instead of its corollary. □

6.13.1 Remark. If M is at least C^1 and the flow is generated by a continuous vectorfield, then S_i $(i = 1, 2)$ admits index pairs that are C^0 manifold with boundary pairs, whence the Čech and singular modules of the Conley indices will be isomorphic. It follows that in (6.17), for $i = 1, 2$, the pairing $\check{\mathfrak{L}}^i$ can be replaced by the pairing $\mathfrak{L}^i$ if $\mathcal{C}\left(\check{f}_!\right)$ and $\mathcal{C}^*\left(\check{f}_!\right)$ are replaced by $\mathcal{C}(f_!) := (i^+_{S_1})^{-1}\mathcal{C}\left(\check{f}_!\right) i^+_{S_2}$ and $\mathcal{C}^*(f_!) := (i^-_{S_1})^{-1}\mathcal{C}^*\left(\check{f}_!\right) i^-_{S_2}$ respectively. It is not difficult to show using Corollary VIII.13.7 of [D] that if f is an orientation preserving local orbital equivalence of S_1 and S_2 then $\mathcal{C}(f_!) = \mathcal{C}\left(f^{-1}\right)_*$ and $\mathcal{C}^*(f_!) = \mathcal{C}^*\left(f^{-1}\right)_*$. Thus, Proposition 6.13 implies Proposition 6.10 when the minimal smoothness conditions are met.

CHAPTER 7

$\mathfrak{L}$ AND $^{\#}\mathfrak{L}$ ON THE CONLEY INDICES OF NORMALLY HYPERBOLIC INVARIANT SUBMANIFOLDS

Throughout this chapter assume M to be of class C^{r+1} and ξ to be a vectorfield on M of class C^r where $r \geq 1$. As before $\dim(M) = m$. For simplicity, all homology is taken with coefficients in R. Also, for each integer $k \geq 0$, let D^k_c denote the closed k-disk of radius c centered at $0 \in \mathbf{R}^k$ and let S^{k-1}_c denote its bounding $k-1$-sphere. In particular, $S^{-1}_c = \partial D^0_c = \emptyset$. As usual, $D^k := D^k_1$ and $S^{k-1} := S^{k-1}_1$. Also, the standard inner product and its associated norm on a Euclidean space are respectively denoted by $\langle \cdot \mid \cdot \rangle$ and $\| \cdot \|$.

A. Summary of Results.

7.1 DEFINITION. Let $\eta = (E, p, X)$ be a vector bundle of rank k ($k \geq 0$) over X paracompact, choose a Euclidean metric $\| \cdot \|$ for E, and let $E_{\geq 1} := \{v \in E : \|v\| \geq 1\}$. The *Thom space of* η, denoted $\mathsf{T}(\eta)$, is the quotient space $E/E_{\geq 1}$ with basepoint $[E_{\geq 1}]$. The definition does not depend on the choice of Euclidean metric: this is obvious for $k = 0$; for $k \geq 1$ see [Hus, Definition 15.1.2]. Note that when $k = 0$, $\mathsf{T}(\eta) \equiv X \amalg \{*\}$ with basepoint $* \notin X$ but is *not* a homeomorph of $X \vee S^0$ where S^0 is identified with $\{x_0\} \cup \{*\}$ for any choice of basepoint $x_0 \in X$ because as pointed spaces they are not homotopy equivalent even though as unpointed spaces they are homeomorphic. The lack of homeomorphism in the pointed case complicates the statement of Theorem 7.6 below.

7.2 DEFINITION. (A) Let S be a compact C^1 submanifold without boundary of M invariant under the action of a C^1 flow ψ in M. Call S a *normally hyperbolic invariant submanifold of* ψ if, and only if, there is a continuous splitting $TM|S = N^u_S \oplus TS \oplus N^s_S$ so that for each $t \in \mathbf{R}$, $T\psi_t$ leaves invariant N^u_S, N^s_S, and TS, and for each $\mathbf{u} \in S$, there exists $\tau \equiv \tau(\mathbf{u}) \in \mathbf{R}$ so that $\|T_{\mathbf{u}}\psi_\tau | N^s_S\| < \min\{\mathrm{m}(T_{\mathbf{u}}\psi_\tau|TS), 1\}$ and $\|T_{\mathbf{u}}\psi_\tau|TS\| > \max\{\mathrm{m}(T_{\mathbf{u}}\psi_\tau|N^u_S), 1\}$ where for a linear map between normed spaces $\mathrm{m}(L) = \inf\{\|Lv\| : \|v\| = 1\}$. Note $\mathrm{m}(L) = \|L^{-1}\|^{-1}$ when L is invertible. This definition is taken from [R] and is equivalent to ψ_t being eventually, relatively 1-normally hyperbolic for each $t \in \mathbf{R}$ as defined in [HPS]. Let us refer to N^s_S and N^u_S as the *expanding and contracting subbundles of the normal bundle of* S.

(B) Let ψ and S be as in (A). The *strong stable manifold at* $\mathbf{u} \in S$, denoted $W^{ss}(\mathbf{u})$, consists of those points $\mathbf{v} \in M$ with the property that the distance between $\psi_t(\mathbf{v})$ and $\psi_t(\mathbf{u})$ goes to zero as $t \to \infty$ at an asymptotic rate that is at least as fast as $\|T_{\mathbf{u}}\psi_\tau|N^s_S\|$. The *strong unstable manifold at* $\mathbf{u}$ is defined similarly by reversing the time direction. For $\mathbf{u}$ a point in an isolated invariant set that is also normally hyperbolic and which admits a block B, the *local strong stable manifold within* B

at $\mathbf{u}$, denoted $W^{ss}(\mathbf{u};B)$, consists of those points in $W^{ss}(\mathbf{u})$ whose positive semi-orbits under the flow lie in B. The *local strong unstable manifold within* B *at* $\mathbf{u}$ is defined analogously, but using negative semi-orbits. For $\lambda = u, s$, $W^\lambda(S)$ is fibered by the family $\{W^{\lambda\lambda}(\mathbf{u}) : \mathbf{u} \in S\}$. When B is a block with corners, transversality at the boundary implies that $W^{\lambda\lambda}(\mathbf{u};B)$ $(\lambda = u, s)$ is a neat submanifold of B; also $\{W^{\lambda\lambda}(\mathbf{u};B) : \mathbf{u} \in S\}$ fibers $W^\lambda(S;B)$.

7.3 DEFINITION. Let S be a closed submanifold of M that is orientable over R and normally hyperbolic relative to the flow generated by a vectorfield on an open neighborhood of S, and assume that N_S^u and N_S^s are R-orientable. Then associated (i) to a choice of R-orientation classes for the expanding and contracting subbundles of the normal bundle, (ii) to a choice of fundamental class for S, and (iii) to the given orientation O^M of M is a sign σ (i.e., either $\sigma = 1 \in R$ or $\sigma = -1 \in R$) that arises in the computation of the intersection pairing on the Conley indices of S. The sign σ is defined as follows.

Set $\ell := \dim(S)$, let $o_S \in H_\ell(S;R)$ denote the chosen fundamental class of S, and let $\nu_S \in H^\ell(S;R)$ denote the generator satisfying $\langle \nu_S, o_S\rangle = 1$. Next, identify the normal bundle of S with a total tubular neighborhood $N_S = (U,p,S)$ of S, and write $N_S = N_S^u \oplus N_S^s$. Under these identifications, for $\lambda = u, s$, let $\mathbf{E}_S^\lambda \subset U$ be the total space of N_S^λ, assume N_S^λ has rank $k_\lambda \geq 0$, and let $\tau^\lambda \in H^{k_\lambda}(\mathbf{E}_S^\lambda, \mathbf{E}_S^\lambda \setminus S; R)$ be an R-orientation of N_S^λ. As the Whitney sum $N_S^u \oplus N_S^s$ is the pull-back by the diagonal map on S of the product bundle $N_S^u \times N_S^s$ and where $\bar\Delta$ is the map on U to the total space of that product bundle induced by the pull-back operation, $\bar\Delta^*(\tau^u \times \tau^s)$ is an orientation class of (U,p,S). Thus, $p^*\nu_S \smile \bar\Delta^*(\tau^u \times \tau^s)$ generates $H^m(U, U\setminus S; R)$ as a consequence of the Thom Isomorphism Theorem. On the other hand, let o_S^U be the fundamental class generating $H_m(U, U \setminus S; R)$ that under the excision induced isomorphism has image $o_S^M \in H_m(M, M \setminus S; R)$ where o_S^M is the fundamental class along S determined by the given orientation O^M of M. Define

$$\sigma \equiv \sigma(\tau^u, \tau^s, o_S, O^M) := \langle p^*\nu_S \smile \bar\Delta^*(\tau^u \times \tau^s), o_S^U\rangle .$$

As defined, σ ostensibly depends on the choice of tubular neighborhood of S. That it does not is a consequence of the fact (e.g., see [H, Theorem 4.5.3]) that any two tubular neighborhoods of S are isotopic in M; details are left to the reader.

7.4 THEOREM. *For S as in Definition 7.3, the following hold:*

(1) *S is an isolated invariant set of the flow generated by ξ;*

(2) *$[\mathcal{C}(S)] = [\mathsf{T}(N_S^u)]$ and $[\mathcal{C}^*(S)] = [\mathsf{T}(N_S^s)]$;*

(3) *$\widetilde{H}_k\mathcal{C}(S;R) \simeq R$ for $k = k_u, k_u + \ell$ and is the zero module for $k < k_u$ and $k > k_u + \ell$; in particular, each R-orientation class τ^u of N_S^u determines a unique generator $\mathbf{o}_{k_u}^u$ of $\widetilde{H}_{k_u}\mathcal{C}(S;R)$, and τ^u and a choice of fundamental class $o_S \in H_\ell(S;R)$ together determine a unique generator $\mathbf{o}_{k_u+\ell}^u$ of $\widetilde{H}_{k_u+\ell}\mathcal{C}(S;R)$;*

(4) *$\widetilde{H}_k\mathcal{C}^*(S;R) \simeq R$ for $k = k_s, k_s + \ell$ and is the zero module for $k < k_s$ and for $k > k_s + \ell$; in particular, each R-orientation class τ^s of N_S^s determines a unique generator $\mathbf{o}_{k_s}^s$ of $\widetilde{H}_{k_s}\mathcal{C}^*(S;R)$, and τ^s and a choice of fundamental class $o_S \in H_\ell(S;R)$ together determine a unique generator $\mathbf{o}_{k_s+\ell}^s$ of $\widetilde{H}_{k_s+\ell}\mathcal{C}^*(S;R)$;*

(5) *given R-orientations* τ^u *of* N_S^u *and* τ^s *of* N_S^s *and given a fundamental class* o_S *of* S*, then, where* $\sigma = \sigma(\tau^u, \tau^s, o_S, O^M)$*, the corresponding generators* $\mathbf{o}_k^\lambda$ $(k = k_\lambda, k_{\lambda+\ell},\ \lambda = u, s)$ *satisfy*

$$\#(\mathbf{o}_{k_u+\ell}^u \mathbin{\widehat{\bullet}} \mathbf{o}_{k_s}^s) = \sigma\,, \qquad \#(\mathbf{o}_{k_u}^u \mathbin{\widehat{\bullet}} \mathbf{o}_{k_s+\ell}^s) = (-1)^{\ell k_u}\sigma$$

and

$$\mathbf{o}_{k_u+\ell}^u \mathbin{\widehat{\bullet}} \mathbf{o}_{k_s+\ell}^s = \sigma o_S\,;$$

(6) *there exists a cofinal family* $\mathfrak{B}(S)$ *of neighborhoods of* S *so that for each* $B \in \mathfrak{B}(S)$*,* B *is a* C^r *isolating block with corners for* S*, and in the sense of Definition 6.7,* $W^\lambda(S; B)$ *geometrically represents* $\mathbf{o}_{k_\lambda+\ell}^\lambda$ *and, for each* $\mathbf{u} \in S$*,* $W^{\lambda\lambda}(\mathbf{u}; B)$ *geometrically represents* $\mathbf{o}_{k_\lambda}^\lambda$ $(\lambda = u, s)$.

As stated in the Introduction, the proof of Theorem 7.4 will be carried out in full only for the special cases of S a hyperbolic critical point and S a hyperbolic closed orbit. These two cases are stated below as Theorems 7.5–6. A proof of the general case will be given elsewhere and depends on the construction of an isolating block B as described in the Introduction that is the total space of the Whitney sum of two disk bundles.

The version of Theorem 7.4 for hyperbolic critical points will be stated based on the following:

HYPOTHESES HCP. *The point* $\mathbf{u}_0$ *is a critical point of* ξ *whose Hessian* $d\xi$ *at* $\mathbf{u}_0$ *has (counting multiplicities)* P *eigenvalues with positive real part and* N *with negative real part where* $P + N = m$.

For the definition of the Hessian of a vectorfield at a critical point as just used see [Ab-R].

As is well-known, $\mathbf{u}_0$ satisfies hypotheses HCP if, and only if, $\mathbf{u}_0$ is a hyperbolic critical point of ξ in which case the singleton $S := \{\mathbf{u}_0\}$ is certainly an R-orientable, normally hyperbolic invariant submanifold where in the notation of Definition 7.3, $k_u = P$, $k_s = N$ and $\ell = 0$. Also, it is immediate from the definitions that the local stable and strong local stable manifolds coincide in the critical point case and similarly for the local unstable manifolds. Since the Thom space of a vector bundle of rank k over a point is obviously a pointed k-sphere, Theorem 7.4 has as corollary the following result where in its statement $* \in S^k$ denotes any chosen basepoint.

7.5 THEOREM. *If* $\mathbf{u}_0 \in M$ *satisfies hypotheses* HCP*, then the conclusions of Theorem 7.4 hold for* $S := \{\mathbf{u}_0\}$ *where* $k_u = P$*,* $k_s = N$*, and* $\ell = 0$*. In particular*

(1) $[\mathcal{C}(S)] = [(S^P, *)]$ *and* $[\mathcal{C}^*(S)] = [(S^N, *)]$*;*

(2) *given R-orientations* τ^u *of* $\mathbf{E}_S^u$ *and* τ^s *of* $\mathbf{E}_S^s$*, the corresponding generators* $\mathbf{o}_P^u$ *of* $\widetilde{H}_P\mathcal{C}(S; R)$ *and* $\mathbf{o}_N^s$ *of* $\widetilde{H}_N\mathcal{C}^*(S; R)$ *satisfy*

$$\#(\mathbf{o}_P^u \mathbin{\widehat{\bullet}} \mathbf{o}_N^s) = \sigma(\tau^u, \tau^s, [\mathbf{u}_0], O^M)$$

where $[\mathbf{u}_0]$ *is the homology class of* $\mathbf{u}_0$ *regarded as a 0-simplex;*

(3) *if* $M = \mathbf{R}^m$*,* $\mathbf{o}_P^u \# \mathbf{o}_N^s$ *generates* $H_m(\mathbf{R}^m, \mathbf{R}^m \setminus \{0\}; R)$.

The manner in which the orientation classes τ^u and τ^s determine the generators $\mathbf{o}_P^u$ and $\mathbf{o}_N^s$ is described below in Lemma 7.13 and the proof of Theorem 7.15.

The version of Theorem 7.4 for hyperbolic closed orbits will be stated based on the following:

HYPOTHESES HCO. *S is a closed orbit of ξ with positive prime period that has (counting multiplicities) P Floquet multipliers with modulus greater than one and N Floquet multipliers with modulus less than one where $P + N = m - 1$.*

As is well-known, S is a hyperbolic closed orbit of ξ if, and only if, hypotheses HCO are satisfied in which case S is an R-orientable, normally hyperbolic invariant submanifold relative to ξ where in the notation of Definition 7.3, $k_u = P$, $k_s = N$, and $\ell = 1$. Thus, Theorem 7.4 has with a bit of work (see Lemma 7.24 and Propositions 7.25–26) the following result as corollary where in its statement, for any integer $k \geq 0$, $S^k\mathbf{RP}^2$ denotes the kth iterated reduced suspension of the projective plane relative to any choice of basepoint $* \in \mathbf{RP}^2$.

7.6 THEOREM. *If S satisfies hypotheses HCO, then the conclusions of Theorem 7.4 hold with $k_u = P$, $k_s = N$, and $\ell = 1$. In particular,*

(1) *either $P = 0$ and $[\mathcal{C}(S)] = [S^1 \amalg \{*\}]$ or $P \geq 1$ and, as N_S^u is or is not orientable over $\mathbf{Z}$, either $[\mathcal{C}(S)] = [S^{P+1} \vee S^P]$ or $[\mathcal{C}(S)] = [S^{P-1}\mathbf{RP}^2]$;*
(2) *either $N = 0$ and $[\mathcal{C}^*(S)] = [S^1 \amalg \{*\}]$ or $N \geq 1$ and, as N_S^s is or is not orientable over $\mathbf{Z}$, either $[\mathcal{C}^*(S)] = [S^{N+1} \vee S^N]$ or $[\mathcal{C}^*(S)] = [S^{N-1}\mathbf{RP}^2]$;*
(3) *given R-orientations τ^u of N_S^u and τ^s of N_S^s and given a fundamental class o_S of S, then, where $\sigma \equiv \sigma(\tau^u, \tau^s, o_S, O^M)$, the corresponding generators $\mathbf{o}_j^u$ of $\widetilde{H}_j\mathcal{C}(S;R)$ $(j = P, P+1)$ and $\mathbf{o}_k^s$ of $\widetilde{H}_k\mathcal{C}^*(S;R)$ $(k = N, N+1)$ satisfy*

$$^{\#}(\mathbf{o}_{P+1}^u \mathbin{\widehat{\bullet}} \mathbf{o}_N^s) = \sigma\,, \qquad ^{\#}(\mathbf{o}_P^u \mathbin{\widehat{\bullet}} \mathbf{o}_{N+1}^s) = (-1)^P\sigma\,,$$

and

$$\mathbf{o}_{P+1}^u \mathbin{\widehat{\bullet}} \mathbf{o}_{N+1}^s = \sigma\, o_S\,.$$

The manner in which the orientation classes τ^u and τ^s determine the generators $\mathbf{o}_P^u$ and $\mathbf{o}_N^s$ and the manner in which τ^u and τ^s together with a fundamental class o_S determine $\mathbf{o}_{P+1}^u$ and $\mathbf{o}_{N+1}^s$ is described below in Lemmas 7.28 and 7.29 and in the proof of Theorem 7.31.

Note that statement (1) of Theorem 7.5 implies that statement (2) of that theorem completely determines $\mathfrak{L}$ and $^{\#}\mathfrak{L}$ for $S = \{\mathbf{u}_0\}$ where $\mathbf{u}_0$ satisfies hypotheses HCP. Similarly, statements (1) and (2) of Theorem 7.6 imply that statement (3) of that theorem completely determines $\mathfrak{L}$ and $^{\#}\mathfrak{L}$ when S satisfies hypotheses HCO.

B. Computational Preliminaries. Made here are the estimates needed to construct isolating blocks under the assumption of hypotheses HCP or HCO and the computations underlying statement (5) of Theorem 7.4 and therefore of statement (2) of Theorem 7.5 and statement (3) of Theorem 7.6.

The first result of this subsection will in effect compute the intersection pairing for hyperbolic critical points and hyperbolic periodic orbits. For other R-orientable, normally hyperbolic invariant submanifolds it only computes the intersection pairing on generating elements of $\widetilde{H}_*\mathcal{C}(S) \otimes \widetilde{H}_*\mathcal{C}^*(S)$ of lowest and highest order; no more can be said without additional knowledge about the homology of S although it will clear from the proof how to use such additional information when available to obtain additional values of the intersection pairing. The result requires hypotheses, stated presently, that motivate the construction of isolating blocks made in Theorems 7.11 and 7.21. In stating them, for reasons that will become clear in the course of stating and proving Theorem 7.21 below, when reference is made to a k-disk bundle, it is not assumed that the structure group is $O(k)$ or a subgroup thereof, but only that the structure group is some subgroup of Homeo($\mathbf{R}^k$) whose elements preserve the Euclidean norm. Also, if η is a k-disk bundle, its associated $k-1$-sphere bundle is denoted by $\dot{\eta}$.

HYPOTHESES AND DEFINITIONS NHIS.

(1) *S is a closed, connected, ℓ-dimensional, R-orientable C^r submanifold of M;*
(2) *$N_S = (U, p, S)$ is a tubular neighborhood of S and admits a continuous splitting $N_S = N_S^u \oplus N_S^s$ where N_S^λ is a k_λ-plane bundle with total space $\mathbf{E}_S^\lambda \subset U$ $(\lambda = u, s)$;*
(3) *$\tau^\lambda \in H^{k_\lambda}(\mathbf{E}_S^\lambda, \mathbf{E}_S^\lambda \setminus S; R)$ is an R-orientation class for N_S^λ $(\lambda = u, s)$;*
(4) *$o_S \in H_\ell(S; R)$ is a fundamental class of S;*
(5) *S is an isolated invariant set of some flow in U, and $\mathcal{N}(S)$, the family of neighborhoods of S, admits a cofinal family $\mathfrak{BW}$ with each $B \in \mathfrak{BW}$ a C^r isolating block with corners having the properties:*
   (a) *B is the total space of a subbundle β of N_S, and the splitting of N_S induces a splitting of $\beta = \omega^u \oplus \omega^s$ with ω^λ a k_λ-disk bundle $(\lambda = u, s)$;*
   (b) *B^+ is the total space of $\dot{\omega}^u \oplus \omega^s$ and B^- is the total space of $\omega^u \oplus \dot{\omega}^s$;*
   (c) *$W^\lambda := W^\lambda(S; B)$ is the total space of ω^λ $(\lambda = u, s)$ and W^u and W^s are topologically transverse at some point $\mathbf{u} \in S$.*

For $\lambda = u, s$, note that W^λ is a manifold with boundary and that ∂W^λ is the total space of $\dot{\omega}^\lambda$. Use $p_\lambda\colon \mathbf{E}_S^\lambda \to S$ to denote the restriction of p to $\mathbf{E}_S^\lambda$, and use $W_\mathbf{u}^\lambda$ and $\partial W_\mathbf{u}^\lambda$ to denote respectively the fiber of ω^λ at $\mathbf{u}$ and the fiber of $\dot{\omega}^\lambda$ at $\mathbf{u}$. Hypotheses NHIS(1)–(5), the Thom Isomorphism Theorem, and the observations that $(W^u, \partial W^u) \subset (B, B^+)$ and that $(W^s, \partial W^s) \subset (B, B^-)$ allow the definition of generating classes for $\widetilde{H}_*\mathcal{C}(S)$ and $\widetilde{H}_*\mathcal{C}^*(S)$ as follows:

(6) for $\lambda = u, s$, let $o_{k_\lambda}^\lambda$ be the generator of $H_{k_\lambda}(W_\mathbf{u}^\lambda, \partial W_\mathbf{u}^\lambda; R)$ satisfying

$$\langle \tau^\lambda | (W_\mathbf{u}^\lambda, \partial W_\mathbf{u}^\lambda), o_{k_\lambda}^\lambda \rangle = 1\,;$$

(7) for $\lambda = u, s$, let $o_{k_\lambda+\ell}^\lambda$ be the generator of $H_{k_\lambda+\ell}(W^\lambda, \partial W^\lambda; R)$ satisfying

$$p_{\lambda *}(\tau^\lambda | (W^\lambda, \partial W^\lambda) \frown o_{k_\lambda+\ell}^\lambda) = o_S\,;$$

(8) let $\mathbf{o}_{k_u}^u$ and $\mathbf{o}_{k_u+\ell}^u$ be the respective canonical images (see Definition 6.7) in $\widetilde{H}_*\mathcal{C}(S)$ of $o_{k_u}^u$ and $o_{k_u+\ell}^u$ and *assume these are independent of* $B \in \mathfrak{BW}$;
(9) let $\mathbf{o}_{k_s}^s$ and $\mathbf{o}_{k_s+\ell}^s$ be the respective canonical images in $\widetilde{H}_*\mathcal{C}^*(S)$ of $o_{k_s}^s$ and $o_{k_s+\ell}^s$ and *assume these are independent of* $B \in \mathfrak{BW}$.

7.7 LEMMA (cf. [D, VIII.11.13 and VIII.13.23]). *Under the assumption of hypotheses and definitions* NHIS *and with* σ *as in Definition 7.3*

$$\#(\mathbf{o}^u_{k_u+\ell} \frown \mathbf{o}^s_{k_s}) = \sigma\,, \qquad \#(\mathbf{o}^u_{k_u} \frown \mathbf{o}^s_{k_s+\ell}) = (-1)^{\ell k_u}\sigma$$

and

$$\mathbf{o}^u_{k_u+\ell} \frown \mathbf{o}^s_{k_s+\ell} = \sigma o_S\,.$$

Proof. The computations are done in three steps: first, relative to a suitable chart, local representatives of the various orientation and fundamental classes are determined; second, formal computations are made which if correct give the desired results; third, the formal computations are justified via a lengthy diagram chase. Throughout, the following facts are used without further explicit comment:

(1) the Kronecker pairing is invariant under isomorphisms on homology and cohomology induced by inclusions and homeomorphisms;
(2) intersection pairings on arbitrary pairs of pairs in a manifold are natural with respect to inclusions and with respect to orientation preserving homeomorphisms; see Lemma 2.10 and [D, Proposition VIII.13.21].

To begin, choose a coordinate chart (Q,ψ) at $\mathbf{u}$ with the following properties: $Q \subset U$, $\psi(Q) = \mathbf{R}^\ell \times \mathbf{R}^{k_u} \times \mathbf{R}^{k_s}$, $\psi(\mathbf{u}) = (0,0,0)$, $\psi(Q\cap S) = \mathbf{R}^\ell \times \{0\} \times \{0\}$, $\psi(\mathbf{E}^u_S \cap Q) = \mathbf{R}^\ell \times \mathbf{R}^{k_u} \times \{0\}$, and $\psi(\mathbf{E}^s_S \cap Q) = \mathbf{R}^\ell \times \{0\} \times \mathbf{R}^{k_s}$, and ψ trivializes the restricted bundles $N_S|(S\cap Q)$ and $N^\lambda_S|(S\cap Q)$ $(\lambda = u,s)$; in particular $p^{-1}(\mathbf{v}) \subset Q$ for each $\mathbf{v} \in Q\cap S$. To satisfy the last property, it may be necessary to first shrink the tubular neighborhood U along those fibers in and near $Q\cap S$. Afterward choose $B \in \mathfrak{BW}$ satisfying $B \subset U$; below, the classes $o^\lambda_{k_\lambda}$ and $o^\lambda_{\ell+k_\lambda}$ $(\lambda = u,s)$ are the ones defined using this B. To facilitate printing set $\mathbf{R}^k_0 := \mathbf{R}^k \setminus \{0\}$ for each integer $k \geq 1$. Below, i_* and i^* always denote inclusion induced maps arising from the inclusion $Q \subset U$.

First, let $\pi_\lambda\colon U \to \mathbf{E}^\lambda_S$ denote the continuous projection arising from the splitting, set $\pi_{Q\,u} := \psi \circ \pi_u|Q \circ \psi^{-1}$, and consider the ladder of isomorphisms

$$\begin{array}{ccccc}
H^{k_u}(U, U\setminus \mathbf{E}^s_S)) & \xrightarrow[\simeq]{i^*} & H^{k_u}(Q, Q\setminus \mathbf{E}^s_S)) & \xrightarrow[\simeq]{\psi^{-1\,*}} & H^{k_u}(\mathbf{R}^\ell \times (\mathbf{R}^{k_u}, \mathbf{R}^{k_u}_0) \times \mathbf{R}^{k_s}) \\
\simeq \uparrow \pi_u^* & & \simeq \uparrow (\pi_u|Q)^* & & \simeq \uparrow \pi^*_{Q\,u} \\
H^{k_u}(\mathbf{E}^u_S, \mathbf{E}^u_S \setminus S) & \xrightarrow[\simeq]{i^*} & H^{k_u}(Q\cap \mathbf{E}^u_S, Q\cap \mathbf{E}^u_S\setminus S) & \xrightarrow[\simeq]{\psi^{-1\,*}} & H^{k_u}(\mathbf{R}^\ell \times (\mathbf{R}^{k_u}, \mathbf{R}^{k_u}_0) \times \{0\})
\end{array}$$

where each vertical arrow is invertible because it is the end map of a strong deformation retraction obtained by contracting each fiber of N^s_S linearly to its zero-point and where i^* in the bottom row is an isomorphism because it restricts an orientation class to an orientation class and the orientation classes here generate as a consequence of the Thom isomorphism theorem because S and $Q\cap S$ are connected. Under the composite isomorphism of the bottom row, the orientation class τ^u corresponds to an element $1 \times \tau'_{k_u} \times 1$ where τ'_{k_u} generates $H^{k_u}(\mathbf{R}^{k_u}, \mathbf{R}^{k_u}_0)$, and since $\pi_{Q\,u}(x,y,z) = (x,y,0)$ for $(x,y,z) \in \mathbf{R}^\ell \times \mathbf{R}^{k_u} \times \mathbf{R}^{k_s}$, it follows that under the composite isomorphism of the top row, $\pi_u^*\tau^u$ corresponds to $1 \times \tau'_{k_u} \times 1 \in$

$H^{k_u}(\mathbf{R}^\ell \times (\mathbf{R}^{k_u}, \mathbf{R}_0^{k_u}) \times \mathbf{R}^{k_s})$. By interchanging the roles of $\mathbf{E}_S^u$ and $\mathbf{E}_S^s$ and of $\mathbf{R}^{k_u}$ and $\mathbf{R}^{k_s}$, one similarly sees that τ^s is represented locally by $1 \times 1 \times \tau''_{k_s} \in H^{k_s}(\mathbf{R}^\ell \times \{0\} \times (\mathbf{R}^{k_s}, \mathbf{R}_0^{k_S}))$ and that $\pi_s^* \tau^s$ is represented locally by $1 \times 1 \times \tau''_{k_s} \in H^{k_s}(\mathbf{R}^\ell \times \mathbf{R}^{k_u} \times (\mathbf{R}^{k_s}, \mathbf{R}_0^{k_S}))$ where τ''_{k_s} generates $H^{k_s}(\mathbf{R}^{k_s}, \mathbf{R}_0^{k_s})$. Because $\bar{\Delta} = (\pi_u, \pi_s) = (\pi_u \times \pi_s) \circ \Delta_U$ where $\Delta_U : U \to U \times U$ is the diagonal map, it follows from the definition of cup products that the orientation class

$$\bar{\Delta}^*(\tau^u \times \tau^s) = \Delta_U^*(\pi_u^* \tau^u \times \pi_s^* \tau^s) = \pi_u^* \tau^u \smile \pi_s^* \tau^s .$$

Then the naturality of cup products implies that under the composite isomorphism

$$H^{k_u+k_s}(U, U\backslash S) \xrightarrow[\simeq]{i^*} H^{k_u+k_s}(Q, Q\backslash S) \xrightarrow[\simeq]{\psi^{-1*}} H^{k_u+k_s}(\mathbf{R}^\ell \times (\mathbf{R}^{k_u}, \mathbf{R}_0^{k_u}) \times (\mathbf{R}^{k_s}, \mathbf{R}_0^{k_s}))$$

$\bar{\Delta}^*(\tau^u \times \tau^s)$ corresponds to $1 \times \tau'_{k_u} \times \tau''_{k_s} = 1 \times \tau'_{k_u} \times 1 \smile 1 \times 1 \times \tau'_{k_s}$.

Next, set $p_Q := \psi|(Q \cap S) \circ p|Q \circ \psi^{-1}$ and consider the ladder of isomorphisms

$$\begin{array}{ccccccc}
H^\ell(U) & \longleftarrow & H^\ell(U, U \setminus p^{-1}(\mathbf{u})) & \xrightarrow{i^*} & H^\ell(Q, Q \setminus p^{-1}(\mathbf{u})) & \xrightarrow[\simeq]{\psi^{-1*}} & H^\ell((\mathbf{R}^\ell, \mathbf{R}_0^\ell) \times \mathbf{R}^{k_u} \times \mathbf{R}^{k_s}) \\
{\scriptstyle p^*}\uparrow{\scriptstyle \simeq} & & {\scriptstyle p^*}\uparrow{\scriptstyle \simeq} & & {\scriptstyle (p|Q)^*}\uparrow{\scriptstyle \simeq} & & {\scriptstyle p_Q^*}\uparrow{\scriptstyle \simeq} \\
H^\ell(S) & \xleftarrow[\simeq]{} & H^\ell(S, S \setminus \{\mathbf{u}\}) & \xrightarrow[\simeq]{i^*} & H^\ell(Q \cap S, Q \cap S \setminus \{\mathbf{u}\}) & \xrightarrow[\simeq]{\psi^{-1*}} & H^\ell((\mathbf{R}^\ell, \mathbf{R}_0^\ell) \times \{0\} \times \{0\})
\end{array}$$

where the unlabeled arrows are inclusion induced. Here, each vertical arrow is an isomorphism as the fiberwise linear contraction of the total space of a vector bundle to its zero section is a strong deformation retraction. Also, i^* in the bottom row is an excision isomorphism, and the bottom unlabeled arrow is invertible because the corresponding homology homomorphism maps fundamental class to fundamental class. Under the composite isomorphism of the bottom row, the class $\nu_S \in H^\ell(S)$ Kronecker dual to the given fundamental class o_S of S corresponds to a class $\tau_\ell \times 1 \times 1$ where τ_ℓ generates $H^\ell(\mathbf{R}^\ell, \mathbf{R}_0^\ell)$. Then under the composite isomorphism of the top row, $p^* \nu_S$ corresponds to $\tau_\ell \times 1 \times 1 \in H^\ell((\mathbf{R}^\ell, \mathbf{R}_0^\ell) \times \mathbf{R}^{k_u} \times \mathbf{R}^{k_s})$ because $p_Q(x, y, z) = (x, 0, 0)$ for $(x, y, z) \in \mathbf{R}^\ell \times \mathbf{R}^{k_u} \times \mathbf{R}^{k_s}$.

We can now compute local representatives for $p^* \nu_S \smile \bar{\Delta}^*(\tau^u \times \tau^s)$ and for the top-dimensional class $p_\lambda^* \nu_S \smile \tau^\lambda$ of N_S^λ $(\lambda = u, s)$. Append an identity homomorphism to the top row of the first ladder considered and compute the cup products of corresponding successive images of $p^* \nu_S$ and $\bar{\Delta}^*(\tau^u \times \tau^s)$ along, respectively, the top rows of the the second and first ladders considered. By naturality of cup products, it follows that under the composite isomorphism

$$H^m(U, U \setminus S) \simeq H^m(U, U \setminus \{\mathbf{u}\}) \xrightarrow[\simeq]{i^*} H^m(Q, Q \setminus \{\mathbf{u}\})$$

$$\xrightarrow[\simeq]{\psi^{-1*}} H^m((\mathbf{R}^\ell, \mathbf{R}_0^\ell) \times (\mathbf{R}^{k_u}, \mathbf{R}_0^{k_u}) \times (\mathbf{R}^{k_s}, \mathbf{R}_0^{k_s}))$$

$p^* \nu_S \smile \bar{\Delta}^*(\tau^u \times \tau^s)$ corresponds to $\tau_\ell \times \tau'_{k_u} \times \tau''_{k_s} = \tau_\ell \times 1 \times 1 \smile 1 \times \tau'_{k_u} \times \tau''_{k_s}$. In similar manner, one shows that the top-dimensional class of N_S^λ, viz., $p_\lambda^* \nu_S \smile \tau^\lambda$, is represented locally in $H^{\ell + k_\lambda}(\psi(Q \cap \mathbf{E}_S^\lambda), \psi(Q \cap \mathbf{E}_S^\lambda \setminus \{\mathbf{u}\}))$ by

$$\tau_\ell \times \tau'_{k_u} \times 1 = \tau_\ell \times 1 \times 1 \smile 1 \times \tau'_{k_u} \times 1 \qquad \text{when } \lambda = u$$

and by

$$\tau_\ell \times 1 \times \tau''_{k_s} = \tau_\ell \times 1 \times 1 \smile 1 \times 1 \times \tau''_{k_s} \qquad \text{when } \lambda = s.$$

The local representative of $p_\lambda^* \nu_S \smile \tau^\lambda$ is used to compute a local representative of $o^\lambda_{\ell+k_\lambda}$ as follows. Because $B \subset U$, also $(W^\lambda, \partial W^\lambda) \subset (\mathbf{E}^\lambda_S, \mathbf{E}^\lambda_S \backslash S)$, and as the former pair consists of the total spaces of a disk-sphere subbundle pair of N^λ_S, the inclusion induced maps on homology and cohomology are isomorphisms. It then follows from the definition of $o^\lambda_{\ell+k_\lambda}$ that its local representative must be Kronecker dual to the local representative of $p_\lambda^* \nu_S \smile \tau^\lambda$, whence in $H_{k_\lambda+\ell}(\psi(Q \cap \mathbf{E}^\lambda_S), \psi(Q \cap \mathbf{E}^\lambda_S \setminus \{\mathbf{u}\}))$ it is represented by

$$(-1)^{\ell k_u} o_\ell \times o'_{k_u} \times [0] \qquad \text{when } \lambda = u$$

and by

$$(-1)^{\ell k_s} o_\ell \times [0] \times o''_{k_s} \qquad \text{when } \lambda = s.$$

Next, choose the orientation $O^{\mathbf{R}^m}$ of $\mathbf{R}^m$ whose fundamental class around 0, call it o_m, locally represents o^U_S as defined in Definition 7.3. To get an explicit expression for o_m, first, note that under the composite isomorphism on homology dual to the composite isomorphism on cohomology given by the bottom row of the second ladder considered, the fundamental class o_S corresponds to an element $o_\ell \times [0] \times [0] \in H_\ell((\mathbf{R}^\ell, \mathbf{R}^\ell_0) \times \{0\} \times \{0\})$ where o_ℓ generates $H_\ell(\mathbf{R}^\ell, \mathbf{R}^\ell_0)$, and note that $\langle \tau_\ell, o_\ell \rangle = \langle \nu_S, o_S \rangle = 1$. Then, because

$$\langle \tau_\ell \times \tau'_{k_u} \times \tau''_{k_s}, o_\ell \times o'_{k_u} \times o''_{k_s} \rangle = (-1)^{\ell k_u + \ell k_s + k_u k_s}$$

it follows with σ defined as in Definition 7.3 that the image of o^U_S under the composite isomorphism (the first unlabeled isomorphism maps fundamental class to fundamental class; the second is an excision isomorphism)

$$H_m(U, U \setminus S) \simeq H_m(U, U \setminus \{\mathbf{u}\}) \simeq H_m(Q, Q \setminus \{\mathbf{u}\}) \xrightarrow[\simeq]{\psi_*} H_m(\psi(Q), \psi(Q \setminus \{\mathbf{u}\}))$$

is $o_m := (-1)^{\ell k_u + \ell k_s + k_u k_s} \sigma o_\ell \times o'_{k_u} \times o''_{k_s}$ which uniquely determines $O^{\mathbf{R}^m}$.

We wish to take advantage of formulae (2.5) in computing the intersection pairing on the various local representatives; however, the use of (2.5) restricts us to computing intersection classes of homology classes that can be represented in open pairs. For the moment, let us ignore this requirement and compute formally ignoring the need to meet the hypotheses under which Poincaré duality holds. Also, we use singular cohomology and cup and cap products rather than their Čech counterparts, and to meet the excisiveness requirements for their use the classes used will be regarded as classes in a larger ambient space obtained by replacing each factor $\{0\}$ with $\mathbf{R}^k$ where $k = \ell$, $k = k_u$, or $k = k_s$ as appropriate. The justification for doing so will be given after the formal computations are completed and follows from fact (2) above and naturality of singular cap and cup products.

First, let us compute $o^u_{k_u+\ell} \mathbin{\widehat{\bullet}} o^s_{k_s+\ell}$. Regard $\sigma 1 \times 1 \times \tau''_{k_s}$ as an element of $H^{k_s}(\mathbf{R}^\ell \times \mathbf{R}^{k_u} \times (\mathbf{R}^{k_s}, \mathbf{R}^{k_s}_0))$. It is straightforward to check that

$$\sigma 1 \times 1 \times \tau''_{k_s} \frown (-1)^{\ell k_u + \ell k_s + k_u k_s} \sigma\, o_\ell \times o'_{k_u} \times o''_{k_s} = (-1)^{\ell k_u} o_\ell \times o'_{k_u} \times [0]$$

whence $\sigma\, 1 \times 1 \times \tau''_{k_s}$ is formally Poincaré dual to $(-1)^{\ell k_u} o_\ell \times o'_{k_u} \times [0]$ when $O^{\mathbf{R}^m}$ is used to orient $\mathbf{R}^m \equiv \mathbf{R}^\ell \times \mathbf{R}^{k_u} \times \mathbf{R}^{k_s}$. Formal use of formula (2.5*b*) then yields the computation (the first equality is only true modulo an isomorphism)

$$\begin{aligned}(-1)^{\ell k_u} o_\ell \times o'_{k_u} \times [0] \,\widehat{\bullet}\, (-1)^{\ell k_s} o_\ell \times [0] \times o''_{k_s} & \\ &= \sigma\, 1 \times 1 \times \tau''_{k_s} \frown (-1)^{\ell k_s} o_\ell \times [0] \times o''_{k_s} \\ &= \sigma\, o_\ell \times [0] \times [0]\,;\end{aligned}$$

from which we conclude that

$$o^u_{k_u+\ell} \,\widehat{\bullet}\, o^s_{k_s+\ell} = \sigma\, o_S\,. \tag{7.1}$$

Similarly, to compute $o^u_{k_u+\ell} \,\widehat{\bullet}\, o^s_{k_s}$ we additionally note that $o^s_{k_s}$ is by definition Kronecker dual to $\tau^s|(W^s_{\mathbf{u}}, W^s_{\mathbf{u}} \setminus \{\mathbf{u}\})$, and the latter is represented locally relative to (Q, ψ) by $1 \times 1 \times \tau''_{k_s}$ which is easily seen to be Kronecker dual to $[0] \times [0] \times o''_{k_s}$. Again, formal use of (2.5*b*) yields (with the same caveat) the computation

$$\begin{aligned}(-1)^{\ell k_u} o_\ell \times o'_{k_u} \times [0] \,\widehat{\bullet}\, [0] \times [0] \times o''_{k_s} &= \sigma\, 1 \times 1 \times \tau''_{k_s} \frown [0] \times [0] \times o''_{k_s} \\ &= \sigma\, [0] \times [0] \times [0]\,;\end{aligned}$$

from which we conclude that

$$\#(o^u_{k_u+\ell} \,\widehat{\bullet}\, o^s_{k_s}) = \sigma\,. \tag{7.2}$$

Finally, since $\tau^u|(W^u_{\mathbf{u}}, W^u_{\mathbf{u}} \setminus \{\mathbf{u}\})$ is represented locally by $1 \times \tau'_{k_u} \times 1$ which is Kronecker dual to $[0] \times o'_{k_u} \times [0]$, the latter is the local representative of $o^u_{k_u}$. Also, it is easily checked that $(-1)^{\ell k_u} \sigma\, \tau_\ell \times 1 \times \tau''_{k_s} \frown o_m = [0] \times o'_{k_u} \times [0]$. Thus again, formal use of (2.5*b*) yields (with caveat as in the two previous cases)

$$\begin{aligned}[0] \times o'_{k_u} \times [0] \,\widehat{\bullet}\, (-1)^{\ell k_s} o_\ell \times [0] \times o''_{k_s} & \\ &= (-1)^{\ell k_u} \sigma\, \tau_\ell \times 1 \times \tau''_{k_s} \frown (-1)^{\ell k_s} o_\ell \times [0] \times o''_{k_s} \\ &= (-1)^{\ell k_u} \sigma\, [0] \times [0] \times [0]\,.\end{aligned}$$

We conclude that

$$\#(o^u_{k_u} \,\widehat{\bullet}\, o^s_{k_s+\ell}) = (-1)^{\ell k_u} \sigma\,. \tag{7.3}$$

It remains to justify the formal computations made. Here, only justification for the computation leading to (7.1) will be made; it is left to the reader to carry out similar justifications for those computations leading to (7.2) and (7.3). Note that the computation *cannot* be justified by invoking invariance of intersection pairings under inclusion directly because the intersection class of an element in $H_{\ell+k_u}((\mathbf{R}^\ell, \mathbf{R}^\ell_0) \times (\mathbf{R}^{k_u}, \mathbf{R}^{k_u}_0) \times \mathbf{R}^{k_s}) \otimes H_{\ell+k_s}((\mathbf{R}^\ell, \mathbf{R}^\ell_0) \times \mathbf{R}^{k_u} \times (\mathbf{R}^{k_s}, \mathbf{R}^{k_s}_0))$ by definition lies in $H_\ell((\mathbf{R}^\ell, \mathbf{R}^\ell_0) \times (\mathbf{R}^{k_u}, \mathbf{R}^{k_u}_0) \times (\mathbf{R}^{k_s}, \mathbf{R}^{k_s}_0)) = \langle 0 \rangle$.

First, it is convenient to work within a relatively compact open subset of $\mathbf{R}^m$ rather than in $\mathbf{R}^m$ itself, and we first show how to do this via invariance of intersection pairings under inclusions. Toward that end, let D^k_ε and S^{k-1}_ε be defined as at the beginning of the chapter, let $B^k_\varepsilon := \operatorname{int}(D^k_\varepsilon)$, and for $0 \le \varepsilon_1 < \varepsilon_2$, let $A^k_{\varepsilon_1\,\varepsilon_2} := B^k_{\varepsilon_2} \setminus D^k_{\varepsilon_1}$. Fix $t > 0$, set $X := B^\ell_t \times B^{k_u}_t \times B^{k_s}_t$, and give X the orientation O^X induced by the orientation $O^{\mathbf{R}^m}$ defined earlier. Choose $q, r, s \in \mathbf{R}$ satisfying $0 < q < r < s < t$. Observe that in $\mathbf{R}^\ell \times \mathbf{R}^{k_u} \times \mathbf{R}^{k_s}$ there exist inclusions of pairs

$$\begin{array}{ccc}
(\mathbf{R}^\ell, \mathbf{R}^\ell \setminus \{0\}) \times (\mathbf{R}^{k_u}, \mathbf{R}^{k_u} \setminus \{0\}) \times \{0\} & & (\mathbf{R}^\ell, \mathbf{R}^\ell \setminus \{0\}) \times \{0\} \times (\mathbf{R}^{k_s}, \mathbf{R}^{k_s} \setminus \{0\}) \\
{\scriptstyle i_1}\big\uparrow & & {\scriptstyle i_2}\big\uparrow \\
(D^\ell_s, S^{\ell-1}_s) \times (D^{k_u}_s, S^{k_u-1}_s) \times \{0\} & \text{and} & (D^\ell_s, S^{\ell-1}_s) \times \{0\} \times (D^{k_s}_s, S^{k_s-1}_s) \\
{\scriptstyle j_1}\big\downarrow & & {\scriptstyle j_2}\big\downarrow \\
(B^\ell_t, A^\ell_{r\,t}) \times (B^{k_u}_t, A^{k_u}_{r\,t}) \times B^{k_s}_r & & (B^\ell_t, A^\ell_{r\,t}) \times B^{k_u}_r \times (B^{k_s}_t, A^{k_s}_{r\,t})
\end{array}$$

each inducing isomorphisms on homology. Also, if we take two singular homology classes, say ξ in the homology of the middle entry in the left column and η from the middle entry in the right column, and form the intersection classes $\xi \frown\!\!\!\!\bullet\; \eta$, $i_{1*}\xi \frown\!\!\!\!\bullet\; j_{1*}\eta$, and $i_{2*}\xi \frown\!\!\!\!\bullet\; j_{2*}\eta$, then the first lies in the Čech homology of the pair $(D^\ell_s, S^\ell_s) \times \{0\} \times \{0\}$, the second in that of the pair $(\mathbf{R}^\ell, \mathbf{R}^\ell \setminus \{0\}) \times \{0\})$, and the third in that of the pair $(B^\ell_t, A^\ell_{r\,t}) \times B^{k_u}_r \times B^{k_s}_r$. Further, since the inclusions

$$(\mathbf{R}^\ell, \mathbf{R}^\ell \setminus \{0\}) \times \{0\}) \xleftarrow{i_3} (D^\ell_s, S^\ell_s) \times \{0\} \times \{0\} \xrightarrow{j_3} (B^\ell_t, A^\ell_{r\,t}) \times B^{k_u}_r \times B^{k_s}_r$$

induce isomorphisms on homology (note that all three pairs have isomorphic Čech and singular homology: the first and third because they are open pairs; the second because it is a pair of Euclidean neighborhood retracts), it follows from the invariance of intersection pairings under inclusions that to compute $i_{1*}\xi \frown\!\!\!\!\bullet\; i_{2*}\eta$ it suffices to compute $j_{1*}\xi \frown\!\!\!\!\bullet\; j_{2*}\eta$. For the latter we are working with pairs of open pairs in X so that the formulae (2.5) can be used. In particular, we may regard the homology classes $(-1)^{\ell k_u} o_\ell \times o'_{k_u} \times [0]$ and $(-1)^{\ell k_s} o_\ell \times [0] \times o''_{k_s}$ as elements of $H_{\ell+k_u}((B^\ell_t, A^\ell_{r\,t}) \times (B^{k_u}_t, A^{k_u}_{r\,t}) \times B^{k_s}_r)$ and $H_{\ell+k_s}((B^\ell_t, A^\ell_{r\,t}) \times B^{k_u}_r \times (B^{k_s}_t, A^{k_s}_{r\,t}))$ respectively.

The justification will now be reduced to diagram chasing, and let us proceed to define various spaces and maps needed to write down the diagrams. Define the pair (K, L), closed relative to X, by the equality

$$(X \setminus L, X \setminus K) := (B^\ell_t, A^\ell_{r\,t}) \times (B^{k_u}_t, A^{k_u}_{r\,t}) \times B^{k_s}_r\,.$$

Set $C := D^\ell_r \times D^{k_u}_r \times D^{k_s}_r$. Then it is easily checked that $K \setminus L = D^\ell_r \times D^{k_u}_r \times B^{k_s}_r \subset C$ and that $(K \cap C, L \cap C) = D^\ell_r \times D^{k_u}_r \times (D^{k_s}_r, S^{k_s-1}_r)$. Also, define

$$Y := B^\ell_t \times B^{k_u}_t \times A^{k_s}_{q\,t}, \qquad Z := A^\ell_{q\,t} \times B^{k_u}_t \times B^{k_s}_t \cup B^\ell_t \times A^{k_u}_{q\,t} \times B^{k_s}_t$$

Next, let us define two homology classes that will be used to define cap product homomorphisms in our diagram; we need an auxiliary diagram. To facilitate printing, set $X_L := X \setminus (L \cap C)$, $X_K := X \setminus (K \cap C)$, $Y_L := Y \setminus (L \cap C)$, and $X_0 := X \setminus \{0\}$. In the commutative diagram of inclusion induced homomorphisms

$$\begin{array}{ccccccc}
H_m(X, X_0) & \xleftarrow{\simeq} & H_m(X, X_K) & \longrightarrow & H_m(X, X_K \cup Y) & \xleftarrow{\simeq} & H_m(X_L, X_K \cup Y_L) \\
\simeq\downarrow & \nwarrow & \downarrow & \swarrow & & \swarrow{\scriptstyle i_{0\,m}} & \\
H_m(\mathbf{R}^m, \mathbf{R}^m_0) & \xleftarrow[k_{0\,m}]{} & H_m(X, Y \cup Z) & & & &
\end{array},$$

in the top row, the rightmost arrow is an excision isomorphism and the leftmost arrow is an isomorphism because it maps fundamental class to fundamental class. Let $o_0^X \in H_m(X, X_0)$ be the fundamental class of X whose image in $H_m(\mathbf{R}^m, \mathbf{R}^m_0)$ under the excision isomorphism that is the leftmost vertical arrow is o_m. Define $\bar{o}_{K\cap C} \in H_m(X_L, X_K \cup Y_L)$ to be the image of o_0^X under the composition of the arrows in the top row with the left-pointing arrows replaced by their inverses. Also, define $\bar{o}_0^X := i_{0\,m}\bar{o}_{K\cap C} \in H_m(X, Y \cup Z)$. Diagram chasing then shows that also $o_m = k_{0\,m}\bar{o}_0^X$. Note too that the underlying inclusion k_0 is a weak deformation retract. Hence, $k_{0\,m}$ is an isomorphism, and $\bar{o}_0^X$ can be characterized as the unique element whose image in $H_m(X, X_0)$ under the inclusion induced oblique arrow is o_0^X. Thus, the definition of $\bar{o}_0^X$ does not depend on K or L; we use this later.

We then have the following diagram of R-module homomorphisms

$$\begin{array}{ccccc}
 & & \check{H}^{k_s}(K, L) & & \\
 & & \downarrow{\scriptstyle e_1^*} & & \\
 & & \check{H}^{k_s}(K \cap C, L \cap C) & & \\
 & & \uparrow{\scriptstyle j^*} & & \\
H^{k_s}(\mathbf{R}^{\ell+k_u} \times (\mathbf{R}^{k_s}, \mathbf{R}^{k_s}_0)) & \longrightarrow & H^{k_s}(X, Y) & \xrightarrow{e_2^*} & H^{k_s}(X \setminus (L \cap C), Y \setminus (L \cap C)) \\
\downarrow{\scriptstyle \frown o_m} & & \downarrow{\scriptstyle \frown \bar{o}_0^X} & & \downarrow{\scriptstyle \frown \bar{o}_{K\cap C}} \\
H_{\ell+k_u}((\mathbf{R}^{\ell+k_u}, \mathbf{R}^{\ell+k_u}_0) \times \mathbf{R}^{k_s}) & \longleftarrow & H_{\ell+k_u}(X, Z) & \longleftarrow & H_{\ell+k_u}(X \setminus (L \cap C), X \setminus (K \cap C)) \\
 & \nwarrow{\scriptstyle k_*} & \uparrow & \swarrow{\scriptstyle e_{3\,*}} & \\
 & & H_{\ell+k_u}(X \setminus L, X \setminus K) & &
\end{array}$$

where all arrows other than the cap product homomorphisms are inclusion induced with e_1^*, e_2^*, and $e_{3\,*}$ excision isomorphisms. The diagram is commutative: the triangles at the bottom are because they consist of inclusion induced morphisms; both rectangles are by naturality of cap products [D, VII.12.6] because $\bar{o}_0^X = i_{0\,m}\bar{o}_{K\cap C}$, because $o_m = k_{0\,m}\bar{o}_0^X$, and because all excisiveness requirements are met because the necessary sets are open.

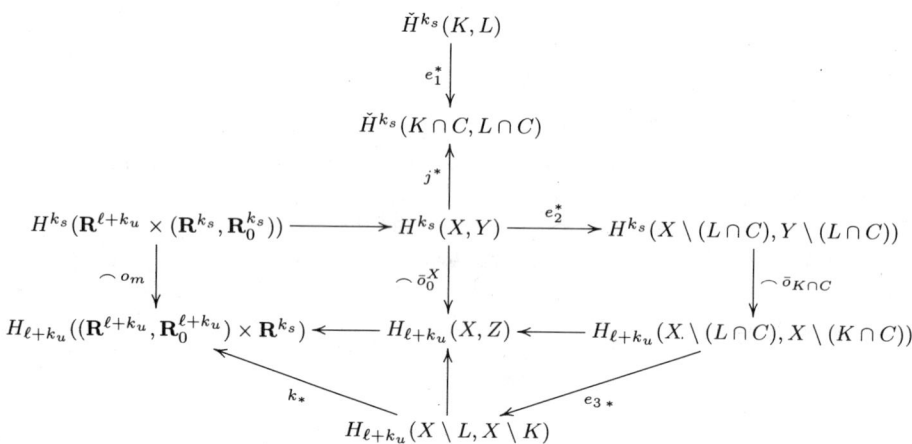

Every arrow in the diagram is an isomorphism. First, note that the inclusion of pairs $(K \cap C, L \cap C) \overset{j}{\hookrightarrow} (X, Y)$ is a deformation retract, whence as the latter pair is a neighborhood pair of the former pair, the canonical map $H^*(X, Y) \to \check{H}^*(K \cap C, L \cap C)$ can be identified with j^* and is an isomorphism. Note, therefore, that the Poincaré duality isomorphism $\frown o^X : \check{H}^{k_s}(K,L) \to H_{\ell+k_u}(X \setminus L, X \setminus K)$ coincides with the composite $e_{3*} \circ \frown \bar{o}_{K\cap C} \circ e_2^* \circ (j^*)^{-1} \circ e_1^*$ (cf. pp. 292–3 and p. 296 of [D]); this implies that $\frown o_{K\cap C}$ is an isomorphism. Also, $\frown o_m$ is an isomorphism because our earlier computation of a formal Poincaré dual of $(-1)^{\ell k_u} o_\ell \times o'_{k_u} \times [0]$ shows that $\frown o_m$ maps cyclic generator to cyclic generator. The inclusions inducing the horizontal arrows in the left rectangle are easily seen to be weak deformation retracts so induce isomorphisms. An elementary diagram chase then shows the remaining arrows to be isomorphisms.

It follows, therefore, that if $\xi \in H_{\ell+k_u}(X \setminus L, X \setminus K)$, then $(\frown o^X)^{-1}\xi$, its Poincaré dual, is uniquely determined by the element $x := (\frown o_m)^{-1} k_* \xi$ and conversely. In particular, this is true for $\xi := (-1)^{\ell k_u} o_\ell \times o'_{k_u} \times [0]$.

Next, we need (but leave it to the reader to write it down) a second diagram entirely analogous to the diagram just constructed but with (K,L) replaced by (K^*, L^*) where the latter pair is defined by

$$(X \setminus L^*, X \setminus K^*) := (B_t^\ell, A_{rt}^\ell) \times B_r^{k_u} \times (B_t^{k_s}, A_{rt}^{k_s})$$

and with Y and Z replaced respectively by Y^* and Z^* defined by

$$Y^* := B_t^\ell \times A_{qt}^{k_u} \times B_t^{k_s}, \qquad Z^* := A_{qt}^\ell \times B_t^{k_u} \times B_t^{k_s} \cup B_t^\ell \times B_t^{k_u} \times A_{qt}^{k_s}.$$

Also, the leftmost vertical arrow gets replaced by the arrow

$$H^{k_u}(\mathbf{R}^\ell \times (\mathbf{R}^{k_u}, \mathbf{R}_0^{k_u}) \times \mathbf{R}^{k_s}) \xrightarrow{\frown o_m} H_{\ell+k_s}((\mathbf{R}^\ell, \mathbf{R}_0^\ell) \times \mathbf{R}^{k_u} \times (\mathbf{R}^{k_s}, \mathbf{R}_0^{k_s})).$$

In direct analogy with the previous case, if $\eta \in H_{\ell+k_s}(X \setminus L^*, X \setminus K^*)$, then its Poincaré dual $(\frown o^X)^{-1}\eta$ is uniquely determined by the element $y := (\frown o_m)^{-1} k'_* \eta$ and conversely where

$$k'_* : H_{\ell+k_s}(X \setminus L^*, X \setminus K^*) \to H_{\ell+k_s}((\mathbf{R}^\ell, \mathbf{R}_0^\ell) \times \mathbf{R}^{k_u} \times (\mathbf{R}^{k_s}, \mathbf{R}_0^{k_s}))$$

is the inclusion induced map. In particular, this is true for $\eta := (-1)^{\ell k_s} o_\ell \times [0] \times o''_{k_s}$; in which case $y = (-1)^{k_u k_s} \sigma 1 \times \tau'_{k_u} \times 1$.

To complete the justification, take the tensor product of the two diagrams, lattice point by lattice point, call the result the source diagram, regard it as one face of a three-dimensional diagram whose opposite face, given momentarily, is a diagram, call it the target diagram, of the same type as those just tensored together where the arrows from the source to target are given by taking cup products on the tensor product of cohomology modules (on the Čech modules the cup product is $\Delta^* \circ \times$), noting that each rectangle with two parallel edges cup products and one edge each in the source and target diagrams is commutative by naturality of cup products, and by defining the arrow on the tensor product of homology modules in the unique way possible so that the three-dimensional diagram is commutative. It will turn out that on the bottom-most tensored pair the arrow to the target diagram is (up to an excision isomorphism labeled e_* below) the intersection pairing $\mathfrak{L}_V$ where

$$V := ((X \setminus L, X \setminus K), (X \setminus L^*, X \setminus K^*)).$$

To facilitate the printing of the target diagram, set $X'' := X \setminus ((K \cap K^*) \cap C)$,

$$X' := X \setminus \big(((L \cap K^*) \cup (K \cap L^*)) \cap C\big) \quad \text{and} \quad Y' := Y \setminus \big(((L \cap K^*) \cup (K \cap L^*)) \cap C\big).$$

Also, define $W := A^{\ell}_{q\,t} \times B^{k_u}_t \times B^{k_s}_t$. Then the target diagram with an extra excision isomorphism added as a bottom tail to get $\mathfrak{L}_V$ exactly is the diagram

$$
\begin{array}{ccccc}
 & & \check{H}^{k_u+k_s}(K \cap K^*, (L \cap K^*) \cup (K \cap L^*)) & & \\
 & & \Big\downarrow {\scriptstyle e''_{1*}} & & \\
 & & \check{H}^{k_u+k_s}((K \cap K^*) \cap C, ((L \cap K^*) \cup (K \cap L^*)) \cap C) & & \\
 & & \Big\uparrow {\scriptstyle j''^*} & & \\
H^{k_u+k_s}(\mathbf{R}^{\ell} \times (\mathbf{R}^{k_u}, \mathbf{R}^{k_u}_0) \times (\mathbf{R}^{k_s}, \mathbf{R}^{k_s}_0)) & \longrightarrow & H^{k_u+k_s}(X, Y \cup Y^*) & \xrightarrow{e''^*_2} & H^{k_u+k_s}(X', Y') \\
\Big\downarrow {\scriptstyle \frown o_m} & & \Big\downarrow {\scriptstyle \frown \bar{o}^X_0} & & \Big\downarrow {\scriptstyle \frown \bar{o}_{K \cap K^* \cap C}} \\
H_{\ell}((\mathbf{R}^{\ell}, \mathbf{R}^{\ell}_0) \times \mathbf{R}^{k_u} \times \mathbf{R}^{k_s}) & \longleftarrow & H_{\ell}(X, W) & \longleftarrow & H_{\ell}(X', X'') \\
 & {\scriptstyle k''_*} \nwarrow & \Big\uparrow & \nearrow {\scriptstyle e''_{3*}} & \\
 & & H_{\ell}(X \setminus ((L \cap K^*) \cup (K \cap L^*)), X \setminus (K \cap K^*)) & & \\
 & & \Big\uparrow {\scriptstyle e_*} & & \\
 & & H_{\ell}(X \setminus (L \cup L^*), X \setminus ((K \cap K^*) \cup (L \cup L^*))) & &
\end{array}
$$

and is seen to be commutative and to have invertible arrows via a proof entirely analogous to that used to construct and analyze the first diagram. Too, because the cup product on $H^{k_s}(\mathbf{R}^{\ell+k_u} \times (\mathbf{R}^{k_s}, \mathbf{R}^{k_s}_0)) \otimes H^{k_u}(\mathbf{R}^{\ell} \times (\mathbf{R}^{k_u}, \mathbf{R}^{k_u}_0) \times \mathbf{R}^{k_s})$ to $H^{k_u+k_s}(\mathbf{R}^{\ell} \times (\mathbf{R}^{k_u}, \mathbf{R}^{k_u}_0) \times (\mathbf{R}^{k_s}, \mathbf{R}^{k_s}_0))$ is an isomorphism—it maps a cyclic generator to a cyclic generator—elementary diagram chasing shows that all arrows in the full three-dimensional diagram are invertible. Let ξ, x, η, and y be as most recently defined. Our last observation yields that $\xi \otimes \eta$ because it generates the module at the bottom-most lattice point of the source diagram has image $\pm o_{\ell} \times [0] \times [0] \in H_{\ell}((B^{\ell}_t, A^{\ell}_{q\,t}) \times B^{k_u}_r \times B^{k_s}_r)$ along any path to that module (the bottom-most lattice point of the target diagram, tail added) in the three-dimensional diagram. Then on the one hand, we computed earlier that $x \frown k'_* \eta = \sigma o_{\ell} \times [0] \times [0]$, and therefore by the associativity identity relating singular cup and cap products,

$$
\begin{aligned}
\sigma o_{\ell} \times [0] \times [0] &= x \frown k'_* \eta = x \frown (y \frown o_m) = (x \smile y) \frown o_m \\
&= \big((\frown o_m)^{-1} k_* \xi \smile (\frown o_m)^{-1} k'_* \eta\big) \frown o_m \,.
\end{aligned}
$$

On the other hand, because traveling down the outer right edge of the source diagram corresponds to $(\frown o^X) \otimes (\frown o^X)$, the tensor product of Poincaré duality isomorphisms, and because traveling down the outer right edge of the target diagram coincides with the Poincaré duality isomorphism $\frown o^X$ followed by e_*^{-1}, it follows from a diagram chase in the three-dimensional diagram that

$$\begin{aligned}\sigma o_\ell \times [0] \times [0] &= k''_*\Big(\big((\frown o^X)^{-1}\xi \smile (\frown o^X)^{-1}\eta\big) \frown o^X\Big)\\ &= k''_* e_*(\xi \mathbin{\widehat{\bullet}}^V \eta)\\ &\equiv k''_* e_* \mathfrak{L}_V(\xi \otimes \eta)\end{aligned}$$

where the second equality is an immediate consequence of the rigorous formulation of formula (2.5a). As already noted, the image of $\xi \otimes \eta$ in the bottom-most module of the target diagram must generate that module; i.e., necessarily, $\mathfrak{L}_V(\xi \otimes \eta) = \pm o_\ell \times [0] \times [0]$, and this and the just given string of equalities yields the desired result because k''_* and e_* are inclusion induced, and therefore, leave signs unchanged.

Note that a rigorous formulation of formula (2.5a) is contained in Proposition 9.18 below—specifically see sequence (9.7)—with the caveat that Proposition 9.18 is formulated for a pair $V \in M^{(2)}_{\Delta o}$ whereas present circumstances dictate that V be a general pair of open pairs that *does not* satisfy the off-diagonal condition. However, the proof of Proposition 9.18 applies equally well to the general case if suitable modifications are made to the lattice points of the diagrams used in its proof, and the reader, with a little thought, should have no difficulty making the necessary changes. For example, because in the general case $(V_0 \cap V_1^*) \cup (V_1 \cap V_0^*) \neq \emptyset$, the last term in sequences (9.7–9) must be changed from $H_{i+j-m}(V_1 \cap V_1^*; G \otimes G')$ to $H_{i+j-m}(V_1 \cap V_1^*, (V_0 \cap V_1^*) \cup (V_1 \cap V_0^*); G \otimes G')$; all other necessary changes can be worked out backwards from this one. □

The next lemma gives a condition for obtaining a Lyapunov function and will be used in the construction of isolating blocks for hyperbolic critical points and hyperbolic closed orbits. It is a simple generalization of well-known estimates; e.g., see [H-S, §§9.1–2]. The following notation and hypotheses are needed for its statement.

Let T^ℓ denote the product of S^1 with itself ℓ times when ℓ is a positive integer. Given $\omega \in \mathbf{R}^\ell_{++}$ and an integer $k \geq 1$ there is a C^∞ universal covering map $/\omega: \mathbf{R}^\ell \times \mathbf{R}^k \to T^\ell \times \mathbf{R}^k$ defined for $(\theta, \mathbf{x}) \in \mathbf{R}^\ell \times \mathbf{R}^k$ by $/\omega(\theta, \mathbf{x}) = (e^{i\theta_1 2\pi/\omega_1}, \ldots, e^{i\theta_\ell 2\pi/\omega_\ell}, \mathbf{x})$. We will say that a function f on $\mathbf{R}^\ell \times \mathbf{R}^k$ to a space Z is multi-periodic in $\theta \in \mathbf{R}^\ell$ with multi-period ω if there exists a function $\tilde{f}: T^\ell \times \mathbf{R}^k \to Z$, necessarily unique, satisfying $\tilde{f} \circ /\omega = f$.

For $\mathbf{u} \in \mathbf{R}^m$, write $\mathbf{u} = (\theta, \mathbf{x}) \in \mathbf{R}^\ell \times \mathbf{R}^{m-\ell} \equiv \mathbf{R}^m$, suppose $\mathbf{E}$ is a subspace of $\mathbf{R}^{m-\ell}$ with orthogonal complement $\mathbf{E}^\perp$ relative to the standard inner product on $\mathbf{R}^{m-\ell}$, let $\pi_E: \mathbf{R}^m \to \mathbf{E}$ and $\pi_{\mathbf{E}^\perp}: \mathbf{R}^m \to \mathbf{E}^\perp$ be the orthogonal projections relative to the standard inner product under the identification $\mathbf{R}^m \equiv \mathbf{R}^\ell \oplus \mathbf{R}^{m-\ell}$, and suppose $A \in L(E, E)$ with the property that the quadratic form Q on $\mathbf{E}$ defined by $Q(y) := \langle Ay \mid y \rangle$ is positive definite. Also, let $K = \{\mathbf{x} \in \mathbf{R}^{m-\ell} : \|\pi_{\mathbf{E}^\perp}(\mathbf{x})\| \leq \|\pi_{\mathbf{E}}(\mathbf{x})\|\}$. Below, $\mathcal{L}_\chi$ denotes Lie differentiation along the trajectories of a vectorfield χ.

7.8 LEMMA. *If χ is a vectorfield on $\mathbf{R}^\ell \times \mathbf{R}^{m-\ell}$ whose local representative $\hat{\chi}$ is multi-periodic in $\theta \in \mathbf{R}^\ell$ with multi-period $\omega \in \mathbf{R}^\ell_{++}$ and satisfies*

$$\pi_E \circ \hat{\chi}(\theta, \mathbf{x}) = A\,\pi_E(\theta, \mathbf{x}) + R_E(\theta, x)$$

where $R_E(\theta,0)=0$ *and* $D_2R_E(\theta,0)=0$ *for each* $\theta\in\mathbf{R}^\ell$*, then*

$$c_1 := \inf\{\langle Ax \mid x\rangle : x\in E\setminus\{0\}\}>0$$

and there exists $c_0>0$ *so that*

$$\mathcal{L}_\chi\|\pi_E(\theta,\mathbf{x})\|^2 \geq c_1\|\pi_E(\theta,\mathbf{x})\|^2 \qquad \text{if } \theta\in\mathbf{R}^\ell \text{ and } \mathbf{x}\in K\cap D^{m-\ell}_{\sqrt{2}c_0}.$$

Proof. For $\mathbf{x}\in\mathbf{R}^{m-\ell}$, write $\mathbf{x}=x_1+x_2$ where $x_1\in\mathbf{E}$ and $x_2\in\mathbf{E}^\perp$, note $c_1=\inf\{\langle Ax_1\,|\,x_1\rangle : \|x_1\|=1\}$, and therefore $c_1>0$ by the positive definiteness of Q and the local compactness of $\mathbf{E}$.

Define $n(\theta,\rho) := \sup\{\|D_2R_\mathbf{E}(\theta,\mathbf{x})\| : \|\mathbf{x}\|\leq\rho\}$ for $\theta\in\mathbf{R}^\ell$ and $\rho>0$. Then n is both lower and upper semi-continuous; hence continuous. The proof of lower semi-continuity follows trivially from the definitions and continuity of $\|D_2R_\mathbf{E}\|$; the proof of upper semi-continuity follows easily from the continuity of $\|D_2R_\mathbf{E}\|$ and the compactness of any closed ball of finite radius in $\mathbf{R}^{m-\ell}$. Details are left to the reader.

Continuity of n and the assumption that $D_2R_\mathbf{E}(\theta,0)=0$ yield for each $\theta\in\mathbf{R}^\ell$, a $\delta(\theta)>0$ so that

$$n(\theta',\rho)<\frac{1}{2\sqrt{2}}c_1 \qquad \text{if } \|\theta-\theta'\|<\delta(\theta) \text{ and } 0<\rho<\delta(\theta).$$

Writing $\omega=(\omega_1,\dots,\omega_\ell)$, let

$$C:=[0,\omega_1]\times\cdots\times[0,\omega_\ell]\subset\mathbf{R}^\ell.$$

By the compactness of C and multi-periodicity of $\hat{\chi}$, hence of $R_\mathbf{E}$, it follows that there exists $c_0>0$ so that

$$0<\rho<\sqrt{2}c_0 \quad \text{implies} \quad n(\theta,\rho)<\frac{1}{2\sqrt{2}}c_1 \qquad \text{for each } \theta\in\mathbf{R}^\ell.$$

Therefore, as $R_E(\theta,0)=0$ for each $\theta\in\mathbf{R}^\ell$, the Cauchy-Schwarz inequality, the Mean Value theorem, and the definition of the cone K together imply that

(7.4)
$$|\langle R_E(\theta,\mathbf{x})\,|\,x_1\rangle| \leq \frac{1}{2\sqrt{2}}c_1\|x_1\|\,\|\mathbf{x}\| \leq \frac{1}{2}c_1\|x_1\|^2 \qquad \text{for } \theta\in\mathbf{R}^\ell,\ \mathbf{x}\in K\cap D^{m-\ell}_{\sqrt{2}c_0}.$$

Then for $(\theta,\mathbf{x})\in\mathbf{R}^\ell\times\mathbf{R}^{m-\ell}$,

$$\begin{aligned}\mathcal{L}_\chi\|\pi_\mathbf{E}(\theta,\mathbf{x})\|^2 &= 2\langle Ax_1+R_E(\theta,\mathbf{x})\,|\,x_1\rangle\\ &\geq 2c_1\|x_1\|^2-2|\langle R_E(\theta,\mathbf{x})\,|\,x_1\rangle|.\end{aligned}$$

Hence, (7.4) implies that

$$\mathcal{L}_\chi\|\pi_\mathbf{E}(\theta,\mathbf{x})\|^2\geq c_1\|x_1\|^2 \qquad \text{if } \theta\in\mathbf{R}^\ell \text{ and } \mathbf{x}\in K\cap D^{m-\ell}_{\sqrt{2}c_0}. \quad \square$$

7.9 PROPOSITION. *If $\hat{\chi} = (\hat{\chi}_1, \hat{\chi}_2, \hat{\chi}_3)$ is a C^1 map of $\mathbf{R}^\ell \times \mathbf{R}^P \times \mathbf{R}^N$ to itself having the form*

$$\begin{aligned} \hat{\chi}_1(\theta, x_1, x_2) &= a + \Theta(\theta, x_1, x_2) \quad \in \mathbf{R}^\ell \\ \hat{\chi}_2(\theta, x_1, x_2) &= Ex_1 + R_1(\theta, x_1, x_2) \in \mathbf{R}^P \qquad \text{for } (\theta, x_1, x_2) \in \mathbf{R}^\ell \times \mathbf{R}^P \times \mathbf{R}^N \\ \hat{\chi}_3(\theta, x_1, x_2) &= Cx_2 + R_2(\theta, x_1, x_2) \in \mathbf{R}^N \end{aligned}$$

where Θ, R_i $(i = 1, 2)$, and the partial derivative $D_j R_i$ $(i = 1, 2,\ j = 2, 3)$ vanish on $\mathbf{R}^\ell \times \{0\} \times \{0\}$, where $E \in L(\mathbf{R}^P, \mathbf{R}^P)$, $C \in L(\mathbf{R}^N, \mathbf{R}^N)$, and the quadratic forms Q_1 and Q_2 defined by $Q_1(x_1) = \langle Ex_1 \mid x_1 \rangle$ for $x_1 \in \mathbf{R}^P$ and $Q_2(x_2) = \langle -Cx_2 \mid x_2 \rangle$ for $x_2 \in \mathbf{R}^N$ are positive definite, and where $\hat{\chi}$ is multi-periodic in $\theta \in \mathbf{R}^\ell$ with multi-period $\omega \in \mathbf{R}^\ell_{++}$, then

$$\tilde{S} := T^\ell \times \{0\} \times \{0\}$$

is an isolated invariant set of the flow generated by the induced vectorfield $\tilde{\chi}$ on $T^\ell \times \mathbf{R}^P \times \mathbf{R}^N$ and for some $c_0 > 0$,

$$0 < c < c_0 \quad \textit{implies} \qquad K_c := T^\ell \times D^P_c \times D^N_c$$

is an isolating block for $\tilde{S}$ with

$$K^+_c = T^\ell \times S^{P-1}_c \times D^N_c \quad \text{and} \quad K^-_c = T^\ell \times D^P_c \times S^{N-1}_c .$$

Proof. First, observe that by hypothesis the vectorfield $\tilde{\chi}$ is tangent to $\tilde{S}$, whence $\tilde{S}$ is a compact invariant set of the flow generated by $\tilde{\chi}$. Thus to finish it must be shown that K_c is an isolating block whose maximal invariant set is $\tilde{S}$.

The inner product on $M := \mathbf{R}^\ell \times \mathbf{R}^P \times \mathbf{R}^N$ is taken to be the direct sum of the standard inner products on $\mathbf{R}^\ell$, $\mathbf{R}^P$, and $\mathbf{R}^N$. Let χ be the vectorfield on M whose local representative is $\hat{\chi}$, set $\mathbf{X} := \mathbf{R}^P \times \mathbf{R}^N$, and define the cones C_1 and C_2 in $\mathbf{X}$ by

$$C_1 := \{(x_1, x_2) \in \mathbf{X} : \|x_2\| \le \|x_1\|\} \quad \text{and} \quad C_2 := \{(x_1, x_2) \in \mathbf{X} : \|x_1\| \le \|x_2\|\}.$$

Also, define $f_1(\theta, x_1, x_2) = \|x_1\|^2$ and define $f_2(\theta, x_1, x_2) = \|x_2\|^2$. Because both are independent of θ, there are corresponding induced functions $\tilde{f}_1$ and $\tilde{f}_2$ on the quotient $T^\ell \times \mathbf{R}^P \times \mathbf{R}^N$ given by $\tilde{f}_i(z, x_1, x_2) = \|x_i\|^2$.

Lemma 7.8 is applied twice: first with $\mathbf{E} := \mathbf{R}^P \times \{0\}$ and A the operator E, second with $\mathbf{E} := \{0\} \times \mathbf{R}^N$ and A the operator $-C$. What results upon observing that $\mathcal{L}_\chi \|\pi_{\mathbf{E}}(\theta, \mathbf{x})\|^2$ is in both cases independent of θ and thereby allowing us to pass to the quotient is that where

$$c_1 := \inf_{\|x_1\|=1} \langle Ex_1 \mid x_1 \rangle \qquad \text{and} \qquad c_2 := \inf_{\|x_2\|=1} \langle -Cx_2 \mid x_2 \rangle$$

c_1 and c_2 are positive, and for some $c_0 > 0$, where $\Sigma_i := T^\ell \times C_i \cap D^{P+N}_{\sqrt{2}c_0}$,

$$(7.5_i) \qquad (-1)^{i-1} \mathcal{L}_{\tilde{\chi}} \tilde{f}_i \ge c_i \tilde{f}_i \quad \text{on } \Sigma_i \qquad (i = 1, 2).$$

Now observe that

$$K_c = \tilde{f}_1^{-1}([0,c]) \cap \tilde{f}_2^{-1}([0,c]),$$

and as a consequence of (7.5_i) $(i = 1,2)$,

$$\mathcal{L}_{\tilde{\chi}}\tilde{f}_1 > 0 \quad \text{on } T^\ell \times S_c^{P-1} \times D_c^N \qquad \text{and} \qquad \mathcal{L}_{\tilde{\chi}}\tilde{f}_2 < 0 \quad \text{on } T^\ell \times D_c^P \times S_c^{N-1}.$$

Similar to observations made in the proof of Proposition 4.4, for $0 < c < c_0$, it follows that K_c is an isolating block relative to the flow generated by $\tilde{\chi}$, that its exit set is

$$K_c^+ = \tilde{f}_1^{-1}(c) \cap K_c = T^\ell \times S_c^{P-1} \times D_c^N,$$

and that its entrance set is

$$K_c^- = \tilde{f}_2^{-1}(c) \cap K_c = T^\ell \times D_c^P \times S_c^{N-1}.$$

Also, note that $\partial K_c = K_c^+ \cup K_c^-$.

It remains to show that K_c isolates $\tilde{S}$. We begin by showing that $\Sigma_{1\,c} := \Sigma_1 \cap K_c$ is positively invariant relative to K_c and that $\Sigma_{2\,c} := \Sigma_2 \cap K_c$ is negatively invariant relative to K_c. To do so, first, observe that $K_c = \Sigma_{1\,c} \cup \Sigma_{2\,c}$ and that $\Sigma_{1\,c} \cap \Sigma_{2\,c} = K_c \cap h^{-1}(0)$ where $h(z, x_1, x_2) = \|x_1\|^2 - \|x_2\|^2$. Hence, to show the desired invariances it will suffice to show that $\mathcal{L}_{\tilde{\chi}}h > 0$ on $h^{-1}(0) \cap K_c \setminus \tilde{S}$. It is easy to compute that

$$\begin{aligned} \mathcal{L}_{\tilde{\chi}}h(z, x_1, x_2) &= 2\langle Ex_1 + R_1(z, x_1, x_2) \mid x_1\rangle - 2\langle Cx_2 + R_2(z, x_1, x_2) \mid |x_2\rangle \\ &\geq 2c_1\|x_1\|^2 + 2c_2\|x_2\|^2 - 2\sum_{i=1,2} \|R_i(z, x_1, x_2)\|\|x_i\|. \end{aligned}$$

As in the proof of the lemma, shrinking c_0 if necessary, it follows that

$$\|R_i(z, x_1, x_2)\| \leq \frac{1}{2\sqrt{2}} \min\{c_1, c_2\}\|\mathbf{x}\| \qquad \text{if } \|\mathbf{x}\| \leq \sqrt{2}c_0.$$

If $h(z, \mathbf{x}) = 0$, then necessarily $\|\mathbf{x}\| = \sqrt{2}\|x_i\|$ for $i = 1, 2$. It follows that

$$\mathcal{L}_{\tilde{\chi}}h(z, x_1, x_2) \geq \min\{c_1, c_2\}\|\mathbf{x}\|^2 \qquad \text{if } \mathbf{x} \in h^{-1}(0) \cap K_c \text{ for } 0 < c < c_0,$$

whence $\mathcal{L}_{\tilde{\chi}}h > 0$ on $K_c \cap h^{-1}(0) \setminus \tilde{S}$ as desired.

Suppose that $\mathbf{u} = (z, x_1, x_2) \in K_c \setminus \tilde{S}$ and that the orbit through $\mathbf{u}$ is complete and contained in K_c. Now, either $\mathbf{u} \in \Sigma_{1\,c}$ or $\mathbf{u} \in \Sigma_{2\,c}$. Assume the former. Then, by the relative positive invariance of $\Sigma_{1\,c}$, the positive semi-orbit $\mathbf{u} \cdot \mathbf{R}^+$ remains in $\Sigma_{1\,c}$. This, however, is impossible, for integrating the inequality of (7.5_1) along this semi-orbit over a positive time interval $[0, t]$ yields that $\tilde{f}_1(\mathbf{u} \cdot t) > e^{c_1 t}$ showing that the positive semi-orbit through $\mathbf{u}$ cannot be contained in K_c. On the other hand, if $\mathbf{u} \in \Sigma_{2\,c}$, then the relative negative invariance of $\Sigma_{2\,c}$ implies that it contains $\mathbf{u} \cdot \mathbf{R}^-$. This too is seen to be impossible, this time by integrating the inequality of (7.5_2) along the negative semi-orbit over an arbitrarily large negative time interval $[t, 0]$. Thus K_c isolates $\tilde{S}$ for $0 < c < c_0$. □

An immediate corollary of the proof of the proposition obtained by taking $\ell = 0$ and /ω the identification $\{0\} \times \mathbf{R}^P \times \mathbf{R}^N \equiv \mathbf{R}^P \times \mathbf{R}^N$ is the following:

7.10 COROLLARY. *If $\hat{\chi} = (\hat{\chi}_1, \hat{\chi}_2)$ is a C^1 map of $\mathbf{R}^P \times \mathbf{R}^N$ to itself having the form*

$$\begin{aligned} \hat{\chi}_1(x_1, x_2) &= Ex_1 + R_1(x_1, x_2) \in \mathbf{R}^P \\ \hat{\chi}_2(x_1, x_2) &= Cx_2 + R_2(x_1, x_2) \in \mathbf{R}^N \end{aligned} \qquad \text{for } (x_1, x_2) \in \mathbf{R}^P \times \mathbf{R}^N$$

where R_i and DR_i vanish at the origin $(0,0) \in \mathbf{R}^P \times \mathbf{R}^N$ and where the quadratic forms Q_1 and Q_2 defined by $Q_1(x_1) = \langle Ex_1 \mid x_1 \rangle$ for $x_1 \in \mathbf{R}^P$ and $Q_2(x_2) = \langle -Cx_2 \mid x_2 \rangle$ for $x_2 \in \mathbf{R}^N$ are positive definite, then $S = \{(0,0)\}$ is an isolated invariant set of the flow generated by the vectorfield χ on $\mathbf{R}^P \times \mathbf{R}^N$ whose local representative is $\hat{\chi}$ and for some $c_0 > 0$, if $0 < c < c_0$, then $B_c := D_c^P \times D_c^N$ is an isolating block for S with

$$B_c^+ = S_c^{P-1} \times D_c^N \quad \text{and} \quad B_c^- = D_c^P \times S_c^{N-1}.$$

C. Results Leading to the Proof of Theorem 7.5. In the theorems of this subsection and the next we will be working with both the flow generated by a given vectorfield and that generated by a linearization along an isolated invariant set S which is then also an isolated invariant set of the flow generated by the linearization. Further, the flows will have common isolating blocks in a sufficiently small neighborhood of the common invariant set. However, the local invariant manifolds within any such block will differ. Hence, to distinguish between the local invariant manifolds of one flow and those of the other, the generating vectorfield will be appended to the notation as a subscript: for example, $W_\xi^u(S;B)$ denotes the local unstable manifold of S within B relative to the flow generated by ξ.

Also, the word "isotopy" will *not* be used in this subsection or the next to mean "ambient isotopy", but will be used according to the following conventions to simplify the statement and exposition of the proofs of the theorems.

7.11 DEFINITION. (A) Two submanifolds with boundary W_0, W_1 of M will be called *isotopic in M* if there exists an isotopy of continuous embeddings from W_0 to W_1, i.e., if there exists continuous $H: W_0 \times [0,1] \to M$ so that writing $H_t(\mathbf{u}) := H(\mathbf{u}, t)$, for each $\mathbf{u} \in W_0$ and $t \in [0,1]$, H_t is a continuous embedding of W_0 into M with $H_0 = 1_{W_0}$ and H_1 a homeomorphism of W_0 onto W_1. Note invariance of domain implies H restricts to an isotopy from ∂W_0 to ∂W_1. Clearly, the notion of isotopy as defined here yields an equivalence relation on submanifolds with boundary of M.

(B) Also of interest to us is the case where for $i = 0, 1$, $(W_i, \partial W_i) \subset (P_1, P_0)$ where (P_1, P_0) is a nested index pair relative to some flow in M. If H is as in part (A) but in fact has image in P_1 with $H_t(\partial W_0) \subset P_0$ for $t \in [0,1]$, then call the pairs $(W_0, \partial W_0)$ and $(W_1, \partial W_1)$ *isotopic in the pair (P_1, P_0)*.

7.12 THEOREM (cf. [C,I.4.3]). *Under the assumption of hypotheses* HCP, *$S = \{\mathbf{u}_0\}$ is an isolated invariant set relative to the flow generated by ξ and also relative to the flow generated by $d\xi$ in some chart neighborhood U of $\mathbf{u}_0$ with chart map ψ satisfying (i) $\psi(U) = \mathbf{R}^P \times \mathbf{R}^N$, (ii) $\psi(\mathbf{u}_0) = (0,0)$, (iii) $T_{\mathbf{u}_0}\psi(\mathbf{E}_S^u) = \mathbf{R}^P \times \{0\}$ and $T_{\mathbf{u}_0}\psi(\mathbf{E}_S^s) = \{0\} \times \mathbf{R}^N$, and there exists a family B_c, $0 < c < c_0$, of C^{r+1},* *codimension zero, compact submanifolds with corners of U exhibiting the following properties:*

(1) *B_c is an isolating block with corners relative to the flows generated by ξ and $d\xi$, and with respect to these two flows B_c has common exit set B_c^+ and common entrance set B_c^- that together satisfy $\partial B_c = B_c^+ \cup B_c^-$;*
(2) *ψ maps B_c onto $D_c^P \times D_c^N$, maps B_c^+ onto $S_c^{P-1} \times D_c^N$, maps B_c^- onto $D_c^P \times S_c^{N-1}$, maps $W_{d\xi}^u(S; B_c)$ onto $D_c^P \times \{0\}$, and maps $W_{d\xi}^s(S; B_c)$ onto $\{0\} \times D_c^N$;*
(3) *B_c/B_c^+ is basepoint homotopic to S^P and B_c/B_c^- to S^N;*
(4) *for $\nu = u, s$, with $W_{\xi\, c}^\nu := W_\xi^\nu(S; B_c)$ and $W_{d\xi\, c}^\nu := W_{d\xi}^\nu(S; B_c)$, there exist isotopies of pairs from $(W_{d\xi\, c}^u, \partial W_{d\xi\, c}^u)$ to $(W_{\xi\, c}^u, \partial W_{\xi\, c}^u)$ in (B_c, B_c^+) and from $(W_{d\xi\, c}^s, \partial W_{d\xi\, c}^s)$ to $(W_{\xi\, c}^s, \partial W_{\xi\, c}^s)$ in (B_c, B_c^-).*

Proof. As is well-known, for all sufficiently small positive ε, $T_{\mathbf{u}_0}M$ admits an ε-almost proper ordered basis for $d\xi$; i.e., if $\varepsilon > 0$ is sufficiently small there exists a basis of $T_{\mathbf{u}_0}M$ so that the matrix of $d\xi$ relative to this basis has the block form

$$\begin{bmatrix} E & 0 \\ 0 & C \end{bmatrix} \tag{7.6}$$

where E is a $P \times P$ expansion matrix in ε-almost proper real Jordan form and C is an $N \times N$ contraction matrix in ε-almost proper real Jordan form. In particular, with E or C written in the form $D + N$ where D is block diagonal and N is lower triangular nilpotent, each non-zero entry of N equals ε with ε positive but sufficiently small so that where $\langle \cdot \mid \cdot \rangle_\varepsilon$ denotes the inner product on $\mathbf{R}^P \oplus \mathbf{R}^N$ making the ε-almost proper basis orthonormal, the quadratic forms Q_1 on $\mathbf{R}^P$ and Q_2 on $\mathbf{R}^N$ defined by

$$Q_1(x_1) = \langle E x_1 \mid x_1 \rangle_\varepsilon \quad \text{for } x_1 \in \mathbf{R}^P$$

and

$$Q_2(x_2) = \langle -C x_2 \mid x_2 \rangle_\varepsilon \quad \text{for } x_2 \in \mathbf{R}^N$$

are positive definite.

Let (U, ψ_0) be a chart at $\mathbf{u}_0$ with ψ_0 mapping U onto $\mathbf{R}^m$ and sending $\mathbf{u}_0$ to the origin, and let $\hat{\xi}_0$ be the local representative of ξ relative to this chart. Then $d\xi$ is represented locally by $D\hat{\xi}_0(0)$ and choosing an ε-almost proper basis $\mathcal{B}_\varepsilon$ of $T_{\mathbf{u}_0}M$ for $d\xi$, the basis is mapped by $T_{\mathbf{u}_0}\psi_0$ to an ε-almost proper basis of $\mathbf{R}^m$ for $D\hat{\xi}_0(0)$. In particular, the matrix of $D\hat{\xi}_0(0)$ relative to this basis is the same as the matrix of $d\xi$ relative to the basis $\mathcal{B}_\varepsilon$. Let L be the linear isomorphism from $\mathbf{R}^m$ to $\mathbf{R}^P \times \mathbf{R}^N$ obtained by mapping each $\mathbf{x} \in \mathbf{R}^m$ to (x_1, x_2) where x_1 consists of the first P coordinates of $\mathbf{x}$ relative to the ordered basis $T_{\mathbf{u}_0}\psi_0(\mathcal{B}_\varepsilon)$ of $\mathbf{R}^m$ and where x_2 consists of the last N coordinates of $\mathbf{x}$ relative to this basis.

Define $\psi = L \circ \psi_0$ and note that the chart (U, ψ) satisfies properties (i)–(iii) in the statement of the theorem. Where $\hat{\xi}$ denotes the local representative of ξ relative to the chart (U, ψ), the hypotheses of Corollary 7.10 are satisfied when $\hat{\chi} := \hat{\xi}$ and also with $\hat{\chi} := D\hat{\xi}(0,0)$, the local representative of $d\xi$ relative to the chart (U, ψ), by construction.

Thus, it follows from Corollary 7.10 that for some positive c_0, if $0 < c < c_0$, then $B_c := \psi^{-1}(D_c^P \times D_c^N)$ is an isolating block relative to the flow generated by ξ and also relative to the flow in U generated by $d\xi$, and for both of these flows B_c isolates S. Also, the exit and entrance sets of B_c as determined by ξ are coincident with those determined by $d\xi$ and where B_c^+ (respectively, B_c^-) denotes the common exit (respectively, entrance) set,

$$B_c^+ = \psi^{-1}(S_c^{P-1} \times D_c^N) \quad \text{and} \quad B_c^- = \psi^{-1}(D_c^P \times S_c^{N-1}).$$

Because D_c^P and D_c^N are contractible it follows when P and N are both positive that B_c/B_c^+ is basepoint homotopic to D_c^P/S_c^{P-1}, a homeomorph of S^P, and that B_c/B_c^- is basepoint homotopic to D_c^N/S_c^{N-1}, a homeomorph of S^N. A separate argument is required to compute B_c/B_c^+ if P is zero or to compute B_c/B_c^- if N is zero. For example, if $P = 0$ and therefore $N = m$, then $D_c^P = \{0\}$ and $S_c^{P-1} = \emptyset$; hence $B_c^+ = \emptyset$ and B_c/B_c^+ is homeomorphic to $D_c^N \amalg \{\emptyset\}$ which contracts to $\{0\} \amalg \{\emptyset\}$, a discrete space with two points and therefore a homeomorph of S^0. Similarly, if $N = 0$, then B_c/B_c^- is homotopic to S^0. Thus, the index spaces associated to B_c have the required homotopy types.

Note it is a simple consequence of the block diagonal form of the matrix representation of $d\xi$ in (7.6) that for $c > 0$ $W_{d\xi}^u(S; B_c) = \psi^{-1}(D_c^P \times \{0\})$ and $W_{d\xi}^s(S; B_c) = \psi^{-1}(\{0\} \times D_c^N)$. Thus, all items in statements (1)–(3) of the present theorem hold.

Finally, let us show that statement (4) holds. With a decrease in c_0 if necessary, standard proofs of the invariant manifold theorems (e.g., see [Ab-R, Lemma 27.2]) show that there is a C^r function $G\colon \mathrm{int}(D_{c_0}^P) \to \mathbf{R}^N$ so that for $0 < c < c_0$ with G_c denoting the restriction of G to D_c^P, the graph of G_c is $\psi(W_\xi^u(S; B_c))$. Note that because $W_\xi^u(S; B_c) \subset B_c$ and $\psi(B_c) = D_c^P \times D_c^N$, in fact G_c has image in D_c^N. Let us regard the graph map associated to G_c as a section of the trivial N-disk bundle $D_c^P \times D_c^N$ over D_c^P so that $\psi(W_{d\xi}^u(S; B_c))$ is the image of the zero section of this bundle. Note that fibers in this bundle over a point in S_c^{P-1}, the exit set of D_c^P relative to the linear flow on $\mathbf{R}^P$ generated by the matrix E, lie entirely in the exit set of B_c. Thus, $\psi(W_\xi^u(S; B_c))$ can be isotoped along fibers of this bundle to the zero section to show that the pair $(W_{\xi\,c}^u, \partial W_{\xi\,c}^u)$ is isotopic to the pair $(W_{d\xi\,c}^u, \partial W_{d\xi\,c}^u)$ in (B_c, B_c^+). In an analogous manner, one shows that the pair $(W_{\xi\,c}^s, \partial W_{\xi\,c}^s)$ is isotopic in (B_c, B_c^-) to the pair $(W_{d\xi\,c}^s, \partial W_{d\xi\,c}^s)$. □

7.13 Lemma. *Under the assumption of hypotheses* HCP *and conclusions of Theorem 7.12, if $\tau^u \in H^P(\mathbf{E}^u, \mathbf{E}^u \setminus \{0\}; R)$ is an R-orientation of the vector space $\mathbf{E}^u$, then for $0 < c < c_0$, there exist unique fundamental classes*

$$o_{\xi\,c}^u \in H_P(W_{\xi\,c}^u, \partial W_{\xi\,c}^u; R) \qquad \text{and} \qquad o_{d\xi\,c}^u \in H_P(W_{d\xi\,c}^u, \partial W_{d\xi\,c}^u; R)$$

that together satisfy

(1) *$o_{\xi\,c}^u$ and $o_{d\xi\,c}^u$ have common image in $H_P(B_c, B_c^+)$;*
(2) *$\langle i_c^{u*}\tau^u, o_{d\xi\,c}^u\rangle = 1$ where $i_c^u\colon (W_{d\xi\,c}^u, \partial W_{d\xi\,c}^u) \to (\mathbf{E}^u, \mathbf{E}^u \setminus \{0\})$ is the embed-*

ding given by the composite

$$(W^u_{d\xi\, c}, \partial W^u_{d\xi\, c}) \xrightarrow{\psi} (D^P_c, S^{P-1}_c) \times \{0\} \subset (\mathbf{R}^P, \mathbf{R}^P \setminus \{0\}) \times \{0\} \xrightarrow{(T_{\mathbf{u}_0}\psi)^{-1}} (\mathbf{E}^u, \mathbf{E}^u \setminus \{0\}).$$

An analogous statement holds for τ^s an R-orientation of the vector space $\mathbf{E}^s$ and the existence of fundamental classes $o^s_{d\xi\, c}$ and $o^s_{\xi\, c}$ for $W^s_{d\xi\, c}$ and $W^s_{\xi\, c}$ respectively satisfying the analogues of conditions (1) *and* (2) *obtained by replacing B^+_c with B^-_c and by replacing $\mathbf{E}^u$ with $\mathbf{E}^s$.*

Proof. Because the inclusions $D^P_c \subset \mathbf{R}^P$ and $S^{P-1}_c \subset \mathbf{R}^P \setminus \{0\}$ are strong deformation retractions, it follows from the Five Lemma and the cohomology sequences of the pairs $(W^u_{d\xi\, c}, \partial W^u_{d\xi\, c})$ and $(\mathbf{E}^u, \mathbf{E}^u \setminus \{0\})$ that i^{u*}_c induces isomorphisms on cohomology. Thus $i^{u*}_c \tau^u$ generates $H^P(W^u_{d\xi\, c}, \partial W^u_{d\xi\, c}; R)$. Hence, for $0 < c < c_0$, choose $o^u_{d\xi\, c}$ to be the unique fundamental class in $H_P(W^u_{d\xi\, c}, \partial W^u_{d\xi\, c}; R)$ dual to $i^{u*}_c \tau^u$ relative to the Kronecker pairing. Thus, $o^u_{d\xi\, c}$ satisfies condition (2). Then condition (i) will be satisfied if, and only if, we choose $o^u_{\xi\, c}$ to be the image of $o^u_{d\xi\, c}$ under the isomorphism induced by the end map of the isotopy from $(W^u_{d\xi\, c}, \partial W^u_{d\xi\, c})$ to $(W^u_{\xi\, c}, \partial W^u_{\xi\, c})$ since the two pairs are isotopic in (B_c, B^+_c).

The analogue for the local stable manifolds is proved in the same manner. □

7.14 DEFINITION. Given τ^u an R-orientation of $\mathbf{E}^u$, for $0 < c < c_0$, let $o^u_{d\xi\, c}$ and $o^u_{\xi\, c}$ be the fundamental classes of $W^u_{d\xi\, c}$ and $W^u_{\xi\, c}$ respectively which together satisfy properties (1) and (2) of Lemma 7.13. Call $\{o^u_{d\xi\, c}\}_{0<c<c_0}$ the *compatible family of fundamental classes determined by the R-orientation τ^u of $\mathbf{E}^u$ for the family of submanifolds* $\mathcal{W}^u_{d\xi} := \{W^u_{d\xi\, c}\}_{0<c<c_0}$, and call $\{o^u_{\xi\, c}\}_{0<c<c_0}$ the *compatible family of fundamental classes determined by the R-orientation τ^u of $\mathbf{E}^u$ for the family of submanifolds* $\mathcal{W}^u_{\xi} := \{W^u_{\xi\, c}\}_{0<c<c_0}$.

Similar designations are made for τ^s an R-orientation of $\mathbf{E}^s$ and the families of fundamental classes $\{o^s_{d\xi\, c}\}_{0<c<c_0}$ and $\{o^s_{\xi\, c}\}_{0<c<c_0}$ determined by Lemma 7.13 for the families of submanifolds $\mathcal{W}^s_{d\xi} := \{W^s_{d\xi\, c}\}_{0<c<c_0}$ and $\mathcal{W}^s_{\xi} := \{W^s_{\xi\, c}\}_{0<c<c_0}$ respectively.

7.15 THEOREM. *Under the assumption of hypotheses* HCP *and conclusions of Theorem 7.12, for $S = \{\mathbf{u}_0\}$,*

(1) $\widetilde{H}_P\mathcal{C}(S; R) \simeq R$ *and* $\widetilde{H}_k\mathcal{C}(S; R) \simeq 0$ *for* $k \neq P$;
(2) *each R-orientation $\tau^u \in H^P(\mathbf{E}^u, \mathbf{E}^u \setminus \{0\}; R)$ determines a unique generator $\mathbf{o}^u_\xi \in \widetilde{H}_P\mathcal{C}(S; R)$ characterized by the property that it is the common canonical image of the members of the compatible family of fundamental classes determined by the R-orientation τ^u of $\mathbf{E}^u$ for the family of submanifolds $\mathcal{W}^u_\xi$;*
(3) $\widetilde{H}_N\mathcal{C}^*(S; R) \simeq R$ *and* $\widetilde{H}_k\mathcal{C}^*(S; R) \simeq 0$ *for* $k \neq N$;

(4) *each R-orientation $\tau^s \in H^N(\mathbf{E}^s, \mathbf{E}^s \setminus \{0\}; R)$ determines a unique generator $\mathbf{o}^s_\xi \in \widetilde{H}_N\mathcal{C}^*(S; R)$ characterized by the property that it is the common canonical image of the members of the compatible family of fundamental classes determined by the R-orientation τ^s of $\mathbf{E}^s$ for the family of submanifolds $\mathcal{W}^s_\xi$;*

(5) $^{\#}(\mathbf{o}^u_\xi \,\widehat{\bullet}\, \mathbf{o}^s_\xi) = \sigma(\tau^u, \tau^s, [\mathbf{u}_0], O^M)$.

Proof. Statements (1) and (3) are an immediate consequence of statement (3) of Theorem 7.12.

To show that two fundamental classes from a compatible family of fundamental classes determined by an R-orientation τ^u of $\mathbf{E}^u$ for the submanifolds $\mathcal{W}^u_\xi$ have common canonical image in $\widetilde{H}_*\mathcal{C}(S; R)$, it suffices to show that under inclusion induced maps the two fundamental classes have common image in the homology of some index space for S because $\widetilde{H}_*\mathcal{C}(S; R)$ is defined as the direct limit of the homology modules of the index spaces in $\mathcal{C}(S)$. Such a procedure will be carried out by (a) showing it true for the compatible family of fundamental classes determined by an R-orientation τ^u of $\mathbf{E}^u$ for the family of submanifolds $\mathcal{W}^u_{d\xi}$, (b) noting that the index space of the common image is also an index space for the flow generated by ξ, and (c) concluding by using property (1) of Lemma 7.13.

Suppose that τ^u is an R-orientation of $\mathbf{E}^u$ and let $\{o^u_{d\xi\, c}\}_{0<c<c_0}$ and $\{o^u_{\xi\, c}\}_{0<c<c_0}$ be the compatible families of fundamental classes determined by τ^u according to the prescription of Lemma 7.13 for the families of submanifolds $\mathcal{W}^u_{d\xi}$ and $\mathcal{W}^u_\xi$ respectively. For $0 < c < c_0$, let k^u_c be the inclusion $(W^u_{\xi\, c}, \partial W^u_{\xi\, c}) \subset (B_c, B^+_c)$ and let $\dot{k}^u_c$ be the inclusion $(W^u_{d\xi\, c}, \partial W^u_{d\xi\, c}) \subset (B_c, B^+_c)$. Now it is immediate from conclusion (2) of Theorem 7.12 that the pair $(W^u_{d\xi\, c}, \partial W^u_{d\xi\, c})$ is a strong deformation retract of the pair (B_c, B^+_c), whence $\dot{k}^u_c$ is a homotopy equivalence for $0 < c < c_0$. Also, because $(W^u_{d\xi\, c}, \partial W^u_{d\xi\, c})$ is isotopic in (B_c, B^+_c) to $(W^u_{\xi\, c}, \partial W^u_{\xi\, c})$, it follows that the latter pair is a weak deformation retract of (B_c, B^+_c); i.e., also k^u_c is a homotopy equivalence for $0 < c < c_0$. Hence, for $0 < c < c_0$, the canonical image of $o^u_{\xi\, c}$ in $\widetilde{H}_P\mathcal{C}(S; R)$, namely, $\bar{b}^+_* k^u_{c\,*} o^u_{\xi\, c}$, generates $\widetilde{H}_P\mathcal{C}(S; R)$. To finish proving statement (2) we must show that the canonical image is the same for each $c \in\,]0, c_0[$. To do so, where $0 < c' < c'' < c_0$, we will use the diagram

$$
\begin{array}{ccccc}
(B_{c'}, B^+_{c'}) & & \xrightarrow{\quad [b^+_{c'}] \quad} & & B_{c'}/B^+_{c'} \\
\Big\downarrow [h] & \searrow & & & \Big\downarrow [\jmath_{c'}] \\
 & & (B_{c''}, CB^+_{c''}) & \longrightarrow & B_{c''}/CB^+_{c''} \\
 & \nearrow & & & \Big\uparrow [\jmath_{c''}] \\
(B_{c''}, B^+_{c''}) & & \xrightarrow{\quad [b^+_{c''}] \quad} & & B_{c''}/B^+_{c''}
\end{array}
\tag{7.7}
$$

of homotopy classes of maps whose construction now begins.

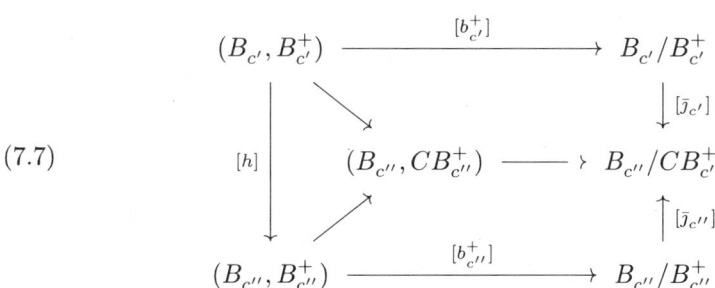

Set $CD^{P,+}_{c''} := \mathrm{cl}\left(D^P_{c''} \setminus D^P_{c'}\right)$ and set $CB^+_{c''} := \psi^{-1}(CD^{P,+}_{c''} \times D^N_{c''})$. Clearly, S has nested index pair $(B_{c''}, CB^+_{c''})$ relative to the flow in U generated by $d\xi$ and also relative to the flow generated by ξ. By construction, for $\lambda = c', c''$, there is an inclusion $j_\lambda\colon (B_\lambda, B_\lambda^+) \subset (B_{c''}, CB_{c''}^+)$ and therefore an inclusion induced map of index pairs $\bar{\jmath}_\lambda\colon B_\lambda/B_\lambda^+ \to B_{c''}/CB_{c''}^+$. The homotopy classes of these inclusions and inclusion induced maps define the slanted arrows and the right vertical arrows of diagram (7.7), and each horizontal arrow there is taken to be the homotopy class of the quotient map from the index pair to its index space.

To define the left vertical arrow in diagram (7.7), we must first for $0 < c, \bar{c} < c_0$ define an isotopy $H^{c\,\bar{c}}$ from $W^u_{d\xi\,c}$ to $W^u_{d\xi\,\bar{c}}$ whose image throughout the isotopy lies in the chart neighborhood U guaranteed by Theorem 7.12 and is mapped by the chart map ψ into $\mathbf{R}^P \times \{0\}$. The definition will use the flow in U generated by $d\xi$ which will be denoted as the map $(\mathbf{u}, t) \mapsto \mathbf{u} * t$. All exit time maps in the definition are taken relative to the flow of $d\xi$. For $0 < c \le \bar{c} < c_0$, define $H^{c\,\bar{c}}$ from $W^u_{d\xi\,c}$ to $W^u_{d\xi\,\bar{c}}$ as follows: for $0 \le t \le 1$,

$$H_t^{c\,\bar{c}}(\mathbf{u}) := \begin{cases} \mathbf{u} & \text{if } \mathbf{u} = \mathbf{u}_0 \\ \mathbf{u} * t\,\tau_{B_{\bar{c}}}(\mathbf{u} * \tau_{B_c}(\mathbf{u})) & \text{if } \mathbf{u} \in W^u_{d\xi\,c} \text{ and } \mathbf{u} \neq \mathbf{u}_0. \end{cases}$$

Note that if $c = \bar{c}$, then $H_t^{c\,c} = 1_{W^u_{d\xi\,c}}$. For $0 < c < \bar{c} < c_0$, define $H^{\bar{c}\,c}$ by

$$H_t^{\bar{c}\,c} := H_{1-t}^{c\,\bar{c}} \circ (H_1^{c\,\bar{c}})^{-1} \qquad \text{for } 0 \le t \le 1.$$

Geometrically, for $0 < c < \bar{c} < c_0$, for $0 \le t \le 1$, $H_t^{c\,\bar{c}}$ leaves the critical point fixed and translates each non-critical orbit segment of $W^u_{d\xi\,c}$ forward by a time (scaled by t) determined by the unique point on the orbit segment in the exit set of B_c and is the time it takes that point to reach the exit set of $B_{\bar{c}}$. Thus, each $H_t^{c\,\bar{c}}$ is one-to-one and maps $W^u_{d\xi\,c}$ into $W^u_{d\xi\,\bar{c}}$, and $H_1^{c\,\bar{c}}$ maps $W^u_{d\xi\,c}$ onto $W^u_{d\xi\,\bar{c}}$. Continuity of $H^{c\,\bar{c}}$ is clear except possibly at points $(\mathbf{u}_0, t)$. To see continuity at such points, let $T_{c\,\bar{c}} = \sup\{\tau_{B_{\bar{c}}}(\mathbf{u}) : \mathbf{u} \in \partial W^u_{d\xi\,c}\}$ and note that $T_{c\,\bar{c}}$ is finite as a consequence of the inequality (7.5_1) as it applies to the proof of Corollary 7.10 and the application of that corollary to the flow in U generated by $d\xi$. Now, if $\mathbf{u}_0 \in W$ open, by continuity of the flow, there exists an open neighborhood V of $\mathbf{u}_0$, $V \subset B_c$, so that $V * [0, T_{c\,\bar{c}}] \subset W$. It follows that if $(\mathbf{u}, t) \in V \times [0,1]$ then $H^{c\,\bar{c}}(\mathbf{u}, t) \in W$. Thus $H^{c\,\bar{c}}$ is an isotopy in $W^u_{d\xi\,\bar{c}}$ from $W^u_{d\xi\,c}$ to $W^u_{d\xi\,\bar{c}}$ for $0 < c \le \bar{c} < c_0$; hence also for $0 < \bar{c} < c < c_0$.

To define the left vertical arrow in diagram (7.7), recall that for $0 < c < c_0$, $\psi(W^u_{d\xi\,c}) = D_c^P \times \{0\}$, let $\pi_1\colon \mathbf{R}^P \times \mathbf{R}^N \to \mathbf{R}^P$ be the projection to the first factor, let $i_{c'}(x) = (x, 0)$ for $x \in D_{c'}^P$, let $\ell_{c'\,c''}$ be the inclusion $D_{c'}^N \subset D_{c''}^N$, define

$$K_t := \psi^{-1} \circ ((\pi_1 \circ \psi \circ H_t^{c'\,c''} \circ \psi^{-1} \circ i_{c'}) \times \ell_{c'\,c''}) \circ \psi \qquad \text{for } 0 \le t \le 1,$$

and set $h := K_1$. Observe that h maps $(B_{c'}, B_{c'}^+)$ to $(B_{c''}, B_{c''}^+)$ and define the left vertical arrow of diagram (7.7) to be the homotopy class of the map h restricted to $(B_{c'}, B_{c'}^+)$.

Diagram (7.7) is commutative. For clearly the upper and lower trapezoids commute being diagrams of the classes of inclusions and quotients, and the left hand triangle commutes because $j_{c''} \circ h$ is homotopic to the inclusion $j_{c'}$ via the homotopy K_t $(0 \le t \le 1)$.

Claim $H_{1*}^{c\bar{c}} o_{d\xi\, c}^u = o_{d\xi\, \bar{c}}^u$ for $0 < c, \bar{c} < c_0$. For the embedding $i_c^u\colon (W_{d\xi\, c}^u, \partial W_{d\xi\, c}^u) \to (\mathbf{E}^u, \mathbf{E}^u \setminus \{0\})$ of Lemma 7.13 is homotopic to the composite $i_{\bar{c}}^u \circ H_1^{c\bar{c}}$ via $H^{c\bar{c}}$. Thus,

$$1 = \langle i_c^{u*} \tau^u, o_{d\xi\, c}^u \rangle = \langle H_1^{c\bar{c}*} i_{\bar{c}}^{u*} \tau^u, o_{d\xi\, c}^u \rangle = \langle i_{\bar{c}}^{u*} \tau^u, H_{1*}^{c\bar{c}} o_{d\xi\, c}^u \rangle ,$$

whence the claim is immediate by definition of $o_{d\xi\, \bar{c}}^u$ as the class dual to $i_{\bar{c}}^{u*} \tau^u$.

Suppose that $0 < c, \bar{c} < c_0$. Let us now show that $o_{\xi\, c}^u$ and $o_{\xi\, \bar{c}}^u$ have common canonical image in $\widetilde{H}_P \mathcal{C}(S;R)$. If $c < \bar{c}$, set $c' := c$ and $c'' := \bar{c}$; otherwise, if $\bar{c} < c$, set $c' := \bar{c}$ and $c'' := c$. Note that $\dot{k}_{c''}^u \circ H_1^{c'\,c''} = h \circ \dot{k}_{c'}^u$. Then, by the claim of the previous paragraph $H_{1*}^{c\bar{c}} o_{d\xi\, c}^u = o_{d\xi\, \bar{c}}^u$; also $H_1^{\bar{c}c} = (H_1^{c\bar{c}})^{-1}$. Hence, if $c = c' < c'' = \bar{c}$, it follows that

$$h_* \dot{k}_{c*}^u o_{d\xi\, c}^u = \dot{k}_{\bar{c}*}^u o_{d\xi\, \bar{c}}^u , \tag{7.8a}$$

but if $\bar{c} = c' < c'' = c$, it follows that

$$h_* \dot{k}_{\bar{c}*}^u o_{d\xi\, \bar{c}}^u = \dot{k}_{c*}^u o_{d\xi\, c}^u . \tag{7.8b}$$

The equations (7.8), the commutativity of diagram (7.7), and the fact—see [K5, Proposition 3.1]—that the homotopy class of an inclusion induced map between index spaces is that of the unique morphism in the Conley index between those index spaces immediately imply that $b_{c*}^+ \dot{k}_{c*}^u o_{d\xi\, c}^u$ and $b_{\bar{c}*}^+ \dot{k}_{\bar{c}*}^u o_{d\xi\, \bar{c}}^u$ have a common successor in $\widetilde{H}_P(B_{c''}/CB_{c''}^+)$ where $c'' = \bar{c}$ if $c < \bar{c}$, but $c'' = c$ if $\bar{c} < c$. Because condition (i) of Lemma 7.13 implies that $\dot{k}_{c*}^u o_{d\xi\, c}^u = k_{c*}^u o_{\xi\, c}^u$ and that $\dot{k}_{\bar{c}*}^u o_{d\xi\, \bar{c}}^u = k_{\bar{c}*}^u o_{\xi\, \bar{c}}^u$, it follows that $b_{c*}^+ k_{c*}^u o_{\xi\, c}^u$ and $b_{\bar{c}*}^+ k_{\bar{c}*}^u o_{\xi\, \bar{c}}^u$ have a common successor in $\widetilde{H}_P(B_{c''}/CB_{c''}^+)$ where $c'' = \bar{c}$ if $c < \bar{c}$, but $c'' = c$ if $\bar{c} < c$. It follows that $o_{\xi\, \bar{c}}^u$ and $o_{\xi\, c}^u$ have common canonical image in $\widetilde{H}_P \mathcal{C}(S;R)$ because as constructed $(B_{c''}, CB_{c''}^+)$ is an index pair for S relative to the flow generated by ξ. This completes the proof of statement (2); statement (4) is proved similarly.

Finally, statement (5) follows immediately from Lemma 7.7 using the flow of $d\xi$. □

7.16 The Proof of Theorem 7.5. Under the assumption of hypotheses HCP, statements (1)–(6) of Theorem 7.4 follow from the results proved above: Statement (1) from Theorem 7.12; statement (2) from conclusion (3) of Theorem 7.12, the connected simple system properties of $\mathcal{C}(S)$ and $\mathcal{C}^*(S)$, and our earlier observation on the Thom space of a vector bundle over a point; and statements (3)–(6) from Theorem 7.15. Note conclusion (3) of Theorem 7.5 follows from conclusion (2) (i.e., statement (5) of Theorem 7.4) and Proposition 2.13. □

D. Results Leading to the Proof of Theorem 7.6. The proof of Theorem 7.6 will be the object of the rest of this section and will be carried out in a lengthy series of steps. Throughout, let us assume that hypotheses HCO hold for the closed orbit S of ξ and let us fix a tubular neighborhood U of S with tubular map p retracting U onto S and with (U, p, S) a vector bundle over S isomorphic to the normal bundle of S. It will also be convenient to use the following notation: if $f\colon V' \to V$ is a local diffeomorphism onto V and ζ is a vectorfield on V, then the formula $\zeta'(z) = (T_z f)^{-1} \circ \zeta(f(z))$ defines a vectorfield on V' and we write $\zeta' = f^*\zeta$; i.e., $f^*\zeta$ is the only vectorfield χ on V' that satisfies the relation $Tf \circ \chi = \zeta \circ f$. If $g: V'' \to V'$ is a local diffeomorphism onto V', then $(f \circ g)^*\zeta = g^* f^* \zeta$ by the chain rule. It is also convenient to use $e^{i\psi}$ as an abbreviation for $(\cos(\psi), \sin(\psi))$ for $\psi \in \mathbf{R}$ and is tautological if one identifies the complex numbers with ordered pairs of reals with $i = (0, 1)$ and the usual multiplication and addition.

Underlying the entire proof of Theorem 7.6 is the following lemma that summarizes the treatment of closed orbits and Floquet theory found in [Ab-R], especially the proof of Theorem 28.1 in which the local stable and unstable manifolds of an elementary closed orbit are constructed. Because there are proofs of existence of a tubular neighborhood of a closed submanifold that are valid when the submanifold and ambient manifold are at best C^2—for example see [H]—it need not be assumed that M is at least C^3 as is done in[Ab-R].

7.17 LEMMA. *There exists a C^r universal covering $\sigma: \mathbf{R} \times \mathbf{R}^P \times \mathbf{R}^N \to U$ with the properties:*

(1) *$\sigma(\mathbf{R} \times \{0\} \times \{0\}) = S$;*

(2) *$\sigma(\theta + 2\tau, x_1, x_2) = \sigma(\theta, x_1, x_2)$ for $(\theta, x_1, x_2) \in \mathbf{R} \times \mathbf{R}^P \times \mathbf{R}^N$; hence, where $/2\tau: \mathbf{R} \times \mathbf{R}^P \times \mathbf{R}^N \to \tilde{U} := S^1 \times \mathbf{R}^P \times \mathbf{R}^N$ is defined by*

$$/2\tau(\theta, x_1, x_2) := (e^{i\theta\pi/\tau}, x_1, x_2)$$

and $\sigma_2: \tilde{U} \to U$ is defined by

$$\sigma_2 := \sigma \circ (/2\tau)^{-1},$$

$\sigma_2: \tilde{U} \to U$ is a well-defined double cover of U;

(3) *where $p_1: \tilde{U} \to S^1$ is projection on the first factor and where $c: S^1 \to S$ is defined by $c(e^{i\theta\pi/\tau}) = \mathbf{u}_0 \cdot \theta$ for some $\mathbf{u}_0 \in S$,*

$$\begin{array}{ccc} \tilde{U} & \xrightarrow{\sigma_2} & U \\ {\scriptstyle p_1}\downarrow & & \downarrow{\scriptstyle p} \\ S^1 & \xrightarrow{c} & S \end{array}$$

is a vector bundle morphism;

(4) *the involutive diffeomorphism $j: \tilde{U} \to \tilde{U}$ determined by the double cover $\sigma_2: \tilde{U} \to U$ (i.e., $j \circ j = 1_{\tilde{U}}$ and $\sigma_2 \circ j = \sigma_2$) has the form*

$$j(e^{i\theta\pi/\tau}, x_1, x_2) = (e^{i(\theta+\tau)\pi/\tau}, L_+(\theta)x_1, L_-(\theta)x_2)$$

where for each $\theta \in \mathbf{R}$, $L_+(\theta)$ and $L_-(\theta)$ are linear automorphisms, the former of $\mathbf{R}^P$, the latter of $\mathbf{R}^N$, with $L_+(\theta)$ (respectively, $L_-(\theta)$) orientation preserving or reversing as N_S^u (respectively, N_S^s) is or is not orientable over $\mathbf{Z}$ and with $L_\pm(\theta + 2\tau) = L_\pm(\theta)$;

(5) *there exist matrices* $E \in \mathrm{Gl}(P;\mathbf{R})$ *and* $C \in \mathrm{Gl}(N;\mathbf{R})$ *in* ε*-almost proper real canonical Jordan form with each eigenvalue of* E *having positive real part and each eigenvalue of* C *having negative real part and with the quadratic forms* $x_1 \mapsto \langle Ex_1 \mid x_1\rangle$ *and* $x_2 \mapsto \langle -Cx_2 \mid x_2\rangle$ *defined for* $x_1 \in \mathbf{R}^P$ *and for* $x_2 \in \mathbf{R}^N$ *being positive definite so that the vectorfield* ξ_σ *on* $\mathbf{R}\times\mathbf{R}^P\times\mathbf{R}^N$ *defined by* $\xi_\sigma := \sigma^*\xi$ *has local representative* $\hat{\xi}_\sigma$ *of the form*

$$\begin{aligned} \hat{\xi}_{\sigma\,1}(\theta,x_1,x_2) &= \quad 1 + \Theta(\theta,x_1,x_2) \in \mathbf{R} \\ \hat{\xi}_{\sigma\,2}(\theta,x_1,x_2) &= Ex_1 + R_1(\theta,x_1,x_2) \in \mathbf{R}^P \qquad \text{for } (\theta,x_1,x_2) \in \mathbf{R}\times\mathbf{R}^P\times\mathbf{R}^N \\ \hat{\xi}_{\sigma\,3}(\theta,x_1,x_2) &= Cx_2 + R_2(\theta,x_1,x_2) \in \mathbf{R}^N \end{aligned}$$

where Θ, R_1 *and* R_2 *are* 2τ*-periodic in* θ *and vanish at all points* $(\theta,0,0) \in \mathbf{R}\times\mathbf{R}^P\times\mathbf{R}^N$ *as do the partial derivatives* D_kR_j, *for* $k = 2,3$ *and* $j = 1,2$.

Using the notation of Lemma 7.17, the following notions occur in the next lemma. A vectorfield χ on $\tilde{U}$ is j-invariant if $j^*\chi = \chi$; a subset $\tilde{C}$ of $\tilde{U}$ is j-invariant if $j(\tilde{C}) = \tilde{C}$; a flow $\tilde{\varphi}$ in $\tilde{C}$ is j-invariant if $j \circ \tilde{\varphi}_t \circ j = \tilde{\varphi}_t$ for each $t \in \mathbf{R}$. Also, if $f\colon V' \to V$ is a covering map, ψ' a flow in V', and ψ a flow in V, then ψ' covers ψ if $f \circ \psi'_t = \psi_t \circ f$ for each $t \in \mathbf{R}$.

7.18 Lemma. *Using the conclusions and notation of Lemma 7.17 if* $\hat{\chi}_\sigma$ *is a self-map of* $\mathbf{R}\times\mathbf{R}^P\times\mathbf{R}^N$ *satisfying*

$$\hat{\chi}_\sigma(\theta+2\tau,x_1,x_2) = \hat{\chi}_\sigma(\theta,x_1,x_2) \qquad \text{for } (\theta,x_1,x_2) \in \mathbf{R}\times\mathbf{R}^P\times\mathbf{R}^N,$$

and if χ_σ *is the vectorfield on* $\mathbf{R}\times\mathbf{R}^P\times\mathbf{R}^N$ *whose local representative is* $\hat{\chi}_\sigma$ *then the following hold:*

(1) *there exists a unique vectorfield* χ_2 *on* $\tilde{U}$ *satisfying* $\chi_\sigma = (/2\tau)^*\chi_2$ *and the flow of* χ_σ *covers the flow of* χ_2;
(2) $\chi_2 = \sigma_2^*\chi$ *for some (necessarily unique) vectorfield* χ *on* U *if, and only if,* $j^*\chi_2 = \chi_2$;
(3) *if* χ_2 *is* j*-invariant, then the flow it generates is* j*-invariant and covers the flow generated by* χ;
(4) *if* χ_2 *is transverse to the codimension one,* C^1 *submanifold* $\tilde{\Sigma}$ *of* $\tilde{U}$, *then* $j^*\chi_2$ *is transverse to the codimension one,* C^1 *submanifold* $j(\tilde{\Sigma})$;
(5) *if* χ_2 *is* j*-invariant and transverse to the codimension one* j*-invariant* C^1 *submanifold* Σ_2, *then* $\Sigma := \sigma_2(\Sigma_2)$ *is a codimension one submanifold of* U *and* χ *is transverse to* Σ;
(6) *if* χ_2 *is* j*-invariant, if* $\tilde{C}$ *is a* j*-invariant isolated invariant set of the flow generated by* χ_2, *and if* $\langle P_1, P_0\rangle$ *is an index pair for* $\tilde{C}$ *relative to that flow, then the definition* $(Q_1,Q_0) := (P_1 \cap j(P_1), Q_1 \cap (P_0 \cap j(P_0)))$ *makes* (Q_1,Q_0) *a nested,* j*-invariant index pair relative to that flow and* $(\sigma_2(Q_1),\sigma_2(Q_0))$ *a nested index pair for* $C := \sigma_2(\tilde{C})$ *relative to the flow generated by* χ; *in particular, if* $\tilde{C}$ *is isolated by the block* K, *then* $N := K \cap j(K)$ *is a* j*-invariant isolating block for* $\tilde{C}$ *with* $N^\pm = (K^\pm \cap j(K)) \cup (j(K^\pm) \cap K)$ *and* $B := \sigma_2(N)$ *is a block isolating* C *relative to the flow generated by* χ *with* $B^\pm = \sigma_2(N^\pm)$ *and with* B *a* C^r *block with corners if* N *is;*

(7) *conversely, if* χ_2 *is* j*-invariant and* B *is an isolating block relative to the flow of* χ *isolating* C, *then* $\sigma_2^{-1}(B)$ *is a* j*-invariant isolating block isolating* $\tilde{C} := \sigma_2^{-1}(C)$ *which is* j*-invariant.*

Proof. In the main, the statements of the lemma follow straightforwardly from the definitions. Note that uniqueness of the integral curve of χ_2 through a given initial point implies that the flow of χ_σ covers that of χ_2, and a similar argument implies that the flow of χ_2 covers that of χ when χ_2 is j-invariant. Also, when χ_2 is j-invariant and K is a block for $\tilde{C}$ as in statement (6), it is easy to prove that $j(K)$ isolates $\tilde{C}$ as a consequence of the j-invariance of the flow. The j-invariance also implies that $\tau_{j(K)}$ and $\tau^*_{j(K)}$ are lower semi-continuous because τ_K and τ^*_K are; upper semi-continuity is a simple consequence of the compactness of $j(K)$. Thus, $j(K)$ is a block whence $K \cap j(K)$ is a j-invariant block since as noted in Definition 1.4(B), the intersection of two blocks is always a block. Proofs of the remaining statements are left as an exercise for the reader. □

Let ξ_σ be the vectorfield defined in statement (5) of Lemma 7.17.

7.19 COROLLARY. *There exists a unique* j*-invariant vectorfield* ξ_2 *on* $\tilde{U}$ *satisfying the relations* $\xi_2 = \sigma_2^*\xi$ *and* $\xi_\sigma = (/2\tau)^*\xi_2$.

Proof. The existence and uniqueness of ξ_2 satisfying $\xi_\sigma = (/2\tau)^*\xi_2$ is immediate from statement (1) of the lemma. As $\xi_\sigma := \sigma^*\xi = (/2\tau)^*\sigma_2^*\xi$, the uniqueness of ξ_2 forces $\xi_2 = \sigma_2^*\xi$. Hence, ξ_2 is j-invariant by statement (2) of the lemma. □

Let $E \in \mathrm{Gl}(P;\mathbf{R})$ and $C \in \mathrm{Gl}(N;\mathbf{R})$ be the matrices of statement (5) of Lemma 7.17 and let $\mathcal{F}\xi_\sigma$ and $\mathcal{F}^\perp\xi_\sigma$ be the vectorfields on $\mathbf{R} \times \mathbf{R}^P \times \mathbf{R}^N$ whose respective local representatives $\mathcal{F}\hat{\xi}_\sigma$ and $\mathcal{F}^\perp\hat{\xi}_\sigma$ are given as follows:

$$\mathcal{F}\hat{\xi}_\sigma(\theta, x_1, x_2) := (1, Ex_1, Cx_2);$$
$$\mathcal{F}^\perp\hat{\xi}_\sigma(\theta, x_1, x_2) := (0, Ex_1, Cx_2).$$

7.20 COROLLARY. *There exists a unique* j*-invariant vectorfield* $\mathcal{F}\xi_2$ *on* $\tilde{U}$ *satisfying* $\mathcal{F}\xi_\sigma = (/2\tau)^*\mathcal{F}\xi_2$ *and a unique vectorfield* $\mathcal{F}\xi$ *on* U *satisfying* $\mathcal{F}\xi_2 = \sigma_2^*\mathcal{F}\xi$ *and* $\mathcal{F}\xi_\sigma = \sigma^*\mathcal{F}\xi$. *An analogous statement holds for the vectorfield* $\mathcal{F}^\perp\xi_\sigma$.

Proof. The existence of a unique $\mathcal{F}\xi_2$ satisfying $\mathcal{F}\xi_\sigma = (/2\tau)^*\mathcal{F}\xi_2$ is immediate from statement (1) of the lemma. Next, note that the involution j lifts to a self-homeomorphism j_σ of $\mathbf{R} \times \mathbf{R}^P \times \mathbf{R}^N$ given explicitly by $j_\sigma(\theta, x_1, x_2) = (\theta + \tau, L_+(\theta)x_1, L_-(\theta)x_2)$ and characterized by the relation $/2\tau \circ j_\sigma = j \circ /2\tau$. From this relation and the fact that $j^*\xi_2 = \xi_2$, it follows that

$$T(/2\tau) \circ \xi_\sigma = T(/2\tau) \circ Tj_\sigma \circ \xi_\sigma \circ j_\sigma .$$

This equality and the uniqueness of Taylor series expansions forces the first-order Taylor polynomials of the local representative $\hat{\xi}_\sigma$ at $(\theta, 0, 0)$ and of the local representative of $Tj_\sigma \circ \xi_\sigma \circ j_\sigma$ at $(\theta, 0, 0)$ to be equal. Because $\mathcal{F}\hat{\xi}_\sigma$ is just the first-order Taylor polynomial of $\hat{\xi}_\sigma$ at $(\theta, 0, 0)$ with the linear term in the first (i.e., θ) component deleted, the just cited equality of first-order Taylor polynomials allows one to compute that

$$T(/2\tau) \circ \mathcal{F}\xi_\sigma = T(/2\tau) \circ Tj_\sigma \circ \mathcal{F}\xi_\sigma \circ j_\sigma .$$

From this and the characterizations $T(/2\tau)\circ\mathcal{F}\xi_\sigma = \mathcal{F}\xi_2\circ/2\tau$ and $/2\tau\circ j_\sigma = j\circ/2\tau$, it readily follows that

$$\mathcal{F}\xi_2\circ/2\tau = Tj\circ\mathcal{F}\xi_2\circ j\circ/2\tau\,;$$

i.e., $\mathcal{F}\xi_2$ is j-invariant. By statement (2) of the lemma, there exists a vectorfield $\mathcal{F}\xi$ on the tubular neighborhood U that is characterized by the equation $\sigma_2^*\mathcal{F}\xi = \mathcal{F}\xi_2$.

The existence of $\mathcal{F}^\perp\xi_2$ is again immediate from statement (1) of the lemma and it is j-invariant as an immediate consequence of the j-invariance of $\mathcal{F}\xi_2$. Hence, that $\mathcal{F}^\perp\xi$ satisfies the required relations is immediate from statement (2) of the lemma. □

In the following theorem we want to distinguish between bundles and fiber bundles especially as regards usage of "subbundle" and "fiber subbundle" and will do so following [Hus] and [St]. A bundle η is a triple, $\eta = (E, p, X)$ where $p\colon E\to X$ is a continuous surjective map: E is called the total space, p the projection, and X the base space of the bundle. For $x\in X$, $p^{-1}(x)$ is the fiber at x. No further structure is implied. Then $\omega := (E', p', X')$ *is a subbundle of* η *if, and only if,* $X' = X$, $E'\subset E$, *and* $p' = p|E'$. Note that there is a category of bundles where a morphism is essentially a fiber preserving map of the total spaces. Hence an equivalence between two bundles is essentially a fiber preserving homeomorphism of their total spaces. See [Hus, p. 14] for details and precise definitions.

A fiber bundle is a bundle with additional structure. The reader is referred to [St,§§2.1–4] for the definition as it is used here of a fiber bundle η with fiber Y and topological structure group G. Given $Z\subset Y$, if the action of G on Y induces an action of G on Z, i.e., if Z is invariant under the action of G on Y, then Steenrod defines—see [St, p. 24]—the subbundle of η determined by the invariant subspace Z of Y, let us denote it by η_Z, that is a fiber bundle with fiber Z and topological structure group $H = G/G_Z$ where $G_Z := \{g\in G : \forall z\in Z,\ g\cdot z = z\}$. Note that G_Z is trivial and $H\equiv G$ if G acts effectively on Z. Call *a fiber bundle* ω *with fiber* Z *and structure group* H *a fiber subbundle of a fiber bundle* η *with fiber* Y *and structure group* G *if, and only if,* $Z\subset Y$, $\omega = \eta_Z$, *and* $H\equiv G/G_Z$. We shall not need to mention explicitly the notion of equivalence of fiber bundles.

The need for the distinction between the notions of subbundle and of fiber subbundle arises because we will necessarily consider k-disk bundles that do not admit $O(k)$ as natural structure group even though the disk bundle is a subbundle of a k-plane bundle. These somewhat anomalous k-disk bundles have as natural structure group a subgroup $\mathfrak{G}_k$ of Homeo($\mathbf{R}^k$) that is an isomorphic image of $\mathrm{Gl}(k;\mathbf{R})$ but not by the inclusion homomorphism. Specifically, for A a linear automorphism of $\mathbf{R}^k$ and where $\|\cdot\|$ denotes the Euclidean norm on $\mathbf{R}^k$, let g_A be the homeomorphism of $\mathbf{R}^k$ with itself defined by

$$g_A(x) := \begin{cases} 0 & \text{if } x = 0 \\ \frac{\|x\|}{\|Ax\|}Ax & \text{if } x\neq 0. \end{cases}$$

Clearly, g_A is norm-preserving and homogeneous of degree one, i.e., $\|g_A(x)\| = \|x\|$ and $g_A(tx) = tg_A(x)$ for $x\in\mathbf{R}^k$ and $t\in\mathbf{R}$. However, g_A is additive if, and only if, A is orthogonal in which case $g_A = A$. It is also straightforward to show that $g_A \circ g_B = g_{AB}$ for A and B both linear automorphisms of $\mathbf{R}^k$. It follows that

$$\mathfrak{G}_k := \{g_A : A \in \mathrm{Gl}(k;\mathbf{R})\}$$

is a group of norm-preserving homeomorphisms of $\mathbf{R}^k$ and is therefore admissible as a structure group for a k-disk bundle. Note that $O(k)$ is a subgroup of $\mathfrak{G}_k$ acting in the usual manner (i.e., matrix multiplication) on elements of $\mathbf{R}^k$ as a consequence of our observation that $g_A = A$ for $A \in O(k)$.

To simplify the statement of the theorem, let us identify N_S, the normal bundle of S, with its realization in the tubular neighborhood U so that each point of U lies on a unique fiber of N_S. Note too that $N_S = N_S^u \oplus N_S^s$. Also, the sphere bundle associated to a disk bundle is denoted below by placing a dot over the symbol used to represent the disk bundle; e.g., if ω is a disk bundle, $\dot\omega$ is its associated sphere bundle.

7.21 THEOREM. *S is a hyperbolic closed orbit of $\mathcal{F}\xi$ and is a circle of degenerate critical points of $\mathcal{F}^\perp\xi$. The tubular neighborhood U of S with tubular map $p\colon U \to S$ contains a family B_c, $0 < c < c_0$, of C^r, codimension zero, compact manifolds with corners exhibiting the following properties:*

(1) *each B_c is an isolating block for S for the flows generated by ξ, $\mathcal{F}\xi$, and $\mathcal{F}^\perp\xi$ and is the total space of a fiber bundle β_c over S with fiber $D_c^P \times D_c^N$ and structure group $\mathfrak{G}_P \times \mathfrak{G}_N$ that is also a subbundle of the tubular neighborhood (U,p,S);*
(2) *$B_c^\pm$ is the total space of a fiber subbundle $\beta_c^\pm$ of β_c with β_c^+ the fiber subbundle determined by $S_c^{P-1} \times D_c^N$ and β_c^- the fiber subbundle determined by $D_c^P \times S_c^{N-1}$;*
(3) *for $0 < c' < c < c_0$, $\beta_{c'}$ is the fiber subbundle of β_c determined by the $\mathfrak{G}_P \times \mathfrak{G}_N$-invariant subset $D_{c'}^P \times D_{c'}^N$ of $D_c^P \times D_c^N$;*
(4) *for $\nu = u, s$, $W^\nu_{\mathcal{F}^\perp\xi}(S;B_c) = W^\nu_{\mathcal{F}\xi}(S;B_c)$ and $W^u_{\mathcal{F}\xi}(S;B_c)$ and $W^s_{\mathcal{F}\xi}(S;B_c)$ are the respective total spaces of fiber subbundles $\omega^u_{\mathcal{F}\xi\,c}$ and $\omega^s_{\mathcal{F}\xi\,c}$ of β_c, the former the P-disk subbundle determined by $D_c^P \times \{0\}$, the latter the N-disk subbundle determined by $\{0\} \times D_c^N$;*
(5) *for $\nu = u, s$, $\omega^\nu_{\mathcal{F}\xi\,c}$ is a subbundle of N_S^ν and is either the 0-disk bundle over S or is bundle equivalent to the unit disk bundle of N_S^ν relative to any norm;*
(6) *the splitting $N_S = N_S^u \oplus N_S^s$ induces splittings*

$$\beta_c \equiv \omega^u_{\mathcal{F}\xi\,c} \oplus \omega^s_{\mathcal{F}\xi\,c}, \quad \beta_c^+ \equiv \dot\omega^u_{\mathcal{F}\xi\,c} \oplus \omega^s_{\mathcal{F}\xi\,c}, \quad \beta_c^- \equiv \omega^u_{\mathcal{F}\xi\,c} \oplus \dot\omega^s_{\mathcal{F}\xi\,c};$$

(7) *$W^u_\xi(S;B_c)$ and $W^s_\xi(S;B_c)$ are the respective total spaces of subbundles $\omega^u_{\xi\,c}$ and $\omega^s_{\xi\,c}$ of β_c, the former a P-disk bundle, the latter an N-disk bundle, and for $\nu = u, s$, with the definitions $W^\nu_{\xi\,c} := W^\nu_\xi(S;B_c)$ and $W^\nu_{\mathcal{F}\xi\,c} := W^\nu_{\mathcal{F}\xi}(S;B_c)$, for $\operatorname{sgn} u = +$ and $\operatorname{sgn} s = -$, the pair $(W^\nu_{\xi\,c}, \partial W^\nu_{\xi\,c})$ is isotopic in $(B_c, B_c^{\operatorname{sgn}\nu})$ to the pair $(W^\nu_{\mathcal{F}\xi\,c}, \partial W^\nu_{\mathcal{F}\xi\,c})$ via an isotopy whose restriction to the fiber pair of $(\omega^\nu_{\xi\,c}, \dot\omega^\nu_{\xi\,c})$ above a point $\mathbf{u} \in S$ is an isotopy in the fiber pair of $(\beta_c, \beta_c^{\operatorname{sgn}\nu})$ above $\mathbf{u}$ to the fiber pair of $(\omega^\nu_{\mathcal{F}\xi\,c}, \dot\omega^\nu_{\mathcal{F}\xi\,c})$ above $\mathbf{u}$;*
(8) *for $\nu = u, s$, for $\mathbf{u} \in S$, $W^{\nu\nu}_\xi(\mathbf{u};B_c)$ coincides with the fiber at $\mathbf{u}$ of $\omega^\nu_{\xi\,c}$;*

(9) *for* $\nu = u, s$, *the pair* $(W^{\nu}_{\mathcal{F}\xi\,c}, \partial W^{\nu}_{\mathcal{F}\xi\,c})$ *is a strong deformation retract of the pair* $(B_c, B_c^{\operatorname{sgn}\nu})$ *and the pair* $(W^{\nu}_{\xi\,c}, \partial W^{\nu}_{\xi\,c})$ *is a weak deformation retract of the pair* $(B_c, B_c^{\operatorname{sgn}\nu})$;

(10) B_c/B_c^+ *is basepoint homotopic to the Thom space of* N^u_S;

(11) B_c/B_c^- *is basepoint homotopic to the Thom space of* N^s_S.

Proof. Let $\mathcal{F}\varphi^{\sigma}$ denote the flow generated by $\mathcal{F}\xi_{\sigma}$. Because $\mathcal{F}\varphi^{\sigma}_t(\theta, x_1, x_2) = (\theta + t, e^{tE}x_1, e^{tC}x_2)$, it is immediate from (5) of Lemma 7.17 and (1) of Lemma 7.18 that the vectorfields $\mathcal{F}\xi_2$ and ξ_2 when restricted to $\tilde{S} := S^1 \times \{0\} \times \{0\} \subset \tilde{U}$ are tangent to $\tilde{S}$, are in fact equal on $\tilde{S}$, and integrate to the same uniform rotation of $\tilde{S}$ making $\tilde{S}$ a closed orbit with fundamental period 2τ for both vectorfields. Further, it is clear that collectively the eigenvalues of E and C form a complete set of Floquet exponents of $\tilde{S}$ as a closed orbit of either $\mathcal{F}\xi_2$ or of ξ_2, whence $\tilde{S}$ is a hyperbolic closed orbit of both vectorfields. As σ_2 is a double cover and since the flow generated by $\mathcal{F}\xi_2$ covers that generated by $\mathcal{F}\xi$, it follows that S is a closed orbit of $\mathcal{F}\xi$ of prime period τ. Thus, where $\mathcal{F}\varphi$ denotes the flow generated by $\mathcal{F}\xi$ and where $\mathcal{F}\varphi^2$ that generated by $\mathcal{F}\xi_2$, if $\tilde{\mathbf{u}} \in \tilde{S}$ and $\mathbf{u} := \sigma_2(\tilde{\mathbf{u}})$, it follows that λ is an eigenvalue of ${}^{\mathbf{C}}T_{\mathbf{u}}\mathcal{F}\varphi_{\tau}$, the complexification of $T_{\mathbf{u}}\mathcal{F}\varphi_{\tau}$, with eigenvector v if, and only if, λ^2 is an eigenvalue of ${}^{\mathbf{C}}T_{\tilde{u}}\mathcal{F}\varphi^2_{2\tau}$ with eigenvector $\tilde{v} = {}^{\mathbf{C}}(T_{\tilde{u}}\sigma_2)^{-1}v$. It follows that S is a hyperbolic closed orbit of $\mathcal{F}\xi$ because it is one for $\mathcal{F}\xi_2$. Also, as $\mathcal{F}^{\perp}\xi_2$ is just $\mathcal{F}\xi_2$ minus the rotational component, clearly each point of $\tilde{S}$ is a degenerate critical point for $\mathcal{F}^{\perp}\xi_2$. It follows that each point of S is a degenerate critical point for $\mathcal{F}^{\perp}\xi$.

From statement (5) of Lemma 7.17 it follows that the hypotheses of Proposition 7.9 are satisfied for $\hat{\chi} := \hat{\xi}_{\sigma}$, for $\hat{\chi} := \mathcal{F}\hat{\xi}_{\sigma}$, and for $\hat{\chi} := \mathcal{F}^{\perp}\hat{\xi}_{\sigma}$. Thus, application of that proposition yields that $\tilde{S} := S^1 \times \{0\} \times \{0\}$ is an isolated invariant set in $\tilde{U}$ relative to the flows generated by ξ_2, $\mathcal{F}\xi_2$, and $\mathcal{F}^{\perp}\xi_2$. Further, the proposition yields $c_0 > 0$ so that if $0 < c < c_0$ then

$$K_c := S^1 \times D^P_c \times D^N_c$$

is an isolating block for $\tilde{S}$ relative to each of these flows that relative to all three has common exit set K_c^+ and common entrance set K_c^- with

$$K_c^+ = S^1 \times S^{P-1}_c \times D^N_c \qquad \text{and} \qquad K_c^- = S^1 \times D^P_c \times S^{N-1}_c .$$

By Lemma 7.18(6), a family of j-invariant blocks for $\tilde{S}$ relative to the flows of ξ_2, $\mathcal{F}\xi_2$, and $\mathcal{F}^{\perp}\xi_2$ is obtained by setting

$$N_c := K_c \cap j(K_c) \qquad \text{for } 0 < c < c_0,$$

and relative to those flows N_c has common exit set N_c^+ and common entrance set N_c^-.

For $0 < c < c_0$, N_c is the total space of a trivializable fiber bundle $\beta_{2\,c}$ over S^1 with fiber $D^P_c \times D^N_c$, and $\beta_{2\,c}$ has fiber subbundles $\beta^+_{2\,c}$ and $\beta^-_{2\,c}$ with respective total spaces N_c^+ and N_c^- and with respective fibers $S^{P-1}_c \times D^N_c$ and $D^P_c \times S^{N-1}_c$. To see that these claims are true, first observe that as D^P_c and D^N_c are Euclidean disks centered at the origin and as L_+ and L_- are 2τ-periodic C^r maps from $\mathbf{R}$ into the space of linear automorphisms of $\mathbf{R}^P$ and $\mathbf{R}^N$ respectively, $L_+(\theta)D_c^P$ and $L_-(\theta)D_c^N$ are, for each $\theta \in \mathbf{R}$, strictly convex topological balls of dimensions P and N respectively centered at the origin. It follows that for each $\theta \in \mathbf{R}$ the intersections

$$D^P_{c\,\theta_+} := D^P_c \cap L_+(\theta+\tau)D^P_c \qquad \text{and} \qquad D^N_{c\,\theta_-} := D^N_c \cap L_-(\theta+\tau)D^N_c \,,$$

are strictly convex topological balls with piecewise smooth boundary (i.e., manifolds with corners) of dimension P and N respectively varying smoothly with θ, and from (4) of Lemma 7.17, the definition of N_c, and (6) of Lemma 7.18, it follows that

$$N_c = \bigcup_{0\le\theta<2\tau} \{e^{i\theta\pi/\tau}\} \times D^P_{c\,\theta_+} \times D^N_{c\,\theta_-} \,, \tag{7.9}$$

that

$$N_c^+ = \bigcup_{0\le\theta<2\tau} \{e^{i\theta\pi/\tau}\} \times \partial D^P_{c\,\theta_+} \times D^N_{c\,\theta_-} \,, \tag{7.10}$$

and that

$$N_c^- = \bigcup_{0\le\theta<2\tau} \{e^{i\theta\pi/\tau}\} \times D^P_{c\,\theta_+} \times \partial D^N_{c\,\theta_-} \,. \tag{7.11}$$

It follows that N_c is a manifold with corners and that each smooth piece of the boundary lies entirely in either N_c^+ or N_c^- and in a local transversal to the flow.

It is clear from (7.9) that N_c is the total space of a bundle $\beta_{2\,c}$ over S_1 whose fiber at $e^{i\theta\pi/\tau}$ is $D^P_{c\,\theta_+} \times D^N_{c\,\theta_-}$. Let $h_{c\,\theta}$ be the homeomorphism of D^P_c onto $D^P_{c\,\theta_+}$ obtained by linearly contracting (when necessary) each radius of D^P_c onto the radius of $D^P_{c\,\theta_+}$ which it contains; i.e., for $s \in [0,1]$ and $x \in S^{P-1}_c$, let $h_{c\,\theta}(sx) := st_{c\,\theta}(x)x$ where $t_{c\,\theta}(x)$ is the unique positive number in $[0,1]$ satisfying $t_{c\,\theta}(x)x \in \partial D^P_{c\,\theta_+}$. Similarly, define a homeomorphism $k_{c\,\theta}$ of D^N_c onto $D^N_{c\,\theta_-}$. The smoothness of $L_\pm$ ensures that the map of $S^1 \times D^P_c \times D^N_c$ onto N_c defined by $(e^{i\theta\pi/\tau}, u_1, u_2) \mapsto (e^{i\theta\pi/\tau}, h_{c\,\theta}(u_1), k_{c\,\theta}(u_2))$ is a homeomorphism and determines an equivalence of $\beta_{2\,c}$ with the trivial fiber bundle over S^1 with fiber $D^P_c \times D^N_c$. It then follows from (7.10) and (7.11) that $N_c^\pm$ is the total space of a fiber subbundle $\beta^\pm_{2\,c}$ of $\beta_{2\,c}$ with $\beta^+_{2\,c}$ the fiber subbundle determined by $S^{P-1}_c \times D^N_c$ and with $\beta^-_{2\,c}$ the fiber subbundle determined by $D^P_c \times S^{N-1}_c$.

Suppose now that $0 < c' < c < c_0$; let us see that $\beta_{2\,c'}$ is the fiber subbundle of $\beta_{2\,c}$ determined by the subset $D^P_{c'} \times D^N_{c'}$ of $D^P_c \times D^N_c$. Clearly it suffices to show that $h_{c'\,\theta}$ is the restriction of $h_{c\,\theta}$ to $D^P_{c'}$ and that $k_{c'\,\theta}$ is the restriction of $k_{c\,\theta}$ to $D^N_{c'}$. Now it follows straightforwardly from the definitions that $h_{c\,\theta}(S^{P-1}_{c'}) = \partial D^P_{c'\,\theta_+}$, whence also $h_{c\,\theta}(D^P_{c'}) = D^P_{c'\,\theta_+}$. In particular, if $x' \in S^{P-1}_{c'}$, then

$$h_{c\,\theta}(x') = \frac{c'}{c} h_{c\,\theta}(\frac{c}{c'}x') = t_{c\,\theta}(\frac{c}{c'}x')x' \in \partial D^P_{c'\,\theta_+} \,.$$

Hence, by definition of $t_{c'\theta}$, necessarily

$$t_{c\theta}(\frac{c}{c'}x') = t_{c'\theta}(x')$$

yielding thereby that $h_{c'\theta}(x') = h_{c\theta}(x')$, which is sufficient to show that $h_{c'\theta}$ is the restriction of $h_{c\theta}$ to $D^P_{c'}$. Similarly, $k_{c'\theta}$ is the restriction of $k_{c\theta}$ to $D^N_{c'}$.

Define $B_c := \sigma_2(N_c)$ for $0 < c < c_0$. By (6) of Lemma 7.18 it is immediate that B_c is an isolating block with corners for S relative to the flows generated by ξ, $\mathcal{F}\xi$, and $\mathcal{F}^{\perp}\xi$. Also, relative to these flows B_c has common exit set B^+_c and common entrance set B^-_c because the analogous statement is true of N_c relative to the flows of ξ_2, $\mathcal{F}\xi_2$, and $\mathcal{F}^{\perp}\xi_2$ which cover the flows of ξ, $\mathcal{F}\xi$, and $\mathcal{F}^{\perp}\xi$ respectively.

A proof of the rest of statement (1) and of statements (2) and (3) are obtained as follows. Because j permutes the fibers of $\beta_{2\,c}$ and because j restricted to $\tilde{S} = S^1 \times \{0\} \times \{0\}$ is just the antipodal map, j can be regarded as defining a bundle self-morphism of $\beta_{2\,c}$ over the antipodal map of S^1. Hence, identifying $\mathbf{u}$ with $j(\mathbf{u})$ for each $\mathbf{u} \in N_c$ and thereby identifying z with $-z$ for each $z \in S^1$, we obtain from $\beta_{2\,c}$ a fiber bundle $\tilde{\beta}_c := (\tilde{B}_c, \tilde{p}_1, \mathbf{RP}^1 \equiv S^1)$ with fiber $D^P \times D^N$ and structure group $\mathfrak{G}_P \times \mathfrak{G}_N$. Specifically, for each $p = [\pm z] \in \mathbf{RP}^1$, there is a connected neighborhood V_p evenly covered relative to the double covering $\tilde{S} \to \mathbf{RP}^1$ induced by j. Let V_0 and V_1 be the disjoint opens mapped by the double covering onto V so that writing $z_\theta := e^{i\theta\pi/\tau}$ for $\theta \in \mathbf{R}$, we have $V_0 = \{z_\theta : \theta \in J_0\}$ for some open interval J_0 and $V_1 = \{z_\theta : \theta \in J_0 + \tau\}$. There are two local trivializations of $\tilde{\beta}_c|V$ obtained by identifying V with V_i $(i = 0, 1)$ and mapping $V_i \times D^P_c \times D^N_c$ onto $\tilde{p}_1^{-1}(V)$ by sending (z_θ, x_1, x_2) to $[(z_\theta, h_\theta(x_1), k_\theta(x_2))]$ where the square brackets denote an equivalence class under the identification of $\mathbf{u}$ and $j(\mathbf{u})$. The transition functions between these two coordinate descriptions are the maps on V into $\mathfrak{G}_P \times \mathfrak{G}_N$ given by

$$[\pm z_\theta] \mapsto g_{L_+(\theta)} \times g_{L_-(\theta)}\,;$$

i.e., in going from $V_0 \times D^P_c \times D^N_c$ to $V_1 \times D^P_c \times D^N_c$, the transition in the fiber at z_θ, $\theta \in J_0$ is given by the homeomorphism $g_{L_+(\theta)} \times g_{L_-(\theta)}$, but in going in the opposite direction, the transition in the fiber at $z_{\theta+\tau}$, $\theta \in J_0$, is the inverse homeomorphism $g_{L_+(\theta+\tau)} \times g_{L_-(\theta+\tau)}$. Thus $\tilde{\beta}_c$ has structure group $\mathfrak{G}_P \times \mathfrak{G}_N$. Also, for $0 < c' < c$, because $\beta_{2\,c'}$ is the fiber subbundle of $\beta_{2\,c}$ determined by $D^P_{c'} \times D^N_{c'}$, it is immediate from the construction that $\tilde{\beta}_{c'}$ is the fiber subbundle of $\tilde{\beta}_c$ determined by $D^P_{c'} \times D^N_{c'}$.

Let $\tilde{\jmath}: \beta_{2\,c} \to \tilde{\beta}_c$ be the bundle map defined by the identification of $\mathbf{u}$ with $j(\mathbf{u})$. As j is the involution of the double cover $\sigma_2: \tilde{U} \to U$, it follows that B_c is homeomorphic to $\tilde{B}_c$ with the homeomorphism being given by $\tilde{\jmath} \circ \sigma_2^{-1}$. It then follows from statement (3) of Lemma 7.17 that this homeomorphism allows us to regard B_c as the total space of a subbundle β_c of the tubular neighborhood, $\beta_c := (B_c, p|B_c, S)$, and also allows us to regard β_c as having fiber $D^P_c \times D^N_c$ with structure group $\mathfrak{G}_P \times \mathfrak{G}_N$. In addition, it follows from this construction of β_c that for $0 < c' < c$ the fiber subbundle of β_c determined by $D^P_{c'} \times D^N_{c'}$ is $\beta_{c'}$ since the same relation holds between $\tilde{\beta}_{c'}$ and $\tilde{\beta}_{c''}$. Also, from the descriptions of $B^\pm_c$ and $N^\pm_c$ above and in particular the descriptions (7.9)–(7.11), it is immediate that B^+_c and B^-_c are the respective total spaces of fiber subbundles β^+_c and β^-_c of β_c, the former the fiber subbundle determined by $S^{P-1}_c \times D^N_c$, the latter that determined by $D^P_c \times S^{N-1}_c$.

To prove statements (4)–(6), let us begin by looking at the local invariant manifolds of $\tilde{S}$ within N_c relative to the flow of $\mathcal{F}\xi_2$, and for simplicity in writing what follows, for $\nu = u, s$, set $W^\nu_{\xi_2\, c} := W^\nu_{\xi_2}(\tilde{S}; N_c)$, set $W^\nu_{\mathcal{F}\xi_2\, c} := W^\nu_{\mathcal{F}\xi_2}(\tilde{S}; N_c)$, and analogously define $W^\nu_{\xi\, c} = W^\nu_{\xi}(S; B_c)$ and $W^\nu_{\mathcal{F}\xi\, c} = W^\nu_{\mathcal{F}\xi}(S; B_c)$. For $\nu = u, s$, as $W^\nu_{\mathcal{F}\xi_2\, c} = W^\nu_{\mathcal{F}\xi_2}(\tilde{S}; K_c) \cap W^\nu \mathcal{F}\xi_2(\tilde{S}; j(K_c))$ and because $W^\nu_{\mathcal{F}\xi_2}(\tilde{S}; j(K_c)) = j(W^\nu_{\mathcal{F}\xi_2}(\tilde{S}; K_c))$ as a consequence of Lemma 7.18(3), it follows from (7.9) and the definition of $\mathcal{F}\xi_2$ that

$$W^u_{\mathcal{F}\xi_2\, c} = \bigcup_{0 \le \theta < 2\tau} \{e^{i\theta\pi/\tau}\} \times D^P_{c\,\theta_+} \times \{0\} \tag{7.12a}$$

and

$$W^s_{\mathcal{F}\xi_2\, c} = \bigcup_{0 \le \theta < 2\tau} \{e^{i\theta\pi/\tau}\} \times \{0\} \times D^N_{c\,\theta_-} \tag{7.12b}$$

so that $W^u_{\mathcal{F}\xi_2\, c}$ is the total space of a P-disk subbundle $\omega^u_{\mathcal{F}\xi_2\, c}$ of $\beta_{2\, c}$ and $W^s_{\mathcal{F}\xi_2\, c}$ is the total space of an N-disk subbundle $\omega^s_{\mathcal{F}\xi_2\, c}$ of $\beta_{2\, c}$ and both are trivializable.

Because $W^\nu_{\mathcal{F}\xi_2\, c}$ is j-invariant for $\nu = u, s$, it follows that the same method used to construct β_c from $\beta_{2\, c}$ when applied to $\omega^\nu_{\mathcal{F}\xi_2\, c}$ yields fiber subbundles $\omega^u_{\mathcal{F}\xi\, c}$ and $\omega^s_{\mathcal{F}\xi\, c}$ of β_c with respective total spaces $W^u_{\mathcal{F}\xi\, c}$ and $W^s_{\mathcal{F}\xi\, c}$ and with $\omega^u_{\mathcal{F}\xi\, c}$ a P-disk bundle and $\omega^s_{\mathcal{F}\xi}$ an N-disk bundle. Also, because $W^\nu_{\mathcal{F}^\perp\xi_2\, c} = W^\nu_{\mathcal{F}\xi\, c}$ for $\nu = u, s$ as a consequence of (7.12) and the definition of $\mathcal{F}^\perp\xi_2$, it follows that $W^\nu_{\mathcal{F}^\perp\xi\, c} = W^\nu_{\mathcal{F}\xi\, c}$ for $\nu = u, s$.

Note too that from (7.9) and (7.12) it is immediate that

$$\beta_{2\, c} = \omega^u_{\mathcal{F}\xi_2\, c} \oplus \omega^s_{\mathcal{F}\xi_2\, c},$$

and from (7.11–13) it follows that

$$\beta^+_{2\, c} = \dot{\omega}^u_{\mathcal{F}\xi_2\, c} \oplus \omega^s_{\mathcal{F}\xi_2\, c} \qquad \text{and} \qquad \beta^-_{2\, c} = \omega^u_{\mathcal{F}\xi_2\, c} \oplus \dot{\omega}^s_{\mathcal{F}\xi_2\, c}.$$

Further each of these splittings is induced by the hyperbolic splitting of the normal bundle N_S as an immediate consequence of the constructions made. In particular, the total space of N^u_S is $\sigma(\mathbf{R} \times \mathbf{R}^P \times \{0\})$, which equals $\sigma_2(S^1 \times \mathbf{R}^P \times \{0\})$, whence from the construction of $\omega^u_{\mathcal{F}\xi\, c}$ it is a subbundle of N^u_S. Similarly, the total space of N^s_S is $\sigma_2(S^1 \times \{0\} \times \mathbf{R}^N)$ and N^s_S has $\omega^s_{\mathcal{F}\xi\, c}$ as subbundle.

We now show that either $\omega^u_{\mathcal{F}\xi\, c}$ is either the 0-disk bundle or is bundle equivalent to the unit disk bundle of N^u_S relative to the norm of some Euclidean metric for N^u_S. An analogous argument establishes the analogous result for $\omega^s_{\mathcal{F}\xi\, c}$ and N^s_S. If $P = 0$, clearly $\omega^u_{\mathcal{F}\xi\, c}$ is the 0-disk bundle over S. Suppose, therefore, $P \ge 1$. Choose a Euclidean metric in N^u_S. As $\sigma_2\colon S^1 \times \mathbf{R}^P \times \{0\} \to N^u_S$ is a double cover, by Lemma 7.17(3) the Euclidean metric pulls back to one in the bundle $N^u_{\tilde{S}} := (S^1 \times \mathbf{R}^P \times \{0\}, p_1, S^1)$ and let $\|\cdot\|_\theta$ denote the resulting norm in the fiber $\mathbf{R}^P \times \{0\} \equiv \mathbf{R}^P$ above the point $e^{i\theta\pi/\tau}$. Note that necessarily this norm is j-invariant; i.e.,

$$\|x\|_\theta = \|L_+(\theta)x\|_{\theta+\tau} \qquad \text{for } x \in \mathbf{R}^P \text{ and } \theta \in \mathbf{R}.$$

Choose $c_1 > 0$ so large that

$$D^P_{c\,\theta_+} \subset \{x \in \mathbf{R}^P : \|x\|_\theta \le c_1\} \qquad \text{for } \theta \in \mathbf{R} \text{ and } 0 < c < c_0.$$

For each $x \in \partial D^P_{c\,\theta_+}$ there exists a unique $t(x) \ge 1$ so that $\|t(x)x\|_\theta = c_1$. Note that $t(x)$ varies continuously with $x \in \partial D^P_{c\,\theta_+}$ and that $t(L_+(\theta)x) = t(x)$. It follows that the formula

$$f(e^{i\theta\pi/\tau}, sx, 0) = (e^{i\theta\pi/\tau}, s\,t(x)x, 0) \qquad \text{for } x \in \partial D^P_{c\,\theta_+} \text{ and } s \in [0,1]$$

defines a j-invariant bundle equivalence of $\omega^u_{\mathcal{F}\xi_2\,c}$ and $\tilde{D}(c_1)$, the disk bundle of radius c_1 in $N^u_{\tilde{S}}$ defined using the norm of the pulled back Euclidean metric, and as a consequence of its j-invariance thereby induces an equivalence of $\omega^u_{\mathcal{F}\xi}$ and $D(c_1)$, the disk bundle of radius c_1 in the chosen Euclidean metric on N^u_S. It follows that $\omega^u_{\mathcal{F}\xi}$ is bundle equivalent to $D(1)$. This completes the proof of statements (4)–(6).

As in the proof of statement (5) of Theorem 7.12, standard proofs of the invariant manifold theorems, e.g., see [Ab-R, Lemma 28.2], imply that decreasing c_0 if necessary there exists a C^r function $G\colon S^1 \times \operatorname{int}(D^P_{c_0}) \to \mathbf{R}^N$ so that for $0 < c < c_0$, if G_c is the restriction of G to $S^1 \times D^P_c$ then the graph of G_c is $W^u_{\xi_2}(\tilde{S}; K_c)$. Further, the proof of the invariant manifold theorems given in [HPS], shows that the graph of the restriction of G_c to $\{e^{i\theta\pi/\tau}\} \times \operatorname{int}(D^P_{c_0})$ is $W^{uu}_{\xi_2}(e^{i\theta\pi/\tau}; K_c)$.

Note that in fact G_c has image in D^N_c since $W^u_{\xi_2}(\tilde{S}; K_c) \subset K_c = S^1 \times D^P_c \times D^N_c$. For $0 < c < c_0$ and $\theta \in \mathbf{R}$, set

$$\hat{W}^u_{\xi_2\,c\,\theta} := \{e^{i\theta\pi/\tau}\} \times D^P_{c\,\theta_+} \qquad \text{and} \qquad \hat{W}^u_{\xi_2\,c} := \bigcup_{\theta \in \mathbf{R}} \hat{W}^u_{\xi_2\,c\,\theta}$$

and define

$$\hat{G}_c := G_c|\hat{W}^u_{\xi_2\,c}, \qquad \text{and} \qquad \hat{G}_{c\,\theta} := G_c|W^u_{\xi_2\,c\,\theta}.$$

Then the graph of $\hat{G}_c$ equals $W^u_{\xi_2\,c}$ and the graph of $\hat{G}_{c\,\theta}$ equals $W^{uu}_{\xi_2}(\bar{z}_\theta, N_c)$ where $\bar{z}_\theta := (z_\theta, 0, 0)$. To prove the statements of the last sentence it suffices, as follows from the observation (7.9), to show that

$$G_c(\{e^{i\theta\pi/\tau}\} \times D^P_{c\,\theta_+}) \subset D^N_{c\,\theta_-} \tag{7.13}$$

and to note that $W^{uu}_{\xi_2}(\bar{z}_\theta, N_c)$ coincides with the local unstable manifold within N_c at $\bar{z}_\theta$ of the diffeomorphism $\varphi^2_{2\tau}$. The inclusion (7.13) follows easily from the equality $W^u_{\xi_2\,c} = W^u_{\xi_2}(\tilde{S}; K_c) \cap j(W^u_{\xi_2}(\tilde{S}; K_c))$ which is proved in the same way that the analogous statement obtained by replacing ξ_2 with $\mathcal{F}\xi_2$ was proved earlier in this demonstration. In particular, the inclusion $W^u_{\xi_2\,c} \subset W^u_{\xi_2}(\tilde{S}; K_c)$ implies that (*i*) $G_c(\{e^{i\theta\pi/\tau}\} \times D^P_{c\,\theta_+})$ is a subset of D^N_c while the inclusion $W^u_{\xi_2\,c} \subset j(W^u_{\xi_2}(\tilde{S}; K_c))$ implies that (*ii*) $G_c(\{e^{i\theta\pi/\tau}\} \times D^P_{c\,\theta_+})$ is a subset of $L_-(\theta+\tau)D^N_c$, and (*i*) and (*ii*) together yield the inclusion (7.13).

Now, by the observation (7.9) we may regard N_c as the total space of an N-disk bundle over $\hat{W}^u_{\xi_2\,c}$, and we may therefore regard the graph map of $\hat{G}_c$ as defining a section of this bundle with image $W^u_{\xi_2\,c}$. Note that the zero section of this bundle is precisely $W^u_{\mathcal{F}\xi_2\,c}$. Thus, the formula

$$Z_t(e^{i\theta\pi/\tau}, x, 0) = (e^{i\theta\pi/\tau}, x, tG_c(e^{i\theta\pi/\tau}, x)) \qquad \text{for } t \in [0,1]$$

defines an isotopy in N_c from $W^u_{\mathcal{F}\xi_2\,c}$ to $W^u_{\xi_2\,c}$ and note that Z_t is j-invariant; i.e., $j \circ Z_t = Z_t \circ j$. Also, it follows from (7.10) that the restriction of the graph map of $\hat{G}_c$ to $\partial \hat{W}^u_{\xi_2\,c} = \bigcup_{\theta\in\mathbf{R}} \{e^{i\theta\pi/\tau}\} \times \partial D^P_{c\,\theta_+}$ has image in N^+_c. Hence, in fact, Z_t is an isotopy from the pair $(W^u_{\mathcal{F}\xi_2\,c}, \partial W^u_{\mathcal{F}\xi_2\,c})$ to the pair $(W^u_{\xi_2\,c}, \partial W^u_{\xi_2\,c})$ in the pair (N_c, N^+_c). Further, throughout the isotopy the image at stage t of the fiber pair of $(\omega^u_{\mathcal{F}\xi_2\,c}, \dot{\omega}^u_{\mathcal{F}\xi_2\,c})$ above any point of S lies within the fiber pair of $(\beta_{2\,c}, \beta^+_{2\,c})$ above that point. It follows that $W^u_{\xi_2\,c}$ is the total space of a trivializable P-disk bundle $\omega^u_{\xi_2\,c}$ that is a subbundle of $\beta_{2\,c}$ with the end map of the isotopy defining a bundle equivalence from $\omega^u_{\mathcal{F}\xi_2\,c}$ to $\omega^u_{\xi_2\,c}$. Once again, application of the method of constructing β_c from $\beta_{2\,c}$ to its j-invariant subbundle $\omega^u_{\xi_2\,c}$ yields that $W^u_{\xi\,c}$ is the total space of a bundle $\omega^u_{\xi\,c}$ with fiber D^P_c, base S, and structure group $\mathfrak{G}_P$, and the fiber of $\omega^u_{\xi\,c}$ at $\mathbf{u} \in S$ coincides with $W^{uu}(\mathbf{u}, B_c)$. The j-invariance of N_c, $W^u_{\xi_2\,c}$, and $W^u_{\mathcal{F}\xi_2\,c}$ imply that $\sigma_2 \circ Z_t \circ \sigma_2^{-1}$ is a well-defined isotopy of the pair $(W^u_{\mathcal{F}\xi\,c}, \partial W^u_{\mathcal{F}\xi\,c})$ to the pair $(W^u_{\xi\,c}, \partial W^u_{\xi\,c})$ in the pair (B_c, B^+_c) in such manner that throughout the isotopy the image of the fiber pair of $(\omega^u_{\mathcal{F}\xi\,c}, \dot{\omega}^u_{\mathcal{F}\xi\,c})$ above a point $\mathbf{u} \in S$ lies in the fiber pair of $(\beta_c\,, \beta^+_c)$ above $\mathbf{u}$ from which it follows that $\omega^u_{\xi\,c}$ has structure group $\mathfrak{G}_P$.

Let us now show that $(W^u_{\mathcal{F}\xi\,c}, \partial W^u_{\mathcal{F}\xi\,c})$ is a strong deformation retract of (B_c, B^+_c) and that $(W^u_{\xi\,c}, \partial W^u_{\xi\,c})$ is a weak deformation retract of (B_c, B^+_c). First, note from (7.12) and (7.9) it follows that the fibers at $e^{i\theta\pi/\tau}$ of $\omega^u_{\mathcal{F}\xi_2\,c}$ and $\beta_{2\,c}$ are respectively $\{e^{i\theta\pi/\tau}\} \times D^P_{c\,\theta_+} \times \{0\}$ and $\{e^{i\theta\pi/\tau}\} \times D^P_{c\,\theta_+} \times D^N_{c\,\theta_-}$. Thus, for $\theta \in \mathbf{R}$, contraction of $D^N_{c\,\theta_-}$ linearly along radii to 0 while leaving $D^P_{c\,\theta_+}$ pointwise fixed, defines a j-invariant strong deformation retraction D_t of the pair (N_c, N^+_c) onto the pair $(W^u_{\mathcal{F}\xi_2\,c}, \partial W^u_{\mathcal{F}\xi_2\,c})$ that at each point of S^1 deforms the fiber of $\beta_{2\,c}$ above that point within itself. Therefore, the composition $\sigma_2 \circ D_t \circ \sigma_2^{-1}$ is a well-defined strong deformation retraction of the pair (B_c, B^+_c) onto $(W^u_{\mathcal{F}\xi\,c}, \partial W^u_{\mathcal{F}\xi\,c})$ that at each point of S deforms the fiber of β_c above that point within itself. As the pair $(W^u_{\xi\,c}, \partial W^u_{\xi\,c})$ is isotopic in (B_c, B^+_c) to the pair $(W^u_{\mathcal{F}\xi\,c}, \partial W^u_{\mathcal{F}\xi\,c})$ and because the latter pair is a strong deformation retract of (B_c, B^+_c), the inclusion $(W^u_{\xi\,c}, \partial W^u_{\xi\,c}) \subset (B_c, B^+_c)$ is a homotopy equivalence as desired.

It is immediate from the previous paragraph that B_c/B^+_c is basepoint homotopic to $W^u_{\mathcal{F}\xi\,c}/\partial W^u_{\mathcal{F}\xi\,c}$ which is basepoint homotopic to $\mathsf{T}(N^u_S)$ as an immediate consequence of statement (5) of the theorem already proved and Definition 7.1.

Arguments analogous to those just given show that $W^s_{\xi\,c}$ and $W^s_{\mathcal{F}\xi\,c}$ are the total spaces of subbundles $\omega^s_{\xi\,c}$ and $\omega^s_{\mathcal{F}\xi\,c}$ of β_c with the latter the fiber subbundle determined by $\{0\} \times D^N_c$ and with the former having $W^{ss}_\xi(\mathbf{u}; B_c)$ as its fiber at $\mathbf{u} \in S$, that the pairs $(W^s_{\xi\,c}, \partial W^s_{\xi\,c})$ and $(W^s_{\mathcal{F}\xi\,c}, \partial W^s_{\mathcal{F}\xi\,c})$ are isotopic in (B_c, B^-_c) via an isotopy whose restriction to the fiber pair of $(\omega^s_{\xi\,c}, \dot{\omega}^s_{\xi\,c})$ above a point $\mathbf{u} \in S$ is an isotopy in the fiber pair of $(\beta_c\,, \beta^-_c)$ above $\mathbf{u}$ to the fiber pair of $(\omega^s_{\mathcal{F}\xi\,c}, \dot{\omega}^s_{\mathcal{F}\xi\,c})$

above $\mathbf{u}$, that $\omega^s_{\mathcal{F}\xi\, c}$ is, as $N = 0$ or as $N \geq 1$, either the 0-disk bundle or equivalent to the unit disk bundle of N^s_S, that therefore $W^s_{\mathcal{F}\xi\, c}/\partial W^s_{\mathcal{F}\xi\, c}$ is homeomorphic to the Thom space $\mathsf{T}(N^s_S)$, that $(W^s_{\mathcal{F}\xi\, c}, \partial W^s_{\mathcal{F}\xi\, c})$ is a strong and $(W^s_{\xi\, c}, \partial W^s_{\xi\, c})$ a weak deformation retract of (B_c, B^-_c), whence the index space B_c/B^-_c is basepoint homotopic to $\mathsf{T}(N^s_S)$. □

The following result will be used below to help identify generators of the homology modules of the Conley indices of a hyperbolic periodic orbit and is corollary to the proofs of Proposition 7.9 and Theorem 7.21. The assumptions and notation are the same as those for Theorem 7.21 and its proof.

7.22 LEMMA. *If $0 < c' < c'' < c_0$, then the total space of the fiber subbundle of $\beta_{c''}$ determined by the $\mathfrak{G}_P \times \mathfrak{G}_N$-invariant subset $\mathrm{cl}\left(D^P_{c''} \setminus D^P_{c'}\right) \times D^N_{c''}$ of $D^P_{c''} \times D^N_{c''}$ is a subset $CB^+_{c''}$ of $B_{c''}$ with the properties*

(1) *$(B_{c''}, CB^+_{c''})$ is a regular index pair for S relative to the flows generated by ξ, $\mathcal{F}\xi$, and $\mathcal{F}^\perp\xi$;*
(2) *if $c' \leq \lambda \leq c''$, then there is an inclusion of nested index pairs $j_\lambda\colon (B_\lambda, B^+_\lambda) \subset (B_{c''}, CB^+_{c''})$.*

Proof. Let $C\beta^+_{c''}$ be the fiber subbundle of of $\beta_{c''}$ determined by $\mathrm{cl}\left(D^P_{c''} \setminus D^P_{c'}\right) \times D^N_{c''}$. Let $CK^+_{c''} := S^1 \times \mathrm{cl}\left(D^P_{c''} \setminus D^P_{c'}\right) \times D^N_{c''}$. Then $\langle K_{c''}, CK^+_{c''}\rangle$ is a regular index pair for $\tilde{S}$ relative to the flows generated by ξ_2, $\mathcal{F}\xi_2$, and $\mathcal{F}^\perp\xi_2$. One only has to verify that $CK^+_{c''}$ is positively invariant relative to $K_{c''}$ with respect to the three flows, for as $K_{c''}$ and $K_{c'}$ are blocks for $\tilde{S}$ relative to all three flows, orbits would then necessarily enter the collar $CK^+_{c''}$ along $K^+_{c'}$ and exit along $K^+_{c''}$. However, the positive invariance relative to all three flows is an immediate consequence of the inequalities (7.5_1) and (7.5_2) in the proof of Proposition 7.9 as they apply to the vectorfields ξ_2, $\mathcal{F}\xi_2$, and $\mathcal{F}^\perp\xi_2$. With the definitions

$$CN^+_{c''} := N_{c''} \cap (CK^+_{c''} \cup j(CK^+_{c''})) \qquad \text{and} \qquad CB^+_{c''} := \sigma_2(CN^+_{c''}),$$

it is immediate from Lemma 7.18(6) that $\langle N_{c''}, CN^+_{c''}\rangle$ and $\langle B_{c''}, CB^+_{c''}\rangle$ are nested index pairs, the former a j-invariant index pair for $\tilde{S}$ with respect to the flows generated by ξ_2, $\mathcal{F}\xi_2$ and $\mathcal{F}^\perp\xi_2$ and the latter an index pair for S with respect to the flows generated by ξ, $\mathcal{F}\xi$, and $\mathcal{F}^\perp\xi$. From the construction of $\beta_{c''}$ in the proof of Theorem 7.21, it follows easily that $CB^+_{c''}$ is the total space of $C\beta^+_{c''}$. The inclusion $(B_\lambda, B^+_\lambda) \subset (B_{c''}, CB^+_{c''})$ holds for $c' \leq \lambda \leq c''$ as an immediate consequence of statements (2) and (3) of Theorem 7.21 and the inclusion $S^{P-1}_\lambda \times D^N_\lambda \subset \mathrm{cl}\left(D^P_{c''} \setminus D^P_{c'}\right) \times D^N_{c''}$, which clearly holds for $c' \leq \lambda \leq c''$. □

Throughout the rest of this section, $\mathcal{C}(S)$ and $\mathcal{C}^*(S)$ will denote the Conley indices of S relative to the flow generated by ξ under the assumption of hypotheses HCO.

7.23 THEOREM. *Under the assumption of the hypotheses* HCO *and the conclusions of Theorem 7.21, the following hold:*

(1) *$[\mathcal{C}(S)] = [\mathsf{T}(N^u_S)]$; in particular, either $P = 0$ and $[\mathcal{C}(S)] = [S^1 \amalg \{*\}]$ or $P \geq 1$ and, as N^u_S is or is not orientable over $\mathbf{Z}$, either $[\mathcal{C}(S)] = [S^{P+1} \vee S^P]$ or $[\mathcal{C}(S)] = [S^{P-1}\mathbf{RP}^2]$;*

(2) $[\mathcal{C}^*(S)] = [\mathsf{T}(N^s_S)]$; *in particular, either* $N = 0$ *and* $[\mathcal{C}^*(S)] = [S^1 \amalg \{*\}]$ *or* $N \geq 1$ *and, as* N^s_S *is or is not orientable over* $\mathbf{Z}$, *either* $[\mathcal{C}^*(S)] = [S^{N+1} \vee S^N]$ *or* $[\mathcal{C}^*(S)] = [S^{N-1}\mathbf{RP}^2]$.

Proof. By Theorem 7.21(9), for $0 < c < c_0$ the index space B_c/B^+_c is basepoint homotopic to $\mathsf{T}(N^u_S)$; hence, by the connected simple system properties of a Conley index, each index space in $\mathcal{C}(S)$ is basepoint homotopic to $\mathsf{T}(N^u_S)$. Similarly, by Theorem 7.21(10), each index space in $\mathcal{C}^*(S)$ is basepoint homotopic to $\mathsf{T}(N^s_S)$.

If $P = 0$, then N^u_S is the vector bundle of rank 0 over S so that $\mathsf{T}(N^u_S) \simeq S^1 \amalg \{*\}$ as a consequence of Definition 7.1. Similarly, when $N = 0$, $\mathsf{T}(N^s_S) \simeq S^1 \amalg \{*\}$. In the remaining cases N^u_S is a P-plane bundle with $P \geq 1$ and N^s_S is an N-plane bundle with $N \geq 1$, whence the theorem is an immediate consequence of Lemma 7.24 following. □

7.24 Lemma. *For each integer* $k \geq 1$, *if* γ *is a* k*-plane bundle over* S^1, *then as* γ *is or is not orientable over* $\mathbf{Z}$ *its Thom space has the pointed homotopy type of* $S^{k+1} \vee S^k$ *or of* $S^{k-1}\mathbf{RP}^2$.

Lemma 7.24 is a simple consequence of the following two propositions. The first of these is well-known and follows quite simply by representing the closed Möbius band as a two-cell with one-pair of edges identified after a half-twist.

7.25 Proposition. *A closed Möbius band with its bounding circle identified to a point and taken as the basepoint is homeomorphic as a pointed space to* $\mathbf{RP}^2$ *with any point of* $\mathbf{RP}^2$ *as basepoint.*

Let $\tilde{S}^2$ denote the quotient space obtained from S^2 by identifying $(0,0,1)$ and $(0,0,-1)$ to a single point taken as the basepoint. $\tilde{S}^2$ is often called the dimpled two-sphere.

7.26 Proposition. $S^2 \vee S^1$ *and* $\tilde{S}^2$ *are basepoint homotopic.*

Proof. The basepoint of S^2 is taken to be its north pole, i.e., the point with Cartesian coordinates $(0,0,1)$, and the basepoint of S^1 is taken to be the point with Cartesian coordinates $(1,0)$. Let $q\colon S^2 \to \tilde{S}^2$ be the quotient map. For $k = 1, 2$, let $j_k\colon S^k \to S^2 \vee S^1$ be the natural embedding. Use will be made of spherical and polar coordinates to define the required homotopy equivalences. Specifically, let $s\colon [0,\pi] \times [0, 2\pi] \to S^2$ be defined by $s(\phi,\theta) := (\sin(\phi)\cos(\theta), \sin(\phi)\sin(\theta), \cos(\phi))$ and let $p\colon [0,2\pi] \to S^1$ be defined by $p(\psi) := (\cos(\psi), \sin(\psi))$. It is also convenient to set $J_i := [\frac{i}{3}, \frac{i+1}{3}]$, $\Phi_i := \pi J_i$, and $\Psi_i := 2\pi J_i$ for $i = 0,1,2$, and to set $I_i := [\frac{i}{2}, \frac{i+1}{2}]$ for $i = 0, 1$.

To define $f\colon \tilde{S}^2 \to S^2 \vee S^1$ it suffices to define $\tilde{f}\colon S^2 \to S^2 \vee S^1$ such that $\tilde{f}(0,0,\pm 1) = *$ where $*$ is the basepoint of $S^2 \vee S^1$, for then take $f = \tilde{f} \circ q^{-1}$. Define $\tilde{f}$ as follows:

$$\tilde{f} \circ s(\phi,\theta) := \begin{cases} j_2 \circ s(3\phi, \theta) & \text{if } \phi \in \Phi_0 \\ j_2 \circ s(2\pi - 3\phi, 0) & \text{if } \phi \in \Phi_1 \\ j_1 \circ p(6\phi - 4\pi) & \text{if } \phi \in \Phi_2 \end{cases}$$

and note that $\tilde{f}$ maps the north and south poles of S^2 to the basepoint of $S^2 \vee S^1$ as required.

The map $g\colon S^2 \vee S^1 \to \tilde{S}^2$ is defined by defining it separately on $S^2 \equiv j_2(S^2)$ and $S^1 \equiv j_1(S^1)$ as follows:

$$g|S^2 := q \qquad \text{and} \qquad g|S^1 \circ p(\psi) = q \circ s(\psi/2, 0)\,.$$

Note that both $g|S^2$ and $g|S^1$ map the basepoint of their domains to the basepoint of $\tilde{S}^2$.

The composite $g \circ f$ is homotopic to $1_{\tilde{S}^2}$ relative to the basepoint via the homotopy H_t defined by

$$H_t \circ q \circ s(\phi, \theta) := \begin{cases} q \circ s(3\phi, \theta) & \text{if } t \in I_0,\ \phi \in \Phi_0 \\ q \circ s(\pi + (1-2t)(\pi - 3\phi), 0) & \text{if } t \in I_0,\ \phi \in \Phi_1 \\ q \circ s(2\pi(3t-1) + 3\phi(1-2t), 0) & \text{if } t \in I_0,\ \phi \in \Phi_2 \\ q \circ s((5-4t)\phi, \theta) & \text{if } t \in I_1,\ \phi \in \Phi_0 \\ q \circ s((2t-1)\phi + (2-2t)\pi, \theta) & \text{if } t \in I_1,\ \phi \in \Phi_1 \cup \Phi_2. \end{cases}$$

The composite $f \circ g$ is homotopic to $1_{S^2 \vee S^1}$ relative to the basepoint via the homotopy K_t defined by

$$K_t|S^2 \circ s(\phi, \theta) := \begin{cases} j_2 \circ s(3\phi, \theta) & \text{if } t \in J_0 \cup J_1,\ \phi \in \Phi_0 \\ j_2 \circ s(2\pi - 3\phi, 0) & \text{if } t \in J_0,\ \phi \in \Phi_1 \\ j_1 \circ p((1-3t)(6\phi - 4\pi)) & \text{if } t \in J_0,\ \phi \in \Phi_2 \\ j_2 \circ s(\pi - (2-3t)(3\phi - \pi), 0) & \text{if } t \in J_1,\ \phi \in \Phi_1 \\ j_2 \circ s((3t-1)\pi, 0) & \text{if } t \in J_1,\ \phi \in \Phi_2 \\ j_2 \circ s((7-6t)\phi, \theta) & \text{if } t \in J_2,\ \phi \in \Phi_0 \\ j_2 \circ s((3t-2)\phi + (3-3t)\pi, \theta) & \text{if } t \in J_2,\ \phi \in \Phi_1 \cup \Phi_2 \end{cases}$$

and

$$K_t|S^1 \circ p(\psi) := \begin{cases} j_2 \circ s((1-2t)\tfrac{3}{2}\psi, 0) & \text{if } t \in I_0,\ \psi \in \Psi_0 \\ j_2 \circ s((1-2t)(2\pi - \tfrac{3}{2}\psi), 0) & \text{if } t \in I_0,\ \psi \in \Psi_1 \\ j_1 \circ p(3\psi - 4\pi) & \text{if } t \in I_0,\ \psi \in \Psi_2 \\ j_1 \circ p((2t-1)\psi) & \text{if } t \in I_1,\ \psi \in \Psi_0 \cup \Psi_1 \\ j_1 \circ p((2-2t)(3\psi - 4\pi) + (2t-1)\psi) & \text{if } t \in I_1,\ \psi \in \Psi_2. \quad \square \end{cases}$$

7.27 Proof of Lemma 7.24. Let ε^ℓ_X denote the trivial ℓ-plane bundle over a space X. As noted in [Sw, p. 229], for η any k-plane bundle over a space X and $x_0 \in X$, the vector bundle $\eta \oplus \varepsilon^\ell_X$ is naturally identified with the product bundle $\eta \times \varepsilon^\ell_{\{x_0\}}$. It follows that

$$\mathsf{T}(\eta \oplus \varepsilon^\ell_X) \simeq \mathsf{T}(\eta) \wedge \mathsf{T}(\varepsilon^\ell_{\{x_0\}}) \simeq \mathsf{T}(\eta) \wedge S^\ell \simeq S^\ell \mathsf{T}(\eta)$$

because the Thom space of a product bundle is the smash product of the Thom spaces of the factors—e.g., see [Sw, Proposition 2.28]—and because the Thom space of an ℓ-plane bundle over a point is obviously homeomorphic to S^ℓ.

Now, as a consequence of the classification theorem for fiber bundles over spheres, [St, Theorem 18.5], for each $k \geq 1$, there are exactly two equivalence classes of k-plane bundles over S^1. One of these is the equivalence class of the trivial k-plane bundle over S^1 which is orientable over $\mathbf{Z}$; the other is the class of the k-plane bundles not orientable over $\mathbf{Z}$ and, in particular, as $k = 1$ or as $k > 1$, is the equivalence class of the canonical line bundle γ_1^1 over $\mathbf{RP}^1 \equiv S^1$ or of the k-plane bundle $\gamma_1^1 \oplus \varepsilon_{S^1}^{k-1}$ over S^1.

First, suppose γ is orientable over $\mathbf{Z}$. If γ is equivalent to the trivial line bundle over S^1, then $\mathsf{T}(\gamma) \simeq S^1 \times D^1 / S^1 \times S^0$ and the latter is clearly a homeomorph of $\tilde{S}^2$, whence the desired result holds for $k = 1$ in the orientable case as a consequence of Proposition 7.26. For $k > 1$, γ must be bundle equivalent to $\varepsilon_{S^1}^k \simeq \varepsilon_{S^1}^1 \oplus \varepsilon_{S^1}^{k-1}$, whence by the observations of the first paragraph of the proof

$$\mathsf{T}(\gamma) \simeq S^{k-1}\mathsf{T}(\varepsilon_{S^1}^1) \simeq S^{k-1}(S^2 \vee S^1) \simeq S^{k+1} \vee S^k$$

yielding the desired result for $k > 1$ in the orientable case.

In the non-orientable case because γ is either equivalent to γ_1^1 when $k = 1$ or to $\gamma_1^1 \oplus \varepsilon_{S^1}^{k-1}$ when $k > 1$ and because $\mathsf{T}(\gamma_1^1)$ is clearly a homeomorph of the closed Möbius band with its bounding circle identified to a point taken as the basepoint, Proposition 7.25 and a computation analogous to that in the orientable case yields that $\mathsf{T}(\gamma) \simeq S^{k-1}\mathbf{RP}^2$ as desired. □

For simplicity in stating the next several lemmas, let us continue to identify $N_S := N_S^u \oplus N_S^s$, the normal bundle of S, with the tubular neighborhood U. Then, where $\mathbf{E}_S^\nu$ is the total space of N_S^ν $(\nu = u, s)$, $\mathbf{E}_S^\nu \subset U$ and $\mathbf{E}_{S\,0}^\nu$, the complement in $\mathbf{E}_S^\nu$ of the zero section of N_S^ν, coincides with $\mathbf{E}_S^\nu \setminus S$. Also, let $W_{\mathcal{F}\xi\,c\,\mathbf{u}}^\nu$ denote the fiber at $\mathbf{u} \in S$ of $\omega_{\mathcal{F}\xi\,c}^\nu$, and where $\dot{\omega}_{\mathcal{F}\xi\,c}^\nu$ is the sphere bundle associated to $\omega_{\mathcal{F}\xi\,c}^\nu$, let $\partial W_{\mathcal{F}\xi\,c\,\mathbf{u}}^\nu$ denote its fiber at $\mathbf{u} \in S$. Similarly, for $\nu = u, s$, $W_{\xi\,c\,u}^\nu$ and $\partial W_{\xi\,c\,\mathbf{u}}^\nu$ denote respectively the fiber at $\mathbf{u}$ of $\omega_{\xi\,c}^\nu$ and the fiber at $\mathbf{u}$ of the associated sphere bundle $\dot{\omega}_{\xi\,c}^\nu$.

7.28 Lemma. *Under the assumption of hypotheses* HCO *and conclusions of Theorem 7.21, if τ^u is an R-orientation of the vector bundle N_S^u, then for $0 < c < c_0$, for each $\mathbf{u} \in S$, there exist unique fundamental classes*

$$o_{\mathcal{F}\xi\,c\,\mathbf{u}}^u \in H_P(W_{\mathcal{F}\xi\,c\,\mathbf{u}}^u, \partial W_{\mathcal{F}\xi\,c\,\mathbf{u}}^u; R) \qquad \text{and} \qquad o_{\xi\,c\,\mathbf{u}}^u \in H_P(W_{\xi\,c\,\mathbf{u}}^u, \partial W_{\xi\,c\,\mathbf{u}}^u; R)$$

that together satisfy

(1) *under the inclusion induced homomorphisms $o_{\mathcal{F}\xi\,c\,\mathbf{u}}^u$ and $o_{\xi\,c\,\mathbf{u}}^u$ have common image in $H_P(B_c, B_c^+; R)$ that generates that R-module;*
(2) *$\langle {i_{c\,\mathbf{u}}^u}^* \tau^u, o_{\mathcal{F}\xi\,c\,\mathbf{u}}^u \rangle = 1$ where $i_{c\,\mathbf{u}}^u \colon (W_{\mathcal{F}\xi\,c\,\mathbf{u}}^u, \partial W_{\mathcal{F}\xi\,c\,\mathbf{u}}^u) \to (\mathbf{E}_S^u, \mathbf{E}_{S\,0}^u)$ is inclusion.*

An analogous statement holds for an R-orientation τ^s of N_S^s and for $0 < c < c_0$, for each $\mathbf{u} \in S$, the existence of fundamental classes $o_{\mathcal{F}\xi\,c\,\mathbf{u}}^s$ and $o_{\xi\,c\,\mathbf{u}}^s$ for the fibers $W_{\mathcal{F}\xi\,c\,\mathbf{u}}^s$ and $W_{\xi\,c\,\mathbf{u}}^s$ respectively satisfying the analogues of conditions (1) and (2) obtained by replacing B_c^+ with B_c^- and by replacing the expanding normal bundle with the contracting one.

Proof. Let $0 < c < c_0$. The inclusion $i^u_c: (W^u_{\mathcal{F}\xi\, c}, \partial W^u_{\mathcal{F}\xi\, c}) \subset (\mathbf{E}^u_S, \mathbf{E}^u_{S\,0})$ induces isomorphisms on cohomology, for it has the factorization

$$(W^u_{\mathcal{F}\xi\, c}, \partial W^u_{\mathcal{F}\xi\, c}) \overset{e_2}{\hookrightarrow} (W^u_{\mathcal{F}\xi\, c}, W^u_{\mathcal{F}\xi\, c} \setminus S) \overset{e_1}{\hookrightarrow} (\mathbf{E}^u_S, \mathbf{E}^u_{S\,0}),$$

and e_1 induces isomorphisms by excision and e_2 induces isomorphisms as a consequence of the Five Lemma and the fact that $\partial W^u_{\mathcal{F}\xi\, c}$, the total space of the sphere bundle $\dot{\omega}^u_{\mathcal{F}\xi\, c}$ associated to the disk bundle $\omega^u_{\mathcal{F}\xi\, c}$, is a strong deformation retract of $W^u_{\mathcal{F}\xi\, c} \setminus S$, the complement of the zero section in the total space of $\omega_{\mathcal{F}\xi\, c}$. Thus, it is immediate that $i^{u\,*}_c \tau^u$ is an orientation class for the P-disk bundle $\omega^u_{\mathcal{F}\xi\, c}$.

The fundamental class $o^u_{\mathcal{F}\xi\, c\mathbf{u}}$ of $W^u_{\mathcal{F}\xi\, c\mathbf{u}}$ ($\mathbf{u} \in S$) can now be determined as follows. Let $\mathbf{u} \in S$. Because the inclusion $i^u_{c\mathbf{u}}$ has the factorization

$$(W^u_{\mathcal{F}\xi\, c\mathbf{u}}, \partial W^u_{\mathcal{F}\xi\, c\mathbf{u}}) \overset{j^u_{c\mathbf{u}}}{\hookrightarrow} (W^u_{\mathcal{F}\xi\, c}, \partial W^u_{\mathcal{F}\xi\, c}) \overset{i^u_c}{\hookrightarrow} (\mathbf{E}^u_S, \mathbf{E}^u_{S\,0}).$$

it is immediate that $i^{u\,*}_{c\mathbf{u}} \tau^u$ generates $H^P(W^u_{\mathcal{F}\xi\, c\mathbf{u}}, \partial W^u_{\mathcal{F}\xi\, c\mathbf{u}}; R)$ as an R-module. Thus, the fundamental class $o^u_{\mathcal{F}\xi\, c\mathbf{u}}$ is uniquely specified by requiring it to be the dual of $i^{u\,*}_{c\mathbf{u}} \tau^u$ relative to the Kronecker pairing; i.e., choose $o^u_{\mathcal{F}\xi\, c\mathbf{u}}$ so that condition (2) is satisfied.

The fundamental class $o^u_{\xi\, c\mathbf{u}}$ of $W^u_{\xi\, c\mathbf{u}}$ ($\mathbf{u} \in S$) is determined from the fundamental class $o^u_{\mathcal{F}\xi\, c\mathbf{u}}$ as follows. Let $\mathbf{u} \in S$. By Theorem 7.21(7), the fiber pair $(W^u_{\mathcal{F}\xi\, c\mathbf{u}}, \partial W^u_{\mathcal{F}\xi\, c\mathbf{u}})$ is isotopic in (B_c, B^+_c) (in fact in the fiber pair over $\mathbf{u}$) to the fiber pair $(W^u_{\xi\, c\mathbf{u}}, \partial W^u_{\xi\, c\mathbf{u}})$. Hence, the image of $o^u_{\mathcal{F}\xi\, c\mathbf{u}}$ under the isomorphism on homology between these fiber pairs induced by the end map of such an isotopy is a fundamental class of $W^u_{\xi\, c\mathbf{u}}$, call it $o^u_{\xi\, c\mathbf{u}}$, which by the manner of its definition has the same image under the homomorphism on homology induced by the inclusion $k^u_{c\mathbf{u}}: (W^u_{\xi\, c\mathbf{u}}, \partial W^u_{\xi\, c\mathbf{u}}) \subset (B_c, B^+_c)$ as does $o^u_{\mathcal{F}\xi\, c\mathbf{u}}$ under the homomorphism on homology induced by the inclusion $\dot{k}^u_{c\mathbf{u}}: (W^u_{\mathcal{F}\xi\, c\mathbf{u}}, \partial W^u_{\mathcal{F}\xi\, c\mathbf{u}}) \subset (B_c, B^+_c)$.

This specifies $o^u_{\xi\, c\mathbf{u}}$ uniquely once we know that $\dot{k}^u_{c\mathbf{u}\,*} o^u_{\mathcal{F}\xi\, c\mathbf{u}}$ is not the zero class in $H_P(B_c, B^+_c; R)$. However, the inclusion $\dot{k}^u_{c\mathbf{u}}$ has the factorization

$$(W^u_{\mathcal{F}\xi\, c\mathbf{u}}, \partial W^u_{\mathcal{F}\xi\, c\mathbf{u}}) \overset{j^u_{c\mathbf{u}}}{\hookrightarrow} (W^u_{\mathcal{F}\xi\, c}, \partial W^u_{\mathcal{F}\xi\, c}) \overset{\dot{k}^u_c}{\hookrightarrow} (B_c, B^+_c),$$

and the inclusion $\dot{k}^u_c$ induces isomorphisms on homology as a consequence of Theorem 7.21(9). Thus, $\dot{k}^u_{c\mathbf{u}\,*} o^u_{\mathcal{F}\xi\, c\mathbf{u}}$ is non-zero if, and only if, $j^u_{c\mathbf{u}\,*} o^u_{\mathcal{F}\xi\, c\mathbf{u}}$ is non-zero. Now not only is $j^u_{c\mathbf{u}\,*} o^u_{\mathcal{F}\xi\, c\mathbf{u}}$ non-zero, in fact it generates $H_P(W^u_{\mathcal{F}\xi\, c}, \partial W^u_{\mathcal{F}\xi\, c}; R)$ because $j^u_{c\mathbf{u}}$ induces an isomorphism on homology in dimension P (the proof will be given momentarily). Hence, $\dot{k}^u_{c\mathbf{u}\,*} o^u_{\mathcal{F}\xi\, c\mathbf{u}}$ generates $H_P(B_c, B^+_c; R)$ as desired.

That $j^u_{c\mathbf{u}}$ induces an isomorphism on homology in dimension P is a consequence of the Thom isomorphism theorem and naturality of cap products. Specifically, the diagram

$$\begin{array}{ccccc}
H_P(W^u_{\mathcal{F}\xi\, c}, \partial W^u_{\mathcal{F}\xi\, c}; R) & \xrightarrow{i^{u\,*}_c \tau^u \frown} & H_0(W^u_{\mathcal{F}\xi\, c}; R) & \xrightarrow{p_*} & H_0(S; R) \\
\uparrow{\scriptstyle j^u_{c\mathbf{u}\,*}} & & \uparrow{\scriptstyle j^u_{c\mathbf{u}\,*}} & & \uparrow \\
H_P(W^u_{\mathcal{F}\xi\, c\mathbf{u}}, \partial W^u_{\mathcal{F}\xi\, c\mathbf{u}}; R) & \xrightarrow{i^{u\,*}_{c\mathbf{u}} \tau^u \frown} & H_0(W^u_{\mathcal{F}\xi\, c\mathbf{u}}; R) & \xrightarrow{p_*} & H_0(\{\mathbf{u}\}; R)
\end{array}$$

is commutative because the vertical arrows are all inclusion induced and the arrows in the bottom row are the restrictions of those in the top row. Specifically, the left-hand rectangle commutes by naturality of cap products. Since the inclusions $S \subset W^u_{\mathcal{F}\xi\, c}$ and $\{\mathbf{u}\} \subset W^u_{\mathcal{F}\xi\, c\mathbf{u}}$ are strong deformation retracts, the horizontal arrows in the right-hand rectangle are isomorphisms. As the rightmost vertical arrow is also an isomorphism, it follows that the middle vertical arrow is too. Hence, also $j^u_{c\mathbf{u}*}$ is an isomorphism in dimension P because the rows of the diagram are isomorphisms by the Thom isomorphism theorem; e.g., see [Sp, Theorem 5.7.10]. □

7.29 LEMMA. *Under the assumption of hypotheses* HCO *and conclusions of Theorem 7.21, if τ^u is an R-orientation of the vector bundle N^u_S and if $o_S \in H_1(S;R)$ is a fundamental class, then for $0 < c < c_0$ there exist unique fundamental classes*

$$o^u_{\mathcal{F}\xi\, c} \in H_{P+1}(W^u_{\mathcal{F}\xi\, c}, \partial W^u_{\mathcal{F}\xi\, c}; R) \qquad \text{and} \qquad o^u_{\xi\, c} \in H_{P+1}(W^u_{\xi\, c}, \partial W^u_{\xi\, c}; R)$$

that together satisfy

(1) *under the inclusion induced isomorphisms $o^u_{\mathcal{F}\xi\, c}$ and $o^u_{\xi\, c}$ have common image in $H_{P+1}(B_c, B^+_c; R)$ generating that R-module;*
(2) *$p_*(i^{u*}_c \tau^u \frown o^u_{\mathcal{F}\xi\, c}) = o_S$ where $i^u_c\colon (W^u_{\mathcal{F}\xi\, c}, \partial W^u_{\mathcal{F}\xi\, c}) \to (\mathbf{E}^u_S, \mathbf{E}^u_{S\,0})$ is inclusion.*

An analogous statement holds for an R-orientation τ^s of N^s_S, a fundamental class o_S of S, and for $0 < c < c_0$, the existence of fundamental classes

$$o^s_{\mathcal{F}\xi\, c} \in H_{P+1}(W^s_{\mathcal{F}\xi\, c}, \partial W^s_{\mathcal{F}\xi\, c}; R) \qquad \text{and} \qquad o^s_{\xi\, c} \in H_{P+1}(W^s_{\xi\, c}, \partial W^s_{\xi\, c}; R)$$

satisfying the analogues of conditions (1) and (2) obtained by replacing B^+_c with B^-_c and by replacing the expanding normal bundle with the contracting one.

Proof. The Thom isomorphism theorem establishes that for $0 < c < c_0$,

$$e \mapsto p_*(i^{u*}_c \tau^u \frown e)$$

defines an isomorphism from $H_{P+1}(W^u_{\mathcal{F}\xi\, c}, \partial W^u_{\mathcal{F}\xi\, c}; R)$ onto $H_1(S)$. Choose $o^u_{\mathcal{F}\xi\, c}$ to be the unique element whose image under this isomorphism is o_S—thus, condition (2) is satisfied—whence it follows that $o^u_{\mathcal{F}\xi\, c}$ is a fundamental class for $W^u_{\mathcal{F}\xi\, c}$. As in the proof of the previous lemma, the fundamental class $o^u_{\xi\, c}$ of $W^u_{\xi\, c}$ is defined as the image under the isomorphism on homology induced by the end map of an isotopy from the pair $(W^u_{\mathcal{F}\xi\, c}, \partial W^u_{\mathcal{F}\xi\, c})$ to the pair $(W^u_{\xi\, c}, \partial W^u_{\xi\, c})$ which at each stage of the isotopy has image in the pair (B_c, B^+_c). Where k^u_c and $\dot{k}^u_c$ are the inclusions

$$k^u_c\colon (W^u_{\xi\, c}, \partial W^u_{\xi\, c}) \subset (B_c, B^+_c) \qquad \text{and} \qquad \dot{k}^u_c\colon (W^u_{\mathcal{F}\xi\, c}, \partial W^u_{\mathcal{F}\xi\, c}) \subset (B_c, B^+_c),$$

it follows that in $H_{P+1}(B_c, B^+_c; R)$,

$$\dot{k}^u_{c*} o^u_{\mathcal{F}\xi\, c} = k^u_{c*} o^u_{\xi\, c}.$$

Further, this common image generates because both $\dot{k}^u_c$ and k^u_c induce isomorphisms on homology by Theorem 7.21(9). □

7.30 DEFINITION. Let τ^u and τ^s be R-orientations of N_S^u and N_S^s respectively and let o_S be a fundamental class of S.

(A) The families of fundamental classes $\{o^u_{\mathcal{F}\xi\, c\, \mathbf{u}} : 0 < c < c_0, \mathbf{u} \in S\}$ and $\{o^u_{\xi\, c\, \mathbf{u}} : 0 < c < c_0, \mathbf{u} \in S\}$ determined by τ^u according to the prescription of Lemma 7.28 are called respectively the *compatible family of fundamental classes determined by τ^u for the fibers of the bundles $\omega^u_{\mathcal{F}\xi\, c}$ $(0 < c < c_0)$* and the *compatible family of fundamental classes determined by τ^u for the fibers of the bundles $\omega^u_{\xi\, c}$* $(0 < c < c_0)$. A similar designation is made for the families of fundamental classes $\{o^s_{\mathcal{F}\xi\, c\, \mathbf{u}} : 0 < c < c_0, \mathbf{u} \in S\}$ and $\{o^s_{\xi\, c\, \mathbf{u}} : 0 < c < c_0, \mathbf{u} \in S\}$ determined by τ^s.

(B) The families of fundamental classes $\{o^u_{\mathcal{F}\xi\, c}\}_{0<c<c_0}$ and $\{o^u_{\xi\, c}\}_{0<c<c_0}$ determined by τ^u and o_S according to the prescription of Lemma 7.29 are called respectively the *compatible family of fundamental classes determined by τ^u and o_S for the family of submanifolds $\mathcal{W}^u_{\mathcal{F}\xi} := \{W^u_{\mathcal{F}\xi\, c}\}_{0<c<c_0}$* and the *compatible family of fundamental classes determined by τ^u and o_S for the family of submanifolds $\mathcal{W}^u_{\xi} := \{W^u_{\xi\, c}\}_{0<c<c_0}$*. A similar designation is made for the family of fundamental classes $\{o^s_{\mathcal{F}\xi\, c}\}_{0<c<c_0}$ for the family of submanifolds $\mathcal{W}^s_{\mathcal{F}\xi} := \{W^s_{\mathcal{F}\xi\, c}\}_{0<c<c_0}$ and for the family of fundamental classes $\{o^s_{\xi\, c}\}_{0<c<c_0}$ for the family of submanifolds $\mathcal{W}^s_{\xi} := \{W^s_{\xi\, c}\}_{0<c<c_0}$.

7.31 THEOREM. *Under the assumption of the hypotheses* HCO *and the conclusions of Theorem 7.21 the following hold:*

(1) *if N^u_S is R-orientable, then*
  (a) *$\widetilde{H}_k \mathcal{C}(S;R) \simeq 0$ for $k \neq P, P+1$;*
  (b) *$\widetilde{H}_P \mathcal{C}(S;R) \simeq R$, and in particular an R-orientation τ^u of N^u_S determines a unique generator $\mathbf{o}^u_{\xi\, P} \in \widetilde{H}_P \mathcal{C}(S;R)$ characterized by the property that it is the common canonical image of all members of the compatible family of fundamental classes determined by τ^u for the fibers of the bundles $\omega^u_{\xi\, c}$ $(0 < c < c_0)$;*
  (c) *$\widetilde{H}_{P+1} \mathcal{C}(S;R) \simeq R$, and in particular an R-orientation τ^u of N^u_S and a fundamental class o_S of S together determine a unique generator $\mathbf{o}^u_{\xi\, P+1} \in \widetilde{H}_{P+1} \mathcal{C}(S;R)$ characterized by the property that it is the common canonical image of all members of the compatible family of fundamental classes determined by τ^u and o_S for the family of submanifolds $\mathcal{W}^u_\xi$;*

(2) *if N^s_S is R-orientable, then*
  (a) *$\widetilde{H}_k \mathcal{C}^*(S;R) \simeq 0$ for $k \neq N, N+1$;*
  (b) *$\widetilde{H}_N \mathcal{C}^*(S;R) \simeq R$, and in particular an R-orientation τ^s of N^s_S determines a unique generator $\mathbf{o}^s_{\xi\, N} \in \widetilde{H}_N \mathcal{C}^*(S;R)$ characterized by the property that it is the common canonical image of all members of the compatible family of fundamental classes determined by τ^s for the fibers of the bundles $\omega^s_{\xi\, c}$ $(0 < c < c_0)$;*
  (c) *$\widetilde{H}_{N+1} \mathcal{C}^*(S;R) \simeq R$, and in particular an R-orientation τ^s of N^s_S and a fundamental class o_S of S together determine a unique generator $\mathbf{o}^s_{\xi\, N+1} \in \widetilde{H}_{N+1} \mathcal{C}^*(S;R)$ characterized by the property that it is the common canonical image of all members of the compatible family of*

fundamental classes determined by τ^s *and* o_S *for the family of submanifolds* $\mathcal{W}^s_\xi$;

(3) *given* R*-orientations* τ^u *of* N^u_S *and* τ^s *of* N^s_S *and given a fundamental class* o_S *of* S*, then where* $\mathbf{o}^u_{\xi\,P}$ *and* $\mathbf{o}^u_{\xi\,P+1}$ *are determined by* τ^u *and* o_S *as in* (1) *and where* $\mathbf{o}^s_{\xi\,N}$ *and* $\mathbf{o}^s_{\xi\,N+1}$ *are determined by* τ^s *and* o_S *as in* (2),

$$^{\#}(\mathbf{o}^u_{\xi\,P+1} \mathbin{\widehat{\bullet}} \mathbf{o}^s_{\xi\,N}) = \sigma\,, \qquad ^{\#}(\mathbf{o}^u_{\xi\,P} \mathbin{\widehat{\bullet}} \mathbf{o}^s_{\xi\,N+1}) = (-1)^P\sigma\,,$$

and

$$\mathbf{o}^u_{\xi\,P+1} \mathbin{\widehat{\bullet}} \mathbf{o}^s_{\xi\,N+1} = \sigma\, o_S\,.$$

Proof. The notation established for the conclusions and proof of Theorem 7.21, Lemmas 7.28 and 7.29, and Definition 7.30 will be used throughout this proof.

Let τ^u and τ^s be R-orientations of N^u_S and N^s_S respectively, let $o_S \in H_1(S;R)$ be a fundamental class, and temporarily fix c, $0 < c < c_0$.

Application of the Thom isomorphism theorem to the P-disk bundle $\omega^u_{\mathcal{F}\xi\,c}$ over S yields

$$H_k(W^u_{\mathcal{F}\xi\,c}, \partial W^u_{\mathcal{F}\xi\,c}; R) \simeq 0 \qquad \text{for } k \neq P, P+1. \tag{7.14}$$

There are two consequences of statement (7.14) with reference to the homology of $\mathcal{C}(S)$ of which the second leads to the definition of the sought for generator $\mathbf{o}^u_{\xi\,P}$. First, Theorem 7.21(8) implies that $H_k(B_c, B_c^+;R) \simeq 0$ for the same range of indices as in (7.14). Hence, also

$$\widetilde{H}_k\mathcal{C}(S;R) \simeq 0\,, \qquad \text{for } k \neq P, P+1.$$

Second, since in particular (7.14) holds for $k = P-1$, by the universal coefficient theorem $H^P(W^u_{\mathcal{F}\xi\,c}, \partial W^u_{\mathcal{F}\xi\,c}; R) \simeq \operatorname{Hom}(H_P(W^u_{\mathcal{F}\xi\,c}, \partial W^u_{\mathcal{F}\xi\,c}; R); R)$, whence there is a unique element $o^u_{P\,c} \in H_P(W^u_{\mathcal{F}\xi\,c}, \partial W^u_{\mathcal{F}\xi\,c}; R)$ satisfying

$$\langle i^{u\,*}_c \tau^u, o^u_{P\,c}\rangle = 1\,.$$

This definition of $o^u_{P\,c}$, the factorization $i^u_{c\mathbf{u}} = i^u_c \circ j^u_{c\mathbf{u}}$, and the characterization of $o^u_{\mathcal{F}\xi\,c\mathbf{u}}$ in statement (2) of Lemma 7.28 imply that

$$j^u_{c\mathbf{u}*} o^u_{\mathcal{F}\xi\,c\mathbf{u}} = o^u_{P\,c} \qquad \text{for each } \mathbf{u} \in S.$$

This equality, the factorization $\dot{k}^u_{c\mathbf{u}} = \dot{k}^u_c \circ j^u_{c\mathbf{u}}$, and statement (1) of Lemma 7.28 imply that the canonical image of $o^u_{\xi\,c\mathbf{u}}$ in $\widetilde{H}_P\mathcal{C}(S;R)$ is independent of $\mathbf{u} \in S$ and cyclically generates $\widetilde{H}_P\mathcal{C}(S;R)$ over R; i.e., where $\bar{b}^+_{c*}\colon H_*(B_c, B_c^+;R) \to \widetilde{H}_*\mathcal{C}(S;R)$ is the canonical isomorphism and where

$$\mathbf{o}^u_{\xi\,P\,c} := \bar{b}^+_{c*}\dot{k}^u_{c*} o^u_{P\,c} = \bar{b}^+_{c*} k^u_{c\mathbf{u}*} o^u_{\xi\,c\mathbf{u}}\,,$$

$r \mapsto r\mathbf{o}^u_{\xi\,P\,c}$ defines an isomorphism $R \simeq \widetilde{H}_P\mathcal{C}(S;R)$. Note that $\mathbf{o}^u_{\xi\,P\,c}$ can be characterized as the unique generator of $\widetilde{H}_P\mathcal{C}(S;R)$ that is the canonical image of the Kronecker dual of $i^{u\,*}_c\tau^u$.

In a similar manner, one obtains a generator $\mathbf{o}^s_{\xi\,N\,c}$ of $\widetilde{H}_N\mathcal{C}^*(S;R)$ that can be characterized as the unique generator of that module that is the canonical image of the Kronecker dual of $i^{s\,*}_c\tau^s$ where i^s_c is the inclusion of pairs $(W^s_{\mathcal{F}\xi\,c}, \partial W^s_{\mathcal{F}\xi\,c}) \subset (\mathbf{E}^s_S, \mathbf{E}^s_{S\,0})$.

For $\nu = u, s$, for $i = P$ when $\nu = u$ but for $i = N$ when $\nu = s$, let $o^\nu_{\xi\,c} \in H_{i+1}(W^\nu_{\xi\,c}, \partial W^\nu_{\xi\,c}; R)$ be the fundamental class determined by τ^ν and o_S according to the prescription of Lemma 7.29. Thus, $k^u_{c\,*}o^u_{\xi\,c}$ generates $H_{P+1}(B_c, B^+_c; R)$ and $k^s_{c\,*}o^s_{\xi\,c}$ generates $H_{N+1}(B_c, B^-_c; R)$. It follows that $\mathbf{o}^u_{\xi\,P+1\,c} := \bar{b}^+_{c\,*}k^u_{c\,*}o^u_{\xi\,c}$ generates $\widetilde{H}_{P+1}\mathcal{C}(S;R)$ and that $\mathbf{o}^s_{\xi\,N+1\,c} := \bar{b}^-_{c\,*}k^s_{c\,*}o^s_{\xi\,N+1\,c}$ generates $\widetilde{H}_{N+1}\mathcal{C}^*(S;R)$.

The generating classes $\mathbf{o}^u_{\xi\,P\,c}$, $\mathbf{o}^u_{\xi\,P+1\,c}$, $\mathbf{o}^s_{\xi\,N\,c}$, and $\mathbf{o}^s_{\xi\,N+1\,c}$ are, in fact, independent of our choice of $c \in\,]0, c_0[$. The proof of this is in the main analogous to the proof of independence of c given in the course of proving Theorem 7.15(2). There are two main structural elements in that proof: the first of these is diagram (7.7); the second is the family of isotopies $H^{c,\bar{c}}_t$ from $W^u_{d\xi\,c}$ to $W^u_{d\xi\,\bar{c}}$ defined for $0 < c, \bar{c} < c_0$ using the flow of $d\xi$. The analogues of these structural elements in the current situation are obtained as follows.

Diagram (7.7) is used as is, but the blocks $B_{c'}$ and $B_{c''}$ where $0 < c' < c'' < c_0$ are now those provided by Theorem 7.21 rather than those provided by Theorem 7.12, and the set $CB^+_{c''}$ is now defined by Lemma 7.22. By that lemma $(B_{c''}, CB^+_{c''})$ is a regular index pair relative to the flows generated by ξ, $\mathcal{F}\xi$, and $\mathcal{F}^\perp\xi$, and the inclusion $j_\lambda\colon (B_\lambda, B^+_\lambda) \to (B_{c''}, CB^+_{c''})$ is defined for $c' \le \lambda \le c''$. As in the proof of Theorem 7.15(2), all arrows in the diagram except the left vertical arrow are the homotopy classes of either inclusions, inclusion induced maps on quotients, or quotient maps. Similarly, to define the left vertical arrow we first must define a family of isotopies $H^{c\,c'}$ from $W^u_{\mathcal{F}\xi\,c}$ to $W^u_{\mathcal{F}\xi\,c'}$ for $0 < c, c' < c_0$.

We proceed to define an isotopy from $W^u_{\mathcal{F}\xi\,c'}$ to $W^u_{\mathcal{F}\xi\,c''}$ whenever $0 < c' \le c'' < c_0$ making use of the fact that $W^u_{\mathcal{F}\xi\,c} = W^u_{\mathcal{F}^\perp\xi\,c}$ for $0 < c < c_0$. In particular, the isotopy is defined using the flow generated by $\mathcal{F}^\perp\xi$ which will be denoted by the map $(\mathbf{u}, t) \mapsto \mathbf{u} * t$, and all exit time maps are computed relative to this flow. For $0 \le t \le 1$ define

$$H^{c'\,c''}_t(\mathbf{u}) := \begin{cases} \mathbf{u} & \text{if } \mathbf{u} \in S \\ \mathbf{u} * t\tau_{B_{c''}}(\mathbf{u} * \tau_{B_{c'}}(\mathbf{u})) & \text{if } \mathbf{u} \in W^u_{\mathcal{F}\xi\,c'} \setminus S. \end{cases}$$

The proof that each $H^{c'\,c''}$ is continuous and defines the required isotopy is essentially the same as in the proof of Theorem 7.15. Note that continuity could *not* be proved using the flow of $\mathcal{F}\xi$ instead of $\mathcal{F}^\perp\xi$.

To define the left vertical arrow $[h]$ in diagram (7.7) in the current context, note that as each point of the tubular neighborhood U lies in a unique fiber of the normal bundle as realized in the tubular neighborhood there are continuous projections $\pi_\nu\colon U \to \mathbf{E}^\nu_S$ for $\nu = u, s$. We write $\mathbf{u} = \mathbf{u}_1 \oplus \mathbf{u}_2$ for the unique $\mathbf{u} \in U$ for which $\mathbf{u}_1 = \pi_u(\mathbf{u})$ and $\mathbf{u}_2 = \pi_s(\mathbf{u})$ whenever $\mathbf{u}_1 \in \mathbf{E}^u_S$, $\mathbf{u}_2 \in \mathbf{E}^s_S$ and $p(\mathbf{u}_1) = p(\mathbf{u}_2)$. Also, note that the restriction of π_ν to B_c has image in $W^\nu_{\mathcal{F}\xi\,c}$ for $\nu = u, s$, for $0 < c < c_0$. Let $\ell_{c'\,c''}$ be the inclusion $\pi_s(B_{c'}) \subset \pi_s(B_{c''})$. Then define $h\colon (B_{c'}, B^+_{c'}) \to (B_{c''}, B^+_{c''})$ by

$$h(\mathbf{u}) = H^{c'\,c''}_1 \circ \pi_u(\mathbf{u}) \oplus \ell_{c'\,c''} \circ \pi_s(\mathbf{u}).$$

The left-hand vertical arrow in diagram (7.7) is $[h]$, the homotopy class of the map h. Note that $j_{c''} \circ h$ is homotopic to $j_{c'}$ via the homotopy $H_t^{c'\,c''} \circ \pi_u \oplus \ell_{c'\,c''} \circ \pi_s$, whence it follows that diagram (7.7) is commutative in the current context.

The proof that $\mathbf{o}^u_{\xi\,P\,c}$ is in fact independent of c is now left to the reader since it is entirely analogous to the computations carried out in the proof of Theorem 7.15(2). Therefore, the canonical image of $o^u_{P\,c}$ is independent of $c \in \,]0, c_0[$, denote it $\mathbf{o}^u_{\xi\,P}$, and generates $\widetilde{H}_P\mathcal{C}(S;R)$. That is, the canonical image of $o^u_{\xi\,c\,\mathbf{u}}$ in $\widetilde{H}_P\mathcal{C}(S;R)$ is independent of $c \in \,]0, c_0[$ and $\mathbf{u} \in S$. In a similar manner, one shows that the canonical image of the class $o^s_{\xi\,c\,\mathbf{u}}$ in $\widetilde{H}_N\mathcal{C}^*(S;R)$ is independent of $c \in \,]0, c_0[$ and $\mathbf{u} \in S$ and generates $\widetilde{H}_N\mathcal{C}^*(S;R)$; let $\mathbf{o}^s_{\xi\,N}$ denote this canonical image.

The proof that $\mathbf{o}^u_{\xi\,P+1\,c}$ is independent of c is also similar to the proof of Theorem 7.15(2); however, we make one additional observation on the classes $o^u_{\mathcal{F}\xi\,c}$, $0 < c < c_0$, before leaving the additional details to the reader. Specifically, to carry out the proof of independence in a manner analogous to that used in the proof of Theorem 7.15 it must be shown that $H^{c\,\bar{c}}_{1\,*} o^u_{\mathcal{F}\xi\,c} = o^u_{\mathcal{F}\xi\,\bar{c}}$. This follows from the definitions of $o^u_{\mathcal{F}\xi\,c}$ and $o^u_{\mathcal{F}\xi\,\bar{c}}$ in terms of the Thom isomorphism and the classes τ^u and o_S according to the prescription of Lemma 7.29 as well as naturality of cap products, the fact that i^u_c is homotopic to $i^u_{\bar{c}} \circ H^{c\,\bar{c}}_1$, and the fact that $H^{c\,\bar{c}}_1$ is the identity on S. These details and the remaining details of the proof of independence of c are left as an exercise for the reader. Once the exercise is carried out, it follows that the canonical image of $o^u_{\xi\,c}$ in $\widetilde{H}_{P+1}\mathcal{C}(S;R)$ is independent of $c \in \,]0, c_0[$, call the canonical image $\mathbf{o}^u_{\xi\,P+1}$, and generates $\widetilde{H}_{P+1}\mathcal{C}(S;R)$. In a similar manner, one shows that the canonical image of the class $o^s_{\xi\,c}$ in $\widetilde{H}_{N+1}\mathcal{C}^*(S;R)$ is independent of $c \in \,]0, c_0[$, call the canonical image $\mathbf{o}^s_{\xi\,N+1}$, and generates $\widetilde{H}_{N+1}\mathcal{C}^*(S;R)$.

Finally, statement (3) holds as an immediate consequence of Lemma 7.7 using the flow of $\mathcal{F}\xi$. □

7.32 The Proof of Theorem 7.6. Under the assumption of hypotheses HCO, conclusions (1)–(6) of Theorem 7.4 follow from the earlier results of this subsection: Conclusion (1) from Theorem 7.21(1), conclusion (2) (and also conclusions (1) and (2) of Theorem 7.6) from Theorem 7.23, and conclusions (3)–(6) (and therefore also conclusion (3) of Theorem 7.6) are restatements of the results of Theorem 7.31 which in the case of conclusion (6) also requires using conclusion (8) of Theorem 7.21. □

CHAPTER 8

PRODUCTS OF INTERSECTION PAIRINGS

For a product of flows on manifolds without boundary, the intersection class pairing on the Conley indices of an isolated invariant set of the product flow is, up to sign, the homology cross-product of the intersection class pairings on the Conley indices of the factor isolated invariant sets when the homology modules of the Conley indices of the factor isolated invariant sets are torsion free relative to the coefficient ring. The analogous result for the intersection number pairing is obtained as a corollary.

Preparatory to these results, we will (1) establish that the Conley index of a product invariant set is up to equivalence of connected simple systems the smash product of the Conley indices of its factors, (2) show that the usual reduced homology cross product (i.e., the homology smash product) is defined for homology classes in the Conley indices of the factors and using (1) yields a class in the homology of the Conley index of the product invariant set, and (3) using (2), prove a Kunneth theorem expressing the graded homology module of the Conley index of a product isolated invariant set in terms of the graded homology modules of the Conley indices of the factor isolated invariant sets.

A. Preliminary Observations and Definitions. For the rest of Chapter 8, S will be an isolated invariant set of the flow on the manifold M and S' will be an isolated invariant set of a flow on a manifold M'. The flow on M' is assumed defined on an open set $\mathfrak{G}'$, $M' \times \{0\} \subset \mathfrak{G}' \subset M' \times \mathbf{R}$, and is given by $(\mathbf{v}, t) \mapsto \mathbf{v} \cdot' t$ for $(\mathbf{v}, t) \in \mathfrak{G}'$. Then there is a product flow on $M \times M'$ given by $\big((\mathbf{u}, \mathbf{v}), t\big) \mapsto (\mathbf{u}, \mathbf{v}) \cdot t := (\mathbf{u} \cdot t, \mathbf{v} \cdot' t)$ defined on the open set $\mathfrak{G}\mathfrak{G}' := \{\big((\mathbf{u}, \mathbf{v}), t\big) : (\mathbf{u}, t) \in \mathfrak{G}$ and $(\mathbf{v}, t) \in \mathfrak{G}'\}$. Also $\langle N_1, N_0\rangle$ and $\langle N_1', N_0'\rangle$ are assumed to be index pairs for S and S', respectively.

In the course of establishing (1) and (2) above, it is convenient to have available the following observation and the definitions immediately following.

8.1 PROPOSITION. *For $\mathcal{X}$ and $\mathcal{Y}$ categories, each functor $F: \mathcal{X} \to \mathcal{Y}$ induces a functor $F_{\mathcal{CSS}}: \mathcal{CSS}(\mathcal{X}) \to \mathcal{CSS}(\mathcal{Y})$ of the same variance as F.*

Proof. To map an object of $\mathcal{CSS}(\mathcal{X})$ to one of $\mathcal{CSS}(\mathcal{Y})$, just apply F to the objects and morphisms of the connected simple system. Similarly, F applied to the diagrams in a map between connected simple systems in $\mathcal{CSS}(\mathcal{X})$ yields one in $\mathcal{CSS}(\mathcal{Y})$ with the variance of $F_{\mathcal{CSS}}$ clearly determined by that of F. □

8.2 DEFINITION. (A) Let $\varinjlim$ denote the functor on $\mathcal{CSS}(\mathcal{G}_R\mathcal{M})$ to $\mathcal{G}_R\mathcal{M}$ obtained by applying Theorem 5.2 to the identity functor on $\mathcal{G}_R\mathcal{M}$. The notation is appropriate because where F is any reduced homology theory on $\mathcal{T}^{*\prime}$, if

$\mathcal{A} \in \operatorname{ob}(\mathcal{CSS}(\mathcal{T}^{*\prime}))$, then $F_{\mathcal{CSS}}(\mathcal{A}) \in \mathcal{CSS}(\mathcal{G}_R\mathcal{M})$ and obviously $\varinjlim F_{\mathcal{CSS}}(\mathcal{A}) = F(\mathcal{A})$ where $F(\mathcal{A})$ is as defined in Definition 1.5(*A*). That is, $\varinjlim F_{\mathcal{CSS}}(\mathcal{A}) := \varinjlim F(A)$ where the limit on the right is over $A \in \mathcal{A}$.

(*B*) Let $\mathcal{A}$, $\mathcal{A}' \in \operatorname{ob}(\mathcal{CSS}(\mathcal{T}^{*\prime}))$. Then $\mathcal{A} \times \mathcal{A}'$ is a connected simple system in $\mathcal{CSS}(\mathcal{T}^{*\prime} \times \mathcal{T}^{*\prime})$. Thus, where $\vee$ and $\wedge$ denote the usual wedge sum and smash product functors on $\mathcal{T}^{*\prime} \times \mathcal{T}^{*\prime}$, Proposition 8.1 yields connected simple systems $\mathcal{A} \vee \mathcal{A}'$ and $\mathcal{A} \wedge \mathcal{A}'$ in $\mathcal{CSS}(\mathcal{T}^{*\prime})$ called respectively the *wedge sum* and *smash product* of $\mathcal{A}$ and $\mathcal{A}'$.

(*C*) For an *R*-module *G*, let $\widetilde{\mathcal{H}}_*(\,\cdot\,; G)$ be the functor on $\mathcal{CSS}(\mathcal{T}^{*\prime})$ to $\mathcal{CSS}(\mathcal{G}_R\mathcal{M})$ induced by $H_*(\,\cdot\,; G)$ according to the prescription of Proposition 8.1. Of particular interest below are the connected simple systems of graded *R*-modules $\widetilde{\mathcal{H}}_*\mathcal{C}(S; G)$, $\widetilde{\mathcal{H}}_*(\mathcal{C}(S) \wedge \mathcal{C}(S'); G)$, and $\widetilde{\mathcal{H}}_*\mathcal{C}(S \times S'; G)$. The last of these is well-defined because $S \times S'$ is an isolated invariant set of the product flow on $M \times M'$; see Proposition 8.3 below.

(*D*) Let *G* and *G′* be *R*-modules and let $\widetilde{\mathcal{H}}_*\mathcal{C}(S; G) \otimes \widetilde{\mathcal{H}}_*\mathcal{C}(S'; G')$ denote the small category each of whose objects is of the form $\widetilde{H}_*(X; G) \otimes \widetilde{H}_*(X'; G')$ where X and X' are index spaces for S and S', respectively, and for X and Y index spaces for S and X' and Y' index spaces for S', let $h^{X\,Y}_{S\,*} \otimes h^{X'\,Y'}_{S'\,*}$ be the unique morphism from $\widetilde{H}_*(X; G) \otimes \widetilde{H}_*(X'; G')$ to $\widetilde{H}_*(Y; G) \otimes \widetilde{H}_*(Y'; G')$ in $\widetilde{\mathcal{H}}_*\mathcal{C}(S; G) \otimes \widetilde{\mathcal{H}}_*\mathcal{C}(S'; G')$. Then $\widetilde{\mathcal{H}}_*\mathcal{C}(S; G) \otimes \widetilde{\mathcal{H}}_*\mathcal{C}(S'; G')$ is a connected simple system in $\mathcal{CSS}(\mathcal{G}_R\mathcal{M})$ since the tensor product of two *R*-module isomorphisms is again an isomorphism—see [Sp, Lemmas 5.1.4 and 5.1.5]. Note that $\varinjlim(\widetilde{\mathcal{H}}_*\mathcal{C}(S; G) \otimes \widetilde{\mathcal{H}}_*\mathcal{C}(S'; G')) = \widetilde{H}_*\mathcal{C}(S; G) \otimes \widetilde{H}_*\mathcal{C}(S'; G')$ because direct limits commute with tensor products.

B. Conley Indices of Product Invariant Sets. Let us observe that if (X, Y) and (X', Y') are topological pairs, then there is a natural homeomorphism of pointed spaces

$$e\colon (X/Y \wedge X'/Y', *) \simeq \left(\frac{X \times X'}{Y \times X' \cup X \times Y'}, *\right)$$

that is an excision; i.e., removing the basepoint from both spaces yields the same set, viz., $X \setminus Y \times X \setminus Y'$. Thus both spaces are one-point compactifications of the same set. This observation is used to establish the natural homeomorphism in the following proposition, a first step in showing (1), and is left as an easy exercise for the reader. Also see [C, §III.6].

8.3 Proposition. *A subset of $M \times M'$ is an isolated invariant set of the product flow if, and only if, it is the Cartesian product of two isolated invariant sets, one from each factor flow. In particular, $S \times S'$ is an isolated invariant set of the product flow and has $\langle N_1 \times N_1', N_0 \times N_1' \cup N_1 \times N_0' \rangle$ as index pair. The index space of this index pair is naturally homeomorphic to the smash product $N_1/N_0 \wedge N_1'/N_0'$ via an excision map. Further, this index pair is regular or has thick exit set if both $\langle N_1, N_0 \rangle$ and $\langle N_1', N_0' \rangle$ are regular or if both have thick exit set.*

Thus in the context of the product flow on $M \times M'$, it makes sense to define the product of the index pairs $\langle N_1, N_0 \rangle$ and $\langle N_1', N_0' \rangle$ by

$$\langle N_1, N_0 \rangle \times \langle N_1', N_0' \rangle := \langle N_1 \times N_1', N_0 \times N_1' \cup N_1 \times N_0' \rangle\,.$$

Note that if both index pairs are nested, then this product coincides with the usual notion of product in the category of topological pairs. To simplify notation in what follows let us assume without loss of generality that both index pairs are nested; for if not replace $\langle N_1, N_0\rangle$ with $(N_1, \cap N_0)$ and $\langle N_1', N_0'\rangle$ with $(N_1', \cap N_0')$.

It is immediate from Proposition 8.3 that the homotopy index (i.e., the homotopy type of any index space) of a product of isolated invariant sets is the smash product of the homotopy indices. This carries over to the full Conley index; i.e., the Conley index of the product is equivalent to the smash product of the indices. *This last statement requires proof,* albeit minimal, because the connected simple systems $\mathcal{C}(S) \wedge \mathcal{C}(S')$ and $\mathcal{C}(S \times S')$ are generally distinct; i.e., in general there are index pairs for $S \times S'$ that are not the product of an index pair for S with one for S'. For example, the product flow on $\mathbf{R}^2$ given by $(x, y) \cdot t := (xe^t, ye^{-t})$ has the origin as isolated invariant set, and any closed disk D_r of radius r centered at the origin is an isolating block for the origin with exit set D_r^+ consisting of those points on the bounding circle of the disk admitting polar coordinates (r, θ) where θ lies either in the interval $[-\pi/4, \pi/4]$ or in the interval $[3\pi/4, 5\pi/4]$. Clearly, the regular index pair (D_r, D_r^+) is not the product of two index pairs for the factor flows; hence, its index space is not homeomorphic to the smash product of any factor index spaces via an excision of the type described above.

8.4 PROPOSITION. *$\mathcal{C}(S) \wedge \mathcal{C}(S')$ is naturally equivalent to $\mathcal{C}(S \times S')$ relative to the morphisms of $\mathcal{CSS}(\mathcal{T}^{*\prime})$, whence $\widetilde{H}_*(\mathcal{C}(S) \wedge \mathcal{C}(S'); G) \simeq \widetilde{H}_*\mathcal{C}(S \times S'; G)$.*

Proof. Define $E: \mathcal{C}(S) \wedge \mathcal{C}(S') \to \mathcal{C}(S \times S')$ a map of connected simple systems as follows. Set $A := N_1/N_0 \wedge N_1'/N_0'$. Then with e the natural homeomorphism of Proposition 8.3, $e(A)$ is an index space in $\mathcal{C}(S \times S')$; hence, the assignment $E_A^{e(A)} := e$ defines a map of connected simple systems $E: \mathcal{C}(S) \wedge \mathcal{C}(S') \to \mathcal{C}(S \times S')$ via the observation used to get equality (5.2) above. This definition is independent of the choice of index spaces N_1/N_0 and N_1'/N_0' because e is natural and makes E an equivalence in $\mathcal{CSS}(\mathcal{T}^{*\prime})$ because e is a homeomorphism. Hence Theorem 5.2 applied to E yields the desired isomorphism on homology. □

8.4.1 Remark. In a similar manner, when S and S' are disjoint isolated invariant sets of the same flow, one can define an equivalence $\mathcal{C}(S \cup S') \simeq \mathcal{C}(S) \vee \mathcal{C}(S')$ using any nested index pair for $S \cup S'$ of the form $(N_1 \cup N_1', N_0 \cup N_0')$ where (N_1, N_0) and (N_1', N_0') are nested index pairs for S and S' respectively with N and N' disjoint. Since $\widetilde{H}_*(N_1/N_0 \vee N_1'/N_0') \simeq \widetilde{H}_*(N_1/N_0) \oplus \widetilde{H}_*(N_1'/N_0')$ and as direct limits commute with direct sums, the factorization $\widetilde{H}_* = \varinjlim \circ \mathcal{H}_*$ then shows $\widetilde{H}_*\mathcal{C}(S \cup S') \simeq \widetilde{H}_*\mathcal{C}(S) \oplus \widetilde{H}_*\mathcal{C}(S')$.

C. A Kunneth Theorem for Conley Indices. The usual exterior product for singular homology, denoted above by $\times$, induces an exterior product on reduced singular homology for pointed spaces, denoted below by $\underline{\wedge}$, as follows. For (X, x_0) and (Y, y_0) pointed spaces and G and G' R-modules, define

$$\underline{\wedge}: \widetilde{H}_*(X; G) \otimes \widetilde{H}_*(Y; G') \to \widetilde{H}_*(X \wedge Y; G \otimes G') \tag{8.1a}$$

as the composition

(8.1b)

$$\widetilde{H}_*(X;G)\otimes\widetilde{H}_*(Y;G') \equiv H_*(X,x_0'G)\otimes H_*(Y,y_0;G') \xrightarrow{\times} H_*(X\times Y, X\vee Y; G\otimes G')$$
$$\xrightarrow{Q_*} H_*(X\wedge Y,*;G\otimes G') \equiv \widetilde{H}_*(X\wedge Y;G\otimes G')$$

where $Q\colon (X\times Y, X\vee Y)\to(X\wedge Y,*)$ is the quotient map which is natural relative to morphisms $f\times g$ in $\mathcal{T}^*$ or $\mathcal{T}^{*\prime}$. The product $\triangle$ is natural; i.e., if $f\colon (X,x_0)\to(X',x_0')$ and $g\colon (Y,y_0)\to(y',y_0')$ are continuous maps of pointed spaces there is the commutative diagram (coefficients suppressed)

(8.2)
$$\begin{array}{ccc} \widetilde{H}_*(X)\otimes\widetilde{H}_*(Y) & \xrightarrow{\triangle} & \widetilde{H}_*(X\wedge Y) \\ {\scriptstyle f_*\otimes g_*}\downarrow & & \downarrow{\scriptstyle (f\wedge g)_*} \\ \widetilde{H}_*(X')\otimes\widetilde{H}_*(Y') & \xrightarrow{\triangle} & \widetilde{H}_*(X'\wedge Y'). \end{array}$$

The commutativity of this diagram is an immediate consequence of the naturality of the usual exterior product for singular homology, i.e., $\times\circ f_*\otimes g_* = (f\times g)_*\circ\times$, and the equality $(f\wedge g)\circ Q^{X\times Y} = Q^{X'\times Y'}\circ(f\times g)$ at the space level where $Q^{X\times Y}\colon (X\times Y, X\vee Y)\to(X\times Y,*)$ and $Q^{X'\times Y'}\colon (X'\times Y, X'\vee Y')\to(X'\wedge Y',*)$ are the quotient maps.

In general, the homomorphism Q_* in (8.1b) need not be one-to-one so information can be lost in passing from the product $\times$ to the product $\triangle$. However, when X and Y are index spaces this is not the case as a consequence of the following.

8.5 Proposition. *For index pairs $\langle N_1,N_0\rangle$ and $\langle N_1',N_0'\rangle$, the quotient map*

$$Q\colon (N_1/N_0\times N_1'/N_0', N_1/N_0\vee N_1'/N_0')\to(N_1/N_0\wedge N_1'/N_0',*)$$

induces isomorphisms on homology for any homology theory.

Proof. Choose $t>0$ large enough so that $U := N_0^{-t}\cap N_1/N_0$ and $U' := N_0'^{-t}\cap N_1'/N_0'$ are closed neighborhoods of $[N_0]$ and $[N_0']$, respectively, in N_1/N_0 and N_1'/N_0', respectively. Let $I:=[0,1]$ and define $f^t\colon N_1/N_0\times I\to N_1/N_0$ by

$$f^t([\mathbf{u}],s) := \begin{cases} [\mathbf{u}\cdot st] & \text{if } \mathbf{u}\cdot[0,st]\subset N_1\setminus N_0, \\ [N_0] & \text{otherwise;} \end{cases}$$

and define $f'^t\colon N_1'/N_0'\times I\to N_1'/N_0'$ analogously. Then f^t and f'^t are continuous basepoint preserving deformations; see [K1, Proposition 3.1] or [C, §III.4.1.B]. Further, the restrictions $D:=f^t|U\times I$ and $D':=f'^t|U'\times I$ contract U and U' to their respective basepoints. Also, by normality, we can choose Urysohn functions $\phi\colon N_1/N_0\to I$ and $\phi'\colon N_1'/N_0'\to I$ so that $\phi([N_0]) = 1 = \phi'([N_0'])$ and $\phi(N_1/N_0\setminus U) = 0 = \phi'(N_1'/N_0'\setminus U')$. It follows from the Corollary to Lemma 3.1 of [K2] that the quotient map (of unpointed spaces)

$$N_1/N_0\times N_1'/N_0'\cup C(N_1/N_0\vee N_1'/N_0')\to N_1/N_0\wedge N_1'/N_0'$$

is a homotopy equivalence. Hence, as in the proof of Proposition 4.2, the Five Lemma implies that the quotient map Q induces isomorphisms on homology. □

To simplify the printing of Proposition 8.6 and Theorem 8.7 below, coefficient modules are suppressed whenever they can be inferred from context. Also, to simplify the statement and proof of the proposition, set $X := (N_1, N_0)$, set $X' := (N_1', N_0')$, set $\overline{X \times X'} := (N_1 \times N_1')/(N_0 \times N_1' \cup N_1 \times N_0')$ and let $n^\times : X \times X' \to (\overline{X \times X'}, *)$ be the quotient map.

8.6 PROPOSITION. *The product* $\triangle$ *of* (8.1) *induces a product*

$$\triangle : \widetilde{H}_* \mathcal{C}(S; G) \otimes \widetilde{H}_* \mathcal{C}(S'; G') \to \widetilde{H}_* \mathcal{C}(S \times S'; G \otimes G') . \tag{8.3}$$

Further, if $\alpha \in H_*(N_1, N_0; G)$ *represents* $\boldsymbol{\alpha} \in \widetilde{H}_*\mathcal{C}(S;G)$ *and* $\alpha' \in H_*(N_1', N_0'; G')$ *represents* $\boldsymbol{\alpha}' \in \widetilde{H}_*\mathcal{C}(S';G')$, *then* $\alpha \times \alpha'$ *represents* $\boldsymbol{\alpha} \triangle \boldsymbol{\alpha}'$; *i.e.,* $n^\times_*(\alpha \times \alpha') \in \widetilde{H}_*(\overline{X \times X'}; G \otimes G')$ *has equivalence class* $\boldsymbol{\alpha} \triangle \boldsymbol{\alpha}' \in \widetilde{H}_*\mathcal{C}(S \times S'; G \otimes G')$.

Proof. Let $\widetilde{\mathcal{H}}_*\mathcal{C}(S)$, $\widetilde{\mathcal{H}}\mathcal{C}(S')$, $\widetilde{\mathcal{H}}\mathcal{C}(S \times S')$, and $\widetilde{\mathcal{H}}_*(\mathcal{C}(S) \wedge \mathcal{C}(S'))$ be the connected simple systems mentioned in Definition 8.2(C), and let $\widetilde{\mathcal{H}}_*\mathcal{C}(S) \otimes \widetilde{\mathcal{H}}_*\mathcal{C}(S')$ be the connected simple system described in Definition 8.2(D).

Where $\langle M_1, M_0 \rangle$ and $\langle M_1', M_0' \rangle$ are also index pairs for S and S', respectively, we have as a special case of diagram (8.2) the commutative diagram

$$\begin{array}{ccc} \widetilde{H}_*(M_1/M_0) \otimes \widetilde{H}_*(M_1'/M_0') & \xrightarrow{\triangle} & \widetilde{H}_*(M_1/M_0 \wedge M_1'/M_0') \\ {\scriptstyle h_{S*} \otimes h_{S'*}} \downarrow & & \downarrow {\scriptstyle (h_S \wedge h_{S'})_*} \\ \widetilde{H}_*(N_1/N_0) \otimes \widetilde{H}_*(N_1'/N_0') & \xrightarrow{\triangle} & \widetilde{H}_*(N_1/N_0 \wedge N_1'/N_0') . \end{array} \tag{8.4}$$

The collection of such diagrams as M_1/M_0 and N_1/N_0 are varied over index spaces for S and as M_1'/M_0' and N_1'/N_0' are varied over index spaces for S' defines a map of connected simple systems from $\widetilde{\mathcal{H}}\mathcal{C}(S) \otimes \widetilde{\mathcal{H}}\mathcal{C}(S')$ to $\widetilde{\mathcal{H}}_*(\mathcal{C}(S) \wedge \mathcal{C}(S'))$ whose image under the functor $\underrightarrow{\mathbf{lim}}: \mathcal{CSS}(\mathcal{G}_R\mathcal{M}) \to \mathcal{G}_R\mathcal{M}$ defined in Definition 8.2(A) is a pairing

$$\tilde{\triangle} : \widetilde{H}_*\mathcal{C}(S;G) \otimes \widetilde{H}_*\mathcal{C}(S';G') \to \widetilde{H}_*(\mathcal{C}(S) \wedge \mathcal{C}(S'); G \otimes G') .$$

Also, application of Proposition 8.1 to the equivalence E of Proposition 8.4, yields an equivalence $\mathcal{E}_* : \widetilde{\mathcal{H}}_*(\mathcal{C}(S) \wedge \mathcal{C}(S')) \simeq \widetilde{\mathcal{H}}_*\mathcal{C}(S \times S')$. Thus, the definition $E_* := \underrightarrow{\mathbf{lim}}(\mathcal{E}_*)$ yields an isomorphism of graded R-modules $E_* : \widetilde{H}_*(\mathcal{C}(S) \wedge \mathcal{C}(S')) \simeq \widetilde{H}_*\mathcal{C}(S \times S')$.

Define the product in (8.3) by $\triangle := E_* \circ \tilde{\triangle}$.

The second assertion of the proposition is equivalent to the equality

$$\boldsymbol{\alpha} \triangle \boldsymbol{\alpha}' = \overline{n^\times}_* \alpha \times \alpha'$$

where utilizing the notation established in Definition 1.5(B), $\overline{n^\times}_* := i_*^{\overline{X \times X'}} n^\times_*$. To prove it we need to present a commutative diagram of graded R-modules, and to help keep it of a reasonable size let us utilize the following notation. In addition to having X, X', and $\overline{X \times X'}$ defined as in the paragraph immediately preceding the statement of the proposition, set $\bar{X} := N_1/N_0$ and $\bar{X}' = N_1'/N_0'$. Also, set

$\bar{n}_* := i_*^{\bar{X}} n_*$ and $\bar{n}'_* = i^{\bar{X}'} n'_*$, and let $e\colon (\bar{X} \wedge \bar{X}', *) \simeq (\overline{X \times X'}, *)$ be the natural homeomorphism of Proposition 8.3. The diagram

$$\begin{array}{ccccc}
H_*(X) \otimes H_*(X') & \xrightarrow{\times} & H_*(X \times X') & \xrightarrow{n_*^{\times}} & \widetilde{H}_*(\overline{X \times X'}) \\
{\scriptstyle n_* \otimes n'_*}\downarrow & & \downarrow{\scriptstyle (n \times n')_*} & & \downarrow{\scriptstyle e_*^{-1}} \\
\widetilde{H}_*(\bar{X}) \otimes \widetilde{H}_*(\bar{X}') & \xrightarrow{\times} & H_*(\bar{X} \times \bar{X}', \bar{X} \vee \bar{X}') & \xrightarrow{Q_*} & \widetilde{H}_*(\bar{X} \wedge \bar{X}')
\end{array}$$

is commutative because naturality of homology cross-products makes the left-hand rectangle commute and commutativity of the diagram of continuous maps underlying the right-hand rectangle makes it commute. Also commutative by definition of the maps of connected simple systems $\tilde{\wedge}$ and E_* is the diagram

$$\begin{array}{ccccc}
\widetilde{H}_*(\bar{X}) \otimes \widetilde{H}_*(\bar{X}') & \xrightarrow{\wedge} & \widetilde{H}_*(\bar{X} \wedge \bar{X}') & \xrightarrow{e_*} & \widetilde{H}_*(\overline{X \times X'}) \\
{\scriptstyle i_*^{\bar{X}} \otimes i_*^{\bar{X}'}}\downarrow & & \downarrow{\scriptstyle i_*^{\bar{X} \wedge \bar{X}'}} & & \downarrow{\scriptstyle i_*^{\overline{X \times X'}}} \\
\widetilde{H}_*\mathcal{C}(S) \otimes \widetilde{H}_*\mathcal{C}(S') & \xrightarrow{\tilde{\wedge}} & \widetilde{H}_*(\mathcal{C}(S) \wedge \mathcal{C}(S')) & \xrightarrow{E_*} & \widetilde{H}_*\mathcal{C}(S \times S').
\end{array}$$

Note that the top left-hand horizontal arrow of this diagram equals the bottom row of the previous diagram. Diagram chasing then yields the desired result. □

In [Mc1], C.K McCord gives a Kunneth Theorem for Conley indices. The context in which his result is stated is more general than that being considered here as it deals with a relative version of the Conley index for a flow and closed subflow on a locally compact metric space. Although the result in the absolute case to be stated presently could be derived from the result in [Mc1] it is no more difficult to give a new proof which can if desired be adapted to the relative case although to give here the requisite definitions would take us too far afield. For the statement of the theorem we drop the assumption that we are dealing with flows on manifolds and instead assume that S and S' are isolated invariant sets of continuous flows on some locally compact Hausdorff spaces.

8.7 THEOREM (Kunneth Theorem for Conley Indices). *For each integer $k \geq 0$, there is a split short exact sequence*

$$(8.5)\quad 0 \to \left[\widetilde{H}_*\mathcal{C}(S;G) \otimes \widetilde{H}_*\mathcal{C}(S';G')\right]_k \xrightarrow{\wedge} \widetilde{H}_k\mathcal{C}(S \times S'; G \otimes G') \to \left[\widetilde{H}_*\mathcal{C}(S;G) * \widetilde{H}_*\mathcal{C}(S';G')\right]_{k-1} \to 0$$

that is natural with respect to morphisms between Conley indices, but the splitting is not natural.

Proof. Let $\langle N_1, N_0 \rangle$ and $\langle N'_1, N'_0 \rangle$ be nested index pairs for S and S' and without loss of generality assume N_1 and N'_1 are isolating neighborhoods of S and S' respectively. For if not, replace N_1 with $\bar{N}_1 := \operatorname{cl}(N_1 \setminus N_0)$ and replace N_0 with $\bar{N}_0 :=$

$N_0 \cap \bar{N}_1$; treat $\langle N_1', N_0' \rangle$ similarly. Set $X := N_1/N_0 \times [N_0']$ and $X' := [N_0] \times N_1'/N_0'$. Then observe that $\{N_1/N_0 \times N_1'/N_0'; X, X'\}$ is an excisive triad. To see this, note that for any $t > 0$, X is a weak deformation retract of $Y := N_1/N_0 \times N_0'^{-t}/N_0'$ and X' is a weak deformation retract of $Y' := N_0^{-t}/N_0 \times N_1'/N_0'$ where the deformations are induced by the flows on the ambient spaces. The observation then follows, since for sufficiently large $t > 0$, the interiors of Y and Y' relative to $X \cup X'$ cover $X \cup X'$ as a consequence of [K1, Proposition 2.9(2)].

Thus, application of the Kunneth Theorem for singular homology yields the short exact sequence

$$(8.6)\quad 0 \longrightarrow \left[\widetilde{H}_*(N_1/N_0) \otimes \widetilde{H}_*(N_1'/N_0')\right]_k \xrightarrow{\times} H_k(N_1/N_0 \times N_1'/N_0', N_1/N_0 \vee N_1'/N_0') \longrightarrow \left[\widetilde{H}_*(N_1/N_0) * \widetilde{H}_*(N_1'/N_0')\right]_{k-1} \longrightarrow 0\,.$$

By Proposition 8.5, $\widetilde{H}_k(N_1/N_0 \wedge N_1'/N_0')$ can be substituted for the middle term in sequence (8.6), whence the exterior product $\times$ becomes the product $\triangle$ of (8.1). Exactness is preserved under direct limits; hence by Theorem 5.2, Definition 8.2, and Propositions 8.3 and 8.4, it follows that the sequence (8.5) is exact where the arrow labeled by $\triangle$ is the product of Proposition 8.6, and the other arrows are the direct limits of the corresponding arrows of the sequence (8.6), i.e., the arrows obtained by applying Theorem 5.2 to the singular homology functor.

Again from Theorem 5.2, the sequence (8.5) is natural relative to morphisms between Conley indices as a consequence of the naturality statement of the standard Kunneth Theorem; it is split, but not naturally, because sequence (8.6) is split non-naturally. Specifically, because the Kunneth Theorem implies that the third arrow from the left in (8.6) has a right inverse (which is not natural), this right inverse defines a non-natural map r: $\left[\widetilde{\mathcal{H}}_*\mathcal{C}(S) * \widetilde{\mathcal{H}}_*\mathcal{C}(S')\right]_{k-1} \to \widetilde{\mathcal{H}}_k\mathcal{C}(S \times S')$ of connected simple systems in $\mathcal{G}_R\mathcal{M}$. Thus, $\varinjlim(r)$ splits (8.5). □

D. Factor and Product Intersection Pairings. In this section, for $i = 1, 2$, assume M_i is a manifold of dimension m_i, oriented over R, and let S_i be an isolated invariant set of a flow in M_i. Thus $S_1 \times S_2$ is an isolated invariant set of the product flow in $M_1 \times M_2$. Also, $M_1 \times M_2$ is given the product orientation over R and G_1, G_1', G_2, and G_2' denote R-modules. The map on homology induced by the isomorphism of coefficient modules $G_1 \otimes G_2 \otimes G_1' \otimes G_2' \simeq G_1 \otimes G_1' \otimes G_2 \otimes G_2'$ obtained by transposition of the second and third factors is denoted by t_*.

The reader is reminded that the statement and proof of the principal result of this section, Theorem 8.10, depends upon the choice of sign convention made in §1.D. The theorem depends on the earlier results of this section and the following two lemmas. For the first, for $i = 1, 2$, let M_i be a manifold without boundary of dimension m_i, assume M_i is oriented over R, and assume $M_1 \times M_2$ is given the product orientation.

8.8 LEMMA. *For $i = 1, 2$, let $\xi_i \in H_{s_i}(V_{1\,i}, V_{0\,i}; G_i)$ and $\eta_i \in H_{t_i}(V_{1\,i}^*, V_{0\,i}^*; G_i')$ where $(V_{1\,i}, V_{0\,i})$ and $(V_{1\,i}^*, V_{0\,i}^*)$ are open topological pairs in M_i. Set $\ell := s_1 + s_2 +$*

$t_1 + t_2 - (m_1 + m_2)$ and set $(X, Y) := (X_1, Y_1) \times (X_2, Y_2)$ where

$$(X_i, Y_i) := \left(V_{1\,i} \cap V_{1\,i}^*, (V_{1\,i} \cap V_{0\,i}^*) \cup (V_{0\,i} \cap V_{1\,i}^*)\right).$$

Then as homology classes in $H_\ell(X, Y; G_1 \otimes G_2 \otimes G_1' \otimes G_2')$,

$$(\xi_1 \times \xi_2) \mathbin{\overset{\frown}{\bullet}} (\eta_1 \times \eta_2) = (-1)^{(m_2 - s_2)(m_1 - t_1)} t_*\left((\xi_1 \mathbin{\overset{\frown}{\bullet}} \eta_1) \times (\xi_2 \mathbin{\overset{\frown}{\bullet}} \eta_2)\right).$$

Also, if the pair of open pairs $((V_{1\,i}, V_{0\,i}), (V_{1\,i}^*, V_{0\,i}^*))$ satisfies the off-diagonal condition $(i = 1, 2)$, then the pair of product pairs

$$\left((V_{1\,1}, V_{0\,1}) \times (V_{1\,2}, V_{0\,2}), (V_{1\,1}^*, V_{0\,1}^*) \times (V_{1\,2}^*, V_{0\,2}^*)\right)$$

also satisfies the off-diagonal condition and $\emptyset = Y_1 = Y_2 = Y$.

Proof. For $i = 1, 2$, let x_i be the Poincaré dual in M_i of ξ_i. From the sign convention as expressed by (1.3) it follows that $(-1)^{m_1(m_2 - s_2)} x_1 \times x_2$ is the Poincaré dual of $\xi_1 \times \xi_2$ relative to the product orientation on $M_1 \times M_2$. Hence (2.5) (also see Proposition 9.18 below) and (1.3) validate the following string of equalities which yields the desired result:

$$\begin{aligned}(\xi_1 \times \xi_2) \mathbin{\overset{\frown}{\bullet}} (\eta_1 \times \eta_2) &= (-1)^{m_1(m_2 - s_2)} (x_1 \times x_2) \frown (\eta_1 \times \eta_2) \\ &= (-1)^{m_1(m_2 - s_2) + (m_2 - s_2)t_1} t_*\left((x_1 \frown \eta_1) \times (x_2 \frown \eta_2)\right) \\ &= (-1)^{(m_2 - s_2)(m_1 + t_1)} t_*\left((\xi_1 \mathbin{\overset{\frown}{\bullet}} \eta_1) \times (\xi_2 \mathbin{\overset{\frown}{\bullet}} \eta_2)\right) \\ &= (-1)^{(m_2 - s_2)(m_1 - t_1)} t_*\left((\xi_1 \mathbin{\overset{\frown}{\bullet}} \eta_1) \times (\xi_2 \mathbin{\overset{\frown}{\bullet}} \eta_2)\right).\end{aligned}$$

The easy verification that $Y_1 = Y_2 = Y = \emptyset$ and that the pair of product pairs satisfies the off-diagonal condition if $((V_{1\,i}, V_{0\,i}), (V_{1\,i}^*, V_{0\,i}^*))$ does for $i = 1, 2$ is left to the reader. □

8.9 LEMMA. *Let*

(8.7)

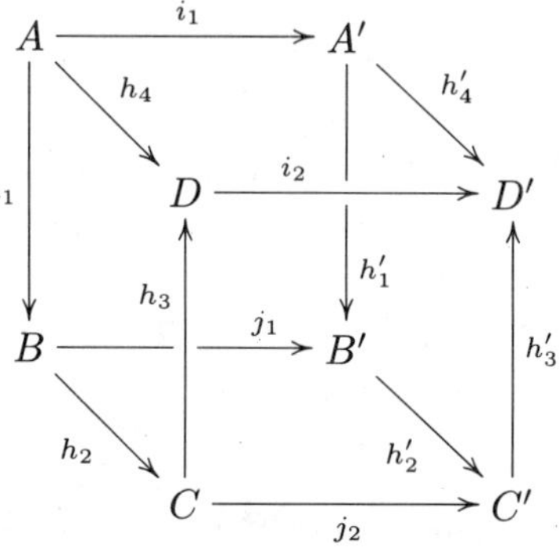

be a diagram of R-module homomorphisms such that $i_2 \circ h_4 = h_4' \circ i_1$, $j_1 \circ h_1 = h_1' \circ i_1$, $j_2 \circ h_2 = h_2' \circ j_1$, and $i_2 \circ h_3 = h_3' \circ j_2$. Then for each $a \in A$ and with $a' := i_1(a)$,

$$i_2(h_4(a) - h_3 \circ h_2 \circ h_1(a)) = h_4'(a') - h_3' \circ h_2' \circ h_1'(a').$$

Proof. The proof is an easy diagram chase left to the reader. □

The reader is reminded that if $\xi_j \in \check{H}_*(S_j)$ then $\xi_1 \times \xi_2 \in \check{H}_*(S_1 \times S_2)$ is defined componentwise from the homology cross-product of singular theory using the cofinal family of product neighborhoods of $S_1 \times S_2$. That is, if U_j is any neighborhood of S_j $(j = 1, 2)$, then

$$(\xi_1 \times \xi_2)_{U_1 \times U_2} := (\xi_1)_{U_1} \times (\xi_2)_{U_2}$$

where the product on the right is of singular homology classes, and if V is any neighborhood of $S_1 \times S_2$, choose U_j a neighborhood of S_j so that $U_1 \times U_2 \subset V$ and define $(\xi_1 \times \xi_2)_V := i_*(\xi_1 \times \xi_2)_{U_1 \times U_2}$ where $i_* \colon H_*(U_1 \times U_2) \to H_*(V)$ is the inclusion induced homomorphism. The definition of $(\xi_1 \times \xi_2)_V$ does not depend on the particular choice of U_j, $j = 1, 2$, by naturality of the singular homology cross-product and because for any space X, $\{\{H_*(U)\}_{U \in \mathcal{N}(X)}, \{i_*^{U'U}\}_{U \leq U' \in \mathcal{N}(X)}\}$ is an inverse system where $i_*^{U'U} \colon H_*(U') \to H_*(U)$ is inclusion induced. It also follows that the Čech homology cross-product is natural relative to inclusion induced homomorphisms.

8.10 THEOREM. *For $i = 1, 2$, let $\boldsymbol{\alpha}_i \in \widetilde{H}_{s_i}\mathcal{C}(S_i; G_i)$, let $\boldsymbol{\gamma}_i \in \widetilde{H}_{t_i}\mathcal{C}^*(S_i; G_i')$, and set $\ell := s_1 + s_2 + t_1 + t_2 - (m_1 + m_2)$. Then in $\check{H}_\ell(S_1 \times S_2; G_1 \otimes G_2 \otimes G_1' \otimes G_2')$,*

$$(\boldsymbol{\alpha}_1 \triangle \boldsymbol{\alpha}_2) \mathbin{\widehat{\bullet}} (\boldsymbol{\gamma}_1 \triangle \boldsymbol{\gamma}_2) = (-1)^{(m_2 - s_2)(m_1 - t_1)} t_*\big((\boldsymbol{\alpha}_1 \mathbin{\widehat{\bullet}} \boldsymbol{\gamma}_1) \times (\boldsymbol{\alpha}_2 \mathbin{\widehat{\bullet}} \boldsymbol{\gamma}_2)\big)\,.$$

Remark. If at least one of $\widetilde{H}_*\mathcal{C}(S)$ and $\widetilde{H}_*\mathcal{C}(S')$ is a torsion free R-module, then $\widetilde{H}_*\mathcal{C}(S) * \widetilde{H}_*\mathcal{C}(S')$, their torsion product, is the trivial graded R-module. Hence, by Theorem 8.7,

$$\triangle \colon \widetilde{H}_*\mathcal{C}(S) \otimes \widetilde{H}_*\mathcal{C}(S') \simeq \widetilde{H}_*\mathcal{C}(S \times S')\ ,$$

and the analogous result holds for the reverse time indices. Thus, in the case when at least one of $\widetilde{H}_*\mathcal{C}(S)$ and $\widetilde{H}_*\mathcal{C}(S')$ is torsion free and when also at least one of $\widetilde{H}_*\mathcal{C}^*(S)$, and $\widetilde{H}_*\mathcal{C}^*(S')$ is torsion free, Theorem 8.10 yields that the intersection pairing for the product invariant set is, up to sign, the product of the intersection pairings for the factor isolated invariant sets.

Proof. Let $\mathfrak{L}^j$ denote the intersection pairing for S_j $(j = 1, 2)$, and let $\mathfrak{L}^\times$ denote the intersection pairing for $S_1 \times S_2$. Also, we use 1_* to denote the identity isomorphism of various graded R-modules, and for any graded R-modules E_* and F_*, we generically use $\tau_* \colon E_* \otimes F_* \to F_* \otimes E_*$ to denote the graded transposition isomorphism given by $\tau_*(e \otimes f) = (-1)^{|e||f|} f \otimes e$. Then where $\bar{s} := s_1 + s_2$, $\bar{t} := t_1 + t_2$, and $\bar{m} := m_1 + m_2$ and where

$$\begin{aligned}
A &:= \widetilde{H}_{s_1}\mathcal{C}(S_1) \otimes \widetilde{H}_{s_2}\mathcal{C}(S_2) \otimes \widetilde{H}_{t_1}\mathcal{C}^*(S_1) \otimes \widetilde{H}_{t_2}\mathcal{C}^*(S_2)\ ,\\
\bar{A} &:= \widetilde{H}_{\bar{s}}\mathcal{C}(S_1 \times S_2) \otimes \widetilde{H}_{\bar{t}}\mathcal{C}^*(S_1 \times S_2)\ ,\\
B &:= \widetilde{H}_{s_1}\mathcal{C}(S_1) \otimes \widetilde{H}_{t_1}\mathcal{C}^*(S_1) \otimes \widetilde{H}_{s_2}\mathcal{C}(S_2) \otimes \widetilde{H}_{t_2}\mathcal{C}^*(S_2)\ ,\\
C &:= \check{H}_{s_1 + t_1 - m_1}(S_1) \otimes \check{H}_{s_2 + t_2 - m_2}(S_2)\,,\\
D &:= \check{H}_{\bar{s} + \bar{t} - \bar{m}}(S_1 \times S_2)\,,
\end{aligned}$$

the statement of the theorem is equivalent to the commutativity of the following diagram:

$$
\begin{array}{ccccc}
A & \xrightarrow{\underline{\triangle}\otimes\underline{\triangle}} & \bar{A} & \xrightarrow{\mathfrak{L}^{\times}} & D \\
{\scriptstyle 1_*\otimes\tau_*\otimes 1_*}\downarrow & & & & \uparrow{\scriptstyle t_*\circ\times} \\
B & & \xrightarrow{(-1)^{(m_2-s_2)(m_1-t_1)}\mathfrak{L}^1\otimes\mathfrak{L}^2} & & C\,.
\end{array}
\tag{8.8}
$$

The proof of commutativity of diagram (8.8) will be carried out by showing that the assumption of non-commutativity leads via Lemma 8.9 to a contradiction of Lemma 8.8. Accordingly, suppose diagram (8.8) does not commute and let h_1, h_2, and h_3 denote respectively the left, bottom, and right arrows of diagram (8.8), and let h_4 denote the composite of the top horizontal arrows in that diagram. Then for some $a \in A$,

$$d := h_4(a) - h_3 \circ h_2 \circ h_1(a) \neq 0 \in \check{H}_{\bar{s}+\bar{t}-\bar{m}}(S_1 \times S_2)\,.$$

By definition of $\check{H}_*(S_1 \times S_2)$, for some neighborhood U of $S_1 \times S_2$, the image of d under the canonical projection to $H_{\bar{s}+\bar{t}-\bar{m}}(U)$, denote it d_U, is not zero, whence for every $U' \in \mathcal{N}(S_1 \times S_2)$, if $U \supset U'$, then the image of d under the canonical projection to $H_{\bar{s}+\bar{t}-\bar{m}}(U')$, denote it $d_{U'}$, is also non-zero. In particular, we can choose $P_j \in \operatorname{ob}(\mathcal{AIP}(S_j))$, $j = 1, 2$, so that $U' := (P_{1\,1} \times P_{1\,2}) \cap (P^*_{1\,1} \times P^*_{1\,2})$ is a neighborhood of $S_1 \times S_2$ contained in U.

As noted in Definition 2.8(C), for $j = 1, 2$, there exists $V_j \in M^{(2)}_{\underline{\Delta}\,o}(P_j)$ satisfying

$$U'' := (V_{1\,1} \times V_{1\,2}) \cap (V^*_{1\,1} \times V^*_{1\,2}) \subset U\,.$$

It follows that the image of d under the inclusion induced homomorphism

$$\check{H}_{\bar{s}+\bar{t}-\bar{m}}(S_1 \times S_2) \xrightarrow{i_2} \check{H}_{\bar{s}+\bar{t}-\bar{m}}((P_{1\,1} \times P_{1\,2}) \cap (P^*_{1\,1} \times P^*_{1\,2}))$$

is non-zero, i.e., $i_2(d) \neq 0$. Also, the projection of $i_2(d)$ into $H_{\bar{s}+\bar{t}-\bar{m}}(U'')$ is non-zero; i.e., where $i'_2 \colon \check{H}_*(U') \to H_*(U'')$ is the natural projection, $i'_2 \circ i_2(d) \neq 0$.

Set

$$
\begin{aligned}
A' &:= H_{s_1}(P_{1\,1}, P_{0\,1}) \otimes H_{s_2}(P_{1\,2}, P_{0\,2}) \otimes H_{t_1}(P^*_{1\,1}, P^*_{0\,1}) \otimes H_{t_2}(P^*_{1\,2}, P^*_{0\,2})\,,\\
\bar{A}' &:= H_{\bar{s}}((P_{1\,1}, P_{0\,1}) \times (P_{1\,2}, P_{0\,2})) \otimes H_{\bar{t}}((P^*_{1\,1}, P^*_{0\,1}) \times (P^*_{1\,2}, P^*_{0\,2}))\,,\\
B' &:= H_{s_1}(P_{1\,1}, P_{0\,1}) \otimes H_{t_1}(P^*_{1\,1}, P^*_{0\,1}) \otimes H_{s_2}(P_{1\,2}, P_{0\,2}) \otimes H_{t_2}(P^*_{1\,2}, P^*_{0\,2})\,,\\
C' &:= \check{H}_{s_1+t_1-m_1}(P_{1\,1} \cap P^*_{1\,1}) \otimes \check{H}_{s_2+t_2-m_2}(P_{1\,2} \cap P^*_{1\,2})\,,\\
D' &:= \check{H}((P_{1\,1} \times P_{1\,2}) \cap (P^*_{1\,1} \times P^*_{1\,2}))\,,
\end{aligned}
$$

and in analogy with the labeling of the arrows of diagram (8.8), in the diagram of homomorphisms

$$
\begin{array}{ccccc}
A' & \xrightarrow{\times\otimes\times} & \bar{A}' & \xrightarrow{\mathfrak{L}^{P_1\times P_2}_{(\bar{s},\bar{t})}} & D' \\
{\scriptstyle 1_*\otimes\tau_*\otimes 1_*}\downarrow & & & & \uparrow{\scriptstyle t_*\circ\times} \\
B & & \xrightarrow{(-1)^{(m_2-s_2)(m_1-t_1)}\mathfrak{L}^{P_1}_{(s_1,t_1)}\otimes\mathfrak{L}^{P_2}_{(s_2,t_2)}} & & C'
\end{array}
\tag{8.9}
$$

let h_1', h_2', and h_3' denote respectively the left, bottom, and right arrows and let h_4' denote the composite of the top horizontal arrows of the diagram.

Diagrams (8.8) and (8.9) correspond to the similarly labeled pair of opposite faces in diagram (8.7) in the statement of Lemma 8.9. Note that i_2 has already been defined, and let the remaining arrows of diagram (8.7) be given as follows: i_1 and j_1 are the inverses of the tensor product of canonical maps from the homology of index pairs to the homology of a Conley index, explicitly,

$$i_1 := (\bar{p}^{1\,+}_* \otimes \bar{p}^{2\,+}_* \otimes \bar{p}^{1\,-}_* \otimes \bar{p}^{2\,-}_*)^{-1}, \qquad j_1 := (\bar{p}^{1\,+}_* \otimes \bar{p}^{1\,-}_* \otimes \bar{p}^{2\,+}_* \otimes \bar{p}^{2\,-}_*)^{-1},$$

and

$$\check{H}_{s_1+t_1-m_1}(S_1) \otimes \check{H}_{s_2+t_2-m_2}(S_2) \xrightarrow{j_2} \check{H}_{s_1+t_1-m_1}(P_{1\,1} \cap P^*_{1\,1}) \otimes \check{H}_{s_2+t_2-m_2}(P_{1\,2} \cap P^*_{1\,2})$$

is the tensor product of the inclusion induced homomorphisms. Note that the commutativity hypotheses of Lemma 8.9 are satisfied as a consequence of Theorem 2.11, Proposition 8.6, and by the naturality of the transposition homomorphism τ_* and of homology cross-products relative to inclusions. Thus where $a' := i_1(a)$, because $i_2(d) \neq 0$, application of Lemma 8.9 yields that

$$0 \neq d' := h_4'(a') - h_3' \circ h_2' \circ h_1'(a') = i_2(d)\,.$$

Next, set

$$\begin{aligned}
A'' &:= H_{s_1}(V_{1\,1}, V_{0\,1}) \otimes H_{s_2}(V_{1\,2}, V_{0\,2}) \otimes H_{t_1}(V^*_{1\,1}, V^*_{0\,1}) \otimes H_{t_2}(V^*_{1\,2}, V^*_{0\,2}),\\
\bar{A}'' &:= H_{\bar{s}}((V_{1\,1}, V_{0\,1}) \times (V_{1\,2}, V_{0\,2})) \otimes H_{\bar{t}}((V^*_{1\,1}, V^*_{0\,1}) \times (V^*_{1\,2}, V^*_{0\,2}))\,,\\
B'' &:= H_{s_1}(V_{1\,1}, V_{0\,1}) \otimes H_{t_1}(V^*_{1\,1}, V^*_{0\,1}) \otimes H_{s_2}(V_{1\,2}, V_{0\,2}) \otimes H_{t_2}(V^*_{1\,2}, V^*_{0\,2}),\\
C'' &:= \check{H}_{s_1+t_1-m_1}(V_{1\,1} \cap V^*_{1\,1}) \otimes \check{H}_{s_2+t_2-m_2}(V_{1\,2} \cap V^*_{1\,2}),\\
D'' &:= \check{H}((V_{1\,1} \times V_{1\,2}) \cap (V^*_{1\,1} \times V^*_{1\,2}))\,,
\end{aligned}$$

and in analogy with the previous two cases, let h_1'', h_2'', and h_3'' denote respectively the left, bottom, and right arrows and let h_4'' denote the composite of the top horizontal arrows of the diagram of R-module homomorphisms

$$(8.10)\qquad \begin{array}{ccccc}
A'' & \xrightarrow{\times\otimes\times} & \bar{A}'' & \xrightarrow{\mathfrak{L}^{V_1 \times V_2}_{P_1 \times P_2\,(\bar{s},\bar{t})}} & D''\\
{\scriptstyle 1_*\otimes\tau_*\otimes 1_*}\big\downarrow & & & & \big\uparrow{\scriptstyle t_*\circ\times}\\
B & & \xrightarrow{(-1)^{(m_2-s_2)(m_1-t_1)}\mathfrak{L}^{V_1}_{P_1\,(s_1,t_1)} \otimes \mathfrak{L}^{V_2}_{P_2\,(s_2,t_2)}} & & C''
\end{array}$$

where $V_1 \times V_2 \in M^{(2)}_{\Delta\,o}(P_1 \times P_2)$ is defined by

$$V_1 \times V_2 := ((V_{1\,1}, V_{0\,1}) \times (V_{1\,2}, V_{0\,2}), (V^*_{1\,1}, V^*_{0\,1}) \times (V^*_{1\,2}, V^*_{0\,2}))\,.$$

Now take diagrams (8.9) and (8.10) as the pair of non-commuting faces in diagram (8.7) with A' taking the role of A, A'' taking the role of A', etc., and with the inclusion induced i_2' defined above corresponding to i_2. Then the remaining arrows, also with a prime appended, are the inclusion induced homomorphisms, i.e., $i_1': A' \to A''$, $j_1': B' \to B''$, and $j_2': C' \to C''$ are the tensor product of inclusion induced homomorphisms between corresponding factors.

Again we can apply Lemma 8.9. The commutativity hypotheses are now satisfied as a consequence of Lemma 2.10 and the naturality of τ_* and of homology cross-products relative to inclusions. Also $i_2'(d') \neq 0$ by choice of P_j and V_j, $j = 1, 2$. This time Lemma 8.9 yields that where $a'' := i_1'(a') = i_1' \circ i_1(a)$

$$0 \neq d'' := h_4''(a'') - h_3'' \circ h_2'' \circ h_1''(a''),$$

whence diagram (8.10) does not commute. However, this contradicts Lemma 8.8 which implies diagram (8.10) does commute for any choice of $V_j \in M^{(2)}_{\not\Delta o}(P_j)$, for $j = 1, 2$. $\square$

8.11 COROLLARY. *For $i = 1, \dots, n$, let S_i be an isolated invariant set of a flow in an m_i-dimensional manifold M_i oriented over R, suppose $\boldsymbol{\alpha}_i \in \widetilde{H}_{s_i}\mathcal{C}(S_i; R)$, $\gamma_i \in \widetilde{H}_{t_i}\mathcal{C}^*(S_i; R)$, and $s_i + t_i = m_i$ for $i = 1, \dots, n$. Also, give $M_1 \times \cdots \times M_n$ the product orientation. Then where $\sigma = \sum_{i=1}^{n-1} \sum_{j=i+1}^{n} s_i t_j$,*

$$^{\#}((\boldsymbol{\alpha}_1 \vartriangle \cdots \vartriangle \boldsymbol{\alpha}_n) \,\widehat{\bullet}\, (\gamma_1 \vartriangle \cdots \vartriangle \gamma_n)) = (-1)^{\sigma} \prod_{\ell=1}^{n} {}^{\#}(\boldsymbol{\alpha}_\ell \,\widehat{\bullet}\, \gamma_\ell)$$

Proof. By induction it suffices to consider the case $n = 2$, but that case follows immediately from Theorem 8.10 and the definition of intersection number because for the exterior product on cohomology $1 \times 1 = 1$, i.e., because the value of the augmentation on the exterior product of two 0-simplexes is the product of the values of the augmentations on the factors. $\square$

CHAPTER 9

THE CAP PRODUCT REPRESENTATION OF $\mathfrak{L}$ AND THE NON-SINGULARITY OF $^{\#}\mathfrak{L}$

The main result of this chapter establishes analogues of the formulae (2.5) for the pairing $\mathfrak{L}$ by representing it in a cofinal family of manifold isolating blocks for an isolated invariant set. An immediate corollary of this representation is the non-singularity of $^{\#}\mathfrak{L}$. The last section, §9.D, briefly examines how the results of this work can be extended to the case where M is a manifold with boundary and $S \cap \partial M \neq \emptyset$.

A. The Cap Product Representation and Corollaries.

9.1 DEFINITION. For each isolated invariant set S of the flow define $\mathfrak{B}(S)$ by $B \in \mathfrak{B}(S)$ if, and only if, B is an isolating block for S and a compact, codimension zero, C^0 submanifold with boundary of M possessing the properties:

(1) $\partial B = B^+ \cup B^-$ and $B^{\pm}$, if non-void, is a codimension zero submanifold with boundary of ∂B;
(2) there exist submanifolds Σ^+ and Σ^- of M and $d > 0$ so that $B^{\pm} \subset \Sigma^{\pm}$, the flow embeds $\Sigma^{\pm} \times \,]-3d, 3d[$ as an open submanifold of M, and $\Sigma^+ \cdot [-d, d] \cap B = \tau_B^{-1}([0, d])$ and $\Sigma^- \cdot [-d, d] \cap B = (\tau_B^*)^{-1}([0, d])$;
(3) if $c_B := B^+ \cap B^- \neq \emptyset$, then $\partial B^+ = c_B = \partial B^-$ and $\partial B^{\pm}$ is bicollared in $\Sigma^{\pm}$.

9.2 PROPOSITION. *If M is C^r $(r \geq 1)$ and the flow in M is generated by a continuous vectorfield, then $\mathfrak{B}(S)$ is cofinal in $\mathcal{N}(S)$, the neighborhoods of S directed by reverse inclusion.*

Proof. If $U \in \mathcal{N}(S)$, then by [WY, Theorem 2.4] there exists $B \subset U$ an isolating block with corners for S; see Definition 4.3. Then $B \in \mathfrak{B}(S)$: property (2) holds because via the implicit function theorem it holds locally at each point of $\Sigma^{\pm}$ and because $B^{\pm}$ is compact; property (3) holds by using the unit disk subbundle of any tubular neighborhood of c_B for the bicollar. □

R. Churchill in [Ch] has constructed isolating blocks for isolated invariant sets of continuous flows in compact metric spaces and in fact his construction can be generalized to flows in locally compact Hausdorff spaces. The construction is quite analogous to that in [WY]. However, the author has not investigated whether or not Churchill's construction in the setting of a continuous flow in a C^0 manifold can be made to yield blocks that are C^0 manifolds with boundary satisfying the criteria defining $\mathfrak{B}(S)$; if so then $\mathfrak{B}(S)$ would be cofinal in $\mathcal{N}(S)$ in the C^0 case also.

For the rest of this chapter, coefficients of homology and cohomology modules are generally suppressed and unless noted otherwise are assumed taken in an R-module G where as usual M is assumed oriented over the PID R. Also, throughout, S is a fixed isolated invariant set of some flow in M.

When $\mathfrak{B}(S) \neq \emptyset$, the Conley indices of S satisfy a duality property: with coefficients taken in any R-module

$$(9.1) \qquad \widetilde{H}^k \mathcal{C}^*(S) \simeq \widetilde{H}_{m-k}\mathcal{C}(S) \quad \text{and} \quad \widetilde{H}^k \mathcal{C}(S) \simeq \widetilde{H}_{m-k}\mathcal{C}^*(S) \qquad \text{for } k \in \mathbf{N}.$$

As mentioned in the Introduction, existence of the isomorphisms of (9.1) (and generalizations that take into account flows on manifolds with boundary and isolated invariant sets that may or may not intersect the boundary—see Theorem 9.20 below) is shown in [Mc2]. The existence of these isomorphisms and generalizations will be referred to as *Poincaré duality of Conley indices.* The implications of the general result in the context of intersection pairings on Conley indices will be discussed briefly in §9.D. So as to be able to present the main result of this chapter, Theorem 9.4 below, the statement of a slight variant of the core lemma of [Mc2] follows but only for the special case of a flow on a manifold without boundary; cf. [Mc2, Lemma 2.3]. In the course of proving Theorem 9.4, a proof of Lemma 9.3 falls out that is quite different than that given in [Mc2]; see Lemma 9.16 below.

9.3 LEMMA. *If $B \in \mathfrak{B}(S)$ and $o_B \in H_m(B, \partial B; R)$ is the fundamental class of B, then the R-module homomorphisms*

$$H^*(B, B^+) \xrightarrow{\frown o_B} H_{m-*}(B, B^-) \qquad \text{and} \qquad H^*(B, B^-) \xrightarrow{\frown o_B} H_{m-*}(B, B^+)$$

are isomorphisms.

Note that the excisiveness conditions needed for the cap product homomorphisms in the above lemma to be well-defined are met because either B^+ and B^- are disjoint and so are both open and closed in ∂B or because they intersect in c_B which is bicollared in ∂B.

Let $\boldsymbol{\alpha} \otimes \boldsymbol{\gamma} \in \widetilde{H}_*\mathcal{C}(S; G) \otimes \widetilde{H}_*\mathcal{C}^*(S; G')$ for some R-modules G and G', assume $B \in \mathfrak{B}(S)$, let $\alpha_B \in H_*(B, B^+; G)$ represent $\boldsymbol{\alpha}$, and let $\gamma_B \in H_*(B, B^-; G')$ represent $\boldsymbol{\gamma}$. By Lemma 9.3 there are unique cohomology classes $x_B \in H^*(B, B^-; G)$ and $y_B \in H^*(B, B^+; G')$ satisfying $\alpha_B = x_B \frown o_B$ and $\gamma_B = y_B \frown o_B$. The main result of this chapter is the following:

9.4 THEOREM. $(\boldsymbol{\alpha} \mathbin{\widehat{\bullet}} \boldsymbol{\gamma})_B = (x_B \smile y_B) \frown o_B = x_B \frown \gamma_B$.

Note that if $B, B' \in \mathfrak{B}(S)$ with $B \supset B'$, then as a consequence of Theorem 2.11 defining $\mathfrak{L}$,

$$i_B^{B'}(\boldsymbol{\alpha} \mathbin{\widehat{\bullet}} \boldsymbol{\gamma})_{B'} = (\boldsymbol{\alpha} \mathbin{\widehat{\bullet}} \boldsymbol{\gamma})_B \qquad \text{for } B, B' \in \mathfrak{B}(S) \text{ and } B \supset B'.$$

Thus, when $\mathfrak{B}(S)$ is cofinal in $\mathcal{N}(S)$, Theorem 9.4 yields by passage to the limit over $\mathfrak{B}(S)$ the following:

9.5 COROLLARY. *If $\mathfrak{B}(S)$ is cofinal in $\mathcal{N}(S)$, then*

$$(\boldsymbol{\alpha} \mathbin{\widehat{\bullet}} \boldsymbol{\gamma}) = \varprojlim (x_B \frown \gamma_B).$$

Perhaps the most important consequence of Theorem 9.4 is the following:

9.6 COROLLARY. *If $\mathfrak{B}(S)$ is non-empty and if $\widetilde{H}_*\mathcal{C}(S;R)$ and $\widetilde{H}_*\mathcal{C}^*(S;R)$ are torsion free, then the intersection number pairing $^{\#}\mathfrak{L}$ defines a non-singular dual pairing of $\widetilde{H}_k\mathcal{C}(S;R)$ and $\widetilde{H}_{m-k}\mathcal{C}^*(S;R)$.*

Proof. Let $B \in \mathfrak{B}(S)$. From Theorem 9.4 it is clear that

$$^{\#}(\boldsymbol{\alpha} \mathbin{\widehat{\bullet}} \gamma) = \langle 1, (x_B \smile y_B) \frown o_B \rangle = (-1)^{k(m-k)} \langle y_B, x_B \frown o_B \rangle .$$

Since $x_B \mapsto x_B \frown o_B$ is an isomorphism of $H^k(B, B^-)$ onto $H_{m-k}(B, B^+)$ and since the Kronecker pairing is a non-singular dual pairing of $H^k(B, B^-)$ and $H_k(B, B^-)$, it follows that

$$\gamma \mapsto {}^{\#}\mathfrak{L}(- \otimes \gamma)$$

is an isomorphism of $\widetilde{H}_k\mathcal{C}^*(S)$ onto $\hom\big[\widetilde{H}_{m-k}\mathcal{C}(S), R\big]$. $\square$

A third corollary is Theorem 8.10 under the additional assumption that for the factor invariant sets S_1 and S_2, the family $\mathfrak{B}(S_i)$ is cofinal in $\mathcal{N}(S_i)$, $i = 1, 2$. One need only perform the computation of Lemma 8.8 with a block-exit set pair $(B_{1\,j}, B_{1\,j}^+)$ replacing $(V_{1\,j}, V_{0\,j})$ and the corresponding block-entrance set pair $(B_{1\,j}, B_{1\,j}^-)$ replacing $(V_{1\,j}^*, V_{0\,j}^*)$. The details of the precise formulation and its proof are left to the reader.

B. Some Technical Propositions on Poincaré Duality Isomorphisms and Čech Cap Products. The proof of Theorem 9.4 is eventually reduced to several diagram chases. However, the construction of the diagrams and the proof that they are commutative requires considerable work. In particular, five technical propositions relating various Poincaré duality isomorphisms will be needed in the proofs of commutativity and these are now listed. Surely all are well-known, but as the author could find no adequate reference, their proofs save for the first which is a quite simple exercise are given in Appendix B. In all of these propositions, any open submanifold of a manifold oriented over R is given the induced orientation over R. Also, whenever X appears in the statement of one of the following five propositions it is assumed to be an m-dimensional manifold oriented over R, and for simplicity coefficients are suppressed, but unless stated otherwise are assumed taken in an R-module G.

9.7 PROPOSITION. *Suppose (K, L) is a closed pair in M with L cobounded in K, and suppose U is open in M and contains the pair (K, L). Then, for $k \in \mathbf{N}$, the diagram*

$$\begin{array}{ccc} & \check{H}^k(K, L) & \\ {\scriptstyle \frown o^U} \swarrow & & \searrow {\scriptstyle \frown o^M} \\ H_{m-k}(U \setminus L, U \setminus K) & \xrightarrow[\simeq]{\text{excision}} & H_{m-k}(M \setminus L, M \setminus K) \end{array}$$

commutes.

The proof, left to the reader, is a straightforward exercise in the definition of the Čech cap product and the use of some obvious excision isomorphisms.

To state the next proposition, note that whenever V is open in a manifold X there is an inclusion induced homomorphism $\check{H}_c^*(V) \to \check{H}_c^*(X)$ obtained by passage to the direct limit over compact $K \subset V$ of the composite homomorphisms

$$\check{H}^*(V, V \setminus K) \xrightarrow{(e_K^*)^{-1}} \check{H}^*(X, X \setminus K) \to \check{H}_c^*(X)$$

where e_K^* is the excision isomorphism and the unlabeled arrow is the canonical map into the direct limit; on page 289 of [D] see statement (6.23) and its proof.

9.8 Proposition. *Suppose V is open in X. Then, where the unlabeled arrows are inclusion induced, for $i \in \mathbf{N}$, the diagram*

$$\begin{array}{ccc} \check{H}_c^i(V) & \longrightarrow & \check{H}_c^i(X) \\ {\scriptstyle \frown o^V}\downarrow & & \downarrow{\scriptstyle \frown o^X} \\ H_{m-i}(V) & \longrightarrow & H_{m-i}(X) \end{array}$$

is commutative.

The next proposition is an analogue of Proposition 9.7 where Čech cohomology is replaced by Čech cohomology with compact supports.

9.9 Proposition. *Let V be open in X, let (K_1, L_1) be a closed (relative to V) pair in V, and let (K_2, L_2) be a closed pair in X. Assume that $(K_1, L_1) \subset (K_2, L_2)$, that this inclusion is an excision, and that $V \setminus L_1 \subset X \setminus L_2$. Then $V \setminus K_1 \subset X \setminus K_2$ and for each $i \in \mathbf{N}$,*

$$\begin{array}{ccc} \check{H}_c^i(K_1 \setminus L_1) & = & \check{H}_c^i(K_2 \setminus L_2) \\ {\scriptstyle \frown o^{V \setminus L_1}}\downarrow & & \downarrow{\scriptstyle \frown o^{X \setminus L_2}} \\ H_{m-i}(V \setminus L_1, V \setminus K_1) & \longrightarrow & H_{m-i}(X \setminus L_2, X \setminus K_2) \end{array}$$

is a commutative diagram of isomorphisms where the unlabeled arrow is an excision isomorphism.

The last two of the five propositions show that standard identities of singular theory relating cap and cross products and cup and cap products carry over to a setting where the cap product is the Čech cap product of a Čech cohomology class and a singular homology class as defined on pp. 292–3 of [D]. The cup product and cross product of Čech cohomology classes are defined in the obvious manner by taking direct limits of singular cup products and singular cross-products of suitable open pairs and therefore require no excisiveness conditions; see [D, §VIII.6.21, p. 288] for some detail. For (K, L) and (K, L') closed pairs in some manifold, the cup product of a Čech class $\check{x} \in \check{H}^*(K, L)$ with a Čech class $\check{y} \in \check{H}_c^*(K, L')$ having compact support can then be defined by passage to the direct limit over elements $\check{x} \smile \check{y}_\omega \in \check{H}(K, L \cup \omega)$ where ω lies in the family of subsets of K that contain L' and are locally compact and cobounded in K and $\check{y}_\omega \mapsto \check{y}$ in the canonical map to the direct limit. So defined $\check{x} \smile \check{y} \in \check{H}_c^*(K, L \cup L')$ because the set of subsets of the form $L \cup \omega$ is cofinal in the family of subsets of K that contain $L \cup L'$ and are locally compact and cobounded in K.

9.10 PROPOSITION. *Suppose K_ℓ is closed in the manifold M_ℓ of dimension m_ℓ and that M_ℓ is R-oriented over K_ℓ with fundamental class o_ℓ along K_ℓ, $\ell = 1, 2$. Then, for $i_1, i_2 \in \mathbf{N}$, the diagram*

$$\begin{array}{ccc} H_{i_1}(M_1, M_1 \setminus K_1) \otimes H_{i_2}(M_2, M_2 \setminus K_2) & \xleftarrow{\frown o_1 \otimes \frown o_2} & \check{H}_c^{m_1-i_1}(K_1) \otimes \check{H}_c^{m_2-i_2}(K_2) \\ \times \downarrow & & \downarrow \times \\ H_{i_1+i_2}(M_1 \times M_2, M_1 \times M_2 \setminus K_1 \times K_2) & \xleftarrow{(-1)^{m_1(m_2-i_2)} \frown o_1 \times o_2} & \check{H}_c^{m_1+m_2-i_1-i_2}(K_1 \times K_2) \end{array}$$

commutes.

9.11 PROPOSITION. *Suppose (K, L) is a closed pair in X. Then the diagram*

$$\begin{array}{ccc} \check{H}(K, L) \otimes \check{H}_c(K) & \xrightarrow{\smile} & \check{H}_c(K, L) \\ 1^* \otimes \frown o^X \downarrow & & \downarrow \frown o^X \\ \check{H}(K, L) \otimes H(X, X \setminus K) & \xrightarrow{\frown} & H(X \setminus L, X \setminus K) \end{array}$$

commutes.

C. Results Leading to the Proof of Theorem 9.4. The first goal of this section is to define an R-module homomorphism

$$D_k \colon \widetilde{H}_k\mathcal{C}(S) \to \widetilde{H}^{m-k}\mathcal{C}^*(S)$$

that will later be seen to be an isomorphism when $\mathfrak{B}(S) \neq \emptyset$ thereby giving a new proof of Poincaré duality of Conley indices for the case $S \cap \partial M = \emptyset$. The following definitions are needed first.

For $k \in \mathbf{N}$, for $P \in \mathrm{ob}(\mathcal{AIP}(S))$ define $D_{P\,k}$ to be the composition of the homomorphisms in the sequence

$$H_k(P_1, P_0) \xrightarrow{i_*} H_k(M \setminus P_0^*, M \setminus P_1^*) \xrightarrow{(\frown o^M)^{-1}} \check{H}^{m-k}(P_1^*, P_0^*)$$

where i_* is inclusion induced and $\frown o^M$ is the relative Poincaré-Alexander duality isomorphism. Also, define $\bar{D}_{P\,k}$ to be the composition of the homomorphisms in the sequence

$$\widetilde{H}_k\mathcal{C}(S) \xrightarrow{(\bar{p}_*^+)^{-1}} H_k(P_1, P_0) \xrightarrow{D_{P\,k}} \check{H}^{m-k}(P_1^*, P_0^*) \xrightarrow{(\bar{p}^{-\,*})^{-1}} \widetilde{H}^{m-k}\mathcal{C}^*(S)\ .$$

9.12 PROPOSITION. $\bar{D}_{P\,k} = \bar{D}_{Q\,k}$ *for all* $P, Q \in \mathrm{ob}(\mathcal{AIP}(S))$.

Proof. Define an equivalence relation D in $\mathrm{ob}(\mathcal{AIP}(S))$ by $P \overset{D}{\sim} Q$ if, and only if, $\bar{D}_{P\,k} = \bar{D}_{Q\,k}$. As already noted in Chapter 2, relative Poincaré-Alexander duality isomorphisms are natural with respect to inclusions. It follows immediately that if $P, Q \in \mathrm{ob}(\mathcal{AIP}(S))$ with $P \hookrightarrow Q$, then $P \overset{D}{\sim} Q$; call this invariance of D under inclusions. The proof then proceeds in analogy with that of Theorem 4.7. That is, (1) as each P in $\mathrm{ob}(\mathcal{AIP}(S))$ includes into an AIP(regular) pair P', by invariance of D under inclusion, each pair of admissible index pairs for S is equivalent to an AIP(regular) pair of index pairs for S; (2) as in the proof of Claim II in the proof of Theorem 4.7, Lemma 4.6 and invariance of D under inclusions show that any two AIP(regular) pairs of index pairs for S are equivalent. $\square$

9.13 DEFINITION. Define $D_k\colon \widetilde{H}_k\mathcal{C}(S) \to \widetilde{\check{H}}^{m-k}\mathcal{C}^*(S)$ to be the unique R-module homomorphism which for each $P \in \mathrm{ob}(\mathcal{AIP}(S))$ equals $\bar{D}_{P\,k}$. Define the R-module homomorphism $D_k^*\colon \widetilde{H}_k\mathcal{C}^*(S) \to \widetilde{\check{H}}^{m-k}\mathcal{C}^*(S)$ analogously by regarding S as an isolated invariant set of the time-reversed flow.

The proof that D_k is an isomorphism will utilize parts of the following lemma. The remaining parts will be used in the construction and proofs of commutativity of several diagrams used to prove Theorem 9.4.

9.14 LEMMA. *Given $B \in \mathfrak{B}(S)$, there exist $P \in \mathrm{ob}(\mathcal{AIP}(S))$, $V \in M^{(2)}_{\mathbb{A}\,o}(P)$, and W, Q, Q^+, Q^- relatively compact, open subsets of M so that $W, Q^\pm \subset Q$ and additionally:*

(1) $P \hookrightarrow ((B, B^+), (B, B^-)) \hookrightarrow ((Q, Q^+), (Q, Q^-))$;
(2) $V_i = W \setminus \mathrm{cl}_W\,(V^*_{1-i})$ *and* $V_i^* = W \setminus \mathrm{cl}_W\,(V_{1-i})$ $(i = 0, 1)$;
(3) $\mathrm{cl}_M\,(V_1 \setminus V_0) \cup \mathrm{cl}_M\,(V_1^* \setminus V_0^*) \subset B^\circ := \mathrm{int}(B)$;
(4) $P_1^* \subset Q \setminus Q^+ \subset P_1^* \cup V_0^*$ *and* $P_1 \subset Q \setminus Q^- \subset P_1 \cup V_0$;
(5) V_1 *and* V_1^* *are weak deformation retracts of* W;
(6) P_i *is a strong deformation retract of* V_i *and of* $W \setminus V^*_{1-i}$ $(i = 0, 1)$;
(7) P_i^* *is a strong deformation retract of* V_i^* *and of* $W \setminus V_{1-i}$ $(i = 0, 1)$;
(8) $W \setminus V_i$ *is a weak deformation retract of* $Q \setminus V_i$ $(i = 0, 1)$;
(9) $W \setminus V_i^*$ *is a weak deformation retract of* $Q \setminus V_i^*$ $(i = 0, 1)$;
(10) B *is a strong deformation retract of* Q *and* $B^\pm$ *is a strong deformation retract of* $Q^\pm$;
(11) B^- *(respectively* B^+*) is a strong deformation retract of* $Q \setminus V_1$ *(respectively* $Q \setminus V_1^*$*)*;
(12) *Each of the pairs* (P_1, P_0), (P_1^*, P_0^*), *and* $(B, B^\pm)$ *has isomorphic Čech and singular cohomology modules (any coefficients).*

Proof. Let us show that we can reduce to the case where $\emptyset \neq c_B := B^+ \cap B^-$. For if c_B is void, there are four possibilities: (i) $B^+ = B^- = \emptyset$, (ii) $B^- = \emptyset$, but $B^+ \neq \emptyset$, (iii) $B^+ = \emptyset$, but $B^- \neq \emptyset$, (iv) B^+ and B^- are both non-void, but are disjoint.

If (i) holds then $\partial B = \emptyset$. Thus, B is open in M as well as being closed and compact; also, necessarily $S = B$. Hence, statements (1)–(12) of the lemma will obviously be satisfied (many vacuously) if $V := P := ((B, B^+), (B, B^-))$, $W := Q := B$, and $Q^\pm := \emptyset$.

Cases (ii) and (iii) are handled analogously; we do only case (ii). If (ii) holds, note that $\partial B = B^+ = \Sigma^+$ so that by definition of $\mathfrak{B}(S)$, the flow determines an open bicollar V_0 of ∂B; say $V_0 = B^+ \cdot]-d, d[$. Then the statements of the lemma are obviously satisfied if $P := ((B, B^+), (\tau_B^{-1}[2d, \infty], B^-))$, $Q := W := V_1 := B \cup V_0$, $V_1^* := W \setminus \mathrm{cl}_W\,(V_0)$, $Q^- := V_0^* := \emptyset$, and $Q^+ := V_0$.

If (iv) holds, then $B^\pm$ coincides with $\Sigma^\pm$ and by hypothesis the flow determines a closed bicollar neighborhood $N^\pm$ of $B^\pm$; say $N^\pm = B^\pm \cdot [-2d, 2d]$ for some $d > 0$. Set $P_1 := B \setminus B^- \cdot [0, 2d[$ and $P_0 := B^+$. Similarly, set $P_1^* := B \setminus B^+ \cdot]-2d, 0]$ and $P_0^* := B^-$. Also, set $V_0 := B^+ \cdot]-d, d[$ and define V_0^* analogously using B^-. Set $Q := W := B \cup V_0 \cup V_0^*$ and define V_1 and V_1^* via the equalities of statement (2) of the lemma. Set $Q^+ := Q \setminus (P_1^* \cup B^- \cdot]-d, 0])$ and define Q^- analogously using

P_1 and B^+. It is easily verified that these choices of P, V, Q, $Q^\pm$, and W fulfill the requirements of the lemma.

Thus, without loss of generality we assume henceforth that $c_B \neq \emptyset$ and note that $c_B = \partial B^\pm$. First some preliminary constructions are needed after which P, W, V, Q and $Q^\pm$ will be constructed in that order.

To begin, where $\Sigma^\pm \supset B^\pm$ is the local section of the flow that exists by the assumption that $B \in \mathfrak{B}(S)$, let $h\colon c_B \times [-1,1] \to \Sigma^-$ be a homeomorphism onto a closed bicollar neighborhood E^- of c_B in Σ^- so that where $U^- := h(c_B \times [0,1[)$ and $\bar{U}^- := h(c_B \times [0,1])$, U^- is an open and $\bar{U}^-$ a closed collar of ∂B^- in B^- with both disjoint from $W^s(S;B)$. Next, with regard to the block B, in condition (2) of Definition 9.1 replace d by $2d$ sufficiently small so that the resulting statement holds true; call it condition (2′). Delete from B those maximal orbit segments in B with one endpoint in U^- and denote the set remaining by $\hat{B}$, an isolating block for S as $\bar{U} \cap W^s(S;B) = \emptyset$. Now, without loss of generality, assume that (i) the flow embeds $\hat{B}^+ \times [-3d, 3d]$ and $\hat{B}^- \times [-3d, 3d]$ onto disjoint closed sets whose intersections with $\hat{B}$ are respectively $\tau_{\hat{B}}^{-1}([0,3d])$ and $\tau_{\hat{B}}^{*\,-1}([0,3d])$ and (ii) $E^- \cdot -d \subset M \setminus \Sigma^+ \cdot [0,d]$ and on $E^- \cdot -d$ the map τ_{Σ^+} is finite, positive, and continuous where

$$\tau_{\Sigma^+}(\mathbf{u}) := \sup\{t \geq 0 : \mathbf{u} \cdot [0,t] \subset M \setminus \Sigma^+\} \qquad \text{for } \mathbf{u} \in E^- \cdot -d.$$

For as $\hat{B}^+$ and $\hat{B}^-$ are disjoint, clearly Definition 9.1 implies that shrinking d as needed will yield (i). Also, $\Sigma^+ \cdot]0,2d] \cap B = \emptyset$; else for some $\mathbf{u} \in \Sigma^+$ and $s \in]0,2d]$, $0 < \bar{s} := s + \tau_B(\mathbf{u} \cdot s) \leq 4d$ and Σ^+ intersects $\Sigma^+ \cdot \bar{s}$ contradicting condition (2′). Then $B^+ \cdot [0,d]$ and $B^- \cdot -d$ are disjoint closed sets; else

$$\emptyset \neq B^+ \cdot]0,2d] \cap B^- \subset \Sigma^+ \cdot]0,2d] \cap B = \emptyset\,.$$

Via normality, continuity of the flow, and compactness of $B^\pm$, it follows that $(B^- \cup E^-) \cdot -d$ and $\operatorname{cl}(\Sigma^+) \cdot [0,d]$ will be disjoint closed sets if the bicollar E^- is shrunk sufficiently and if Σ^+ is replaced with a relatively compact, open relative to Σ^+ neighborhood of B^+. The inclusion in (ii) then follows immediately. As $(c_B \cdot -d) \cdot d$ lies in the open $\Sigma^+ \cdot]-\varepsilon, \varepsilon[$ for $0 < \varepsilon < 6d$, it follows that shrinking the bicollar E^-, as needed, will ensure that τ_{Σ^+} is finite, positive, and continuous on $E^- \cdot -d$.

Define $P \in \operatorname{ob}(\mathcal{AIP}(S))$ by setting

$$\begin{aligned} P_1 &:= \tau_{\hat{B}}^{*\,-1}[2d,\infty]\,, & P_0 &:= \hat{B}^+\,,\\ P_1^* &:= \tau_{\hat{B}}^{-1}[2d,\infty]\,, & P_0^* &:= \hat{B}^-\,. \end{aligned}$$

Note that $P_1 = \hat{B} \setminus \hat{B}^- \cdot [0,2d[$ and $P_1^* = \hat{B} \setminus \hat{B}^+ \cdot]-2d,0]$, whence P_1 and P_1^* are squeezes of $\hat{B}$, the former along the entrance set, the latter along the exit set. Thus, (P_1, P_0) is a (block, exit set)-pair and (P_1^*, P_0^*) a (block, entrance set)-pair. Therefore, both are regular index pairs and by construction P satisfies the off-diagonal condition; hence, $P \in \operatorname{ob}(\mathcal{AIP}(S))$.

Define W and V as follows. First, define $Y^- \subset U^-$ by $Y^- := h(\partial B^- \times]\frac{1}{2},1[)$ and let Y be the union of the maximal orbit segments in B with one endpoint in Y^-. Next, set $Y^+ := Y \cap B^+$ and set $Z^\pm := Y^\pm \cup \hat{B}^\pm$. Then $Z^\pm$ is an open neighborhood of $\hat{B}^{\pm}$ relative to $\Sigma^{\pm}$. It follows that $Z^{\pm}\cdot\,]-d,d[$ is an open neighborhood of $\hat{B}^{\pm}$ relative to M. Define

$$W := \hat{B} \cup Y \cup Z^{-}\cdot\,]-d,0[\,\cup Z^{+}\cdot\,]0,d[\,.$$

It follows easily that W is an open neighborhood of $\hat{B}$ relative to M (and therefore of P_1 and P_1^*) and that W has compact closure. Next, define $V \in M^{(2)}_{\mathbb{A}o}(P)$ by

$$\begin{aligned} V_1 &:= W \setminus Z^{-}\cdot\,]-d,d]\,, & V_0 &:= Z^{+}\cdot\,]-d,d[\,,\\ V_1^* &:= W \setminus Z^{+}\cdot\,]-d,d]\,, & V_0^* &:= Z^{-}\cdot\,]-d,d[\,. \end{aligned}$$

Define

$$Q := B\cup B^{+}\cdot[0,d[\,\cup B^{-}\cdot\,]-d,0]\cup\bigcup\{\mathbf{u}\cdot\,]-d,\tau_{\Sigma^+}(\mathbf{u}\cdot -d)+d[\,:\mathbf{u}\in h(c_B\times\,]-1,0])\}\,.$$

To define Q^+ and Q^-, first define

$$\tilde{P}_1 := P_1 \cup P_0\cdot[0,d[\,, \qquad \text{and} \qquad \tilde{P}_1^* := P_1^* \cup P_0^*\cdot\,]-d,0]$$

and then set

$$Q^+ := Q \setminus \tilde{P}_1^* \qquad \text{and} \qquad Q^- := Q \setminus \tilde{P}_1\,.$$

It is clear from their construction that $P \in \mathrm{ob}(\mathcal{AIP}(S))$, that $V \in M^{(2)}_{\mathbb{A}o}(P)$, and that W, Q, Q^+, and Q^- are relatively compact, open subsets of M such that $W, Q^{\pm} \subset Q$ and statements (1)–(4) are satisfied. It remains to describe the requisite weak and strong deformation retractions of statements (5)–(11) and show the isomorphisms of statement (12).

A weak deformation retraction of W into V_1 is given by

$$(\mathbf{u},s) \mapsto \begin{cases} \mathbf{u}\cdot s\min\{2d,\tau_B(\mathbf{u})\} & \text{if } \mathbf{u}\in (W\cap B)\cup Z^{-}\cdot\,]-d,0],\ s\in[0,1]\\ \mathbf{u} & \text{if } \mathbf{u}\in Z^{+}\cdot[0,d[\,,\ s\in[0,1]\end{cases}$$

where $\tau_B(\mathbf{u}\cdot r) := \tau_B(\mathbf{u}) - r$ for $\mathbf{u}\in Z^-$ and $r\in\,]-d,0]$. A weak deformation retraction of W into V_1^* is defined similarly.

Strong deformation retractions of V_0 onto P_0 and of $W\setminus V_1$ onto P_0^* will be defined presently; strong deformation retractions of V_0^* onto P_0^* and of $W\setminus V_1^*$ onto P_0 can be defined similarly and are left for the reader. Because of the way $Y^{\pm}$ is defined in terms of the collar neighborhood of ∂B^-, where $\bar{Y}^- := h(\partial B^-\times\,]\frac{1}{2},1])$, there is a strong deformation retraction $\hat{D}^{\pm}$ of $Z^{\pm}$ onto $\hat{B}^{\pm}$ with the deformations $\hat{D}^-$ and $\hat{D}^+$ satisfying the relation

$$\hat{D}^-(\mathbf{u},s)\cdot\tau_B(\hat{D}^-(\mathbf{u},s)) = \hat{D}^+(\mathbf{u}\cdot\tau_B(\mathbf{u}),s) \qquad \text{for } \mathbf{u}\in\bar{Y}^-,\ s\in[0,1].$$

Note too that $W\setminus V_1 = Z^-\cdot\,]-d,d]$. Then define a strong deformation retraction D^+ of V_0 onto P_0 and a strong deformation retraction D^- of $W\setminus V_1$ onto P_0^* by the formula

$$D^{\pm}(\mathbf{u}\cdot t,s) := \hat{D}^{\pm}(\mathbf{u},s)\cdot(1-s)t$$

where for D^+ the formula is defined for each $\mathbf{u} \in Z^+$, $t \in \,]-d, d[$, and $s \in [0,1]$, but for D^- the formula is for each $\mathbf{u} \in Z^-$, $t \in \,]-d, d]$, and $s \in [0,1]$.

The definition of strong deformation retractions of V_1 onto P_1, of V_1^* onto P_1^*, of $W \setminus V_0$ onto P_1^*, and of $W \setminus V_0^*$ onto P_1 are more involved. Definitions of the deformation retractions of V_1 onto P_1 and of $W \setminus V_0$ onto P_1^* will be made. The others are defined analogously. The deformations will be given piecewise. Accordingly, define

$$V_+ := Z^+ \cdot [0, d[\,, \quad V_- := Z^- \cdot \,]d, 2d]\,,$$

and

$$\bar{Y}_{2d} := \bigcup \{\mathbf{u} \cdot [2d, \tau_B(\mathbf{u})] : \mathbf{u} \in \bar{Y}^-\}.$$

Note that V_+, V_-, $\bar{Y}_{2d}$, and P_1 are all subsets of V_1, each is closed relative to V_1, and the union of these four sets equals V_1. Also for each $\mathbf{u} \in \bar{Y}_{2d}$, there are unique $\mathbf{y} \equiv \mathbf{y}(\mathbf{u}) \in \bar{Y}^-$ and $r \equiv r(\mathbf{u}) \in [0,1]$, both varying continuously with $\mathbf{u}$, so that $\mathbf{u} = \mathbf{y} \cdot (2d + r\tau_B(\mathbf{y} \cdot 2d))$, and for each $\mathbf{u} \in V_-$, there are unique $\mathbf{z} \equiv \mathbf{z}(\mathbf{u}) \in Z^-$ and $t \equiv t(\mathbf{u}) \in \,]d, 2d]$, both varying continuously with $\mathbf{u}$, so that $\mathbf{u} = \mathbf{z} \cdot t$. Then define

$$D(\mathbf{u}, s) := \begin{cases} \mathbf{u} & \text{for } \mathbf{u} \in P_1,\ s \in [0,1]; \\ D^+(\mathbf{u}, s) & \text{for } \mathbf{u} \in V_+,\ s \in [0,1]; \\ \hat{D}^-(\mathbf{y}, s) \cdot \big[2d + r\tau_B(\hat{D}^-(\mathbf{y}, s) \cdot 2d)\big] & \text{for } \mathbf{u} \in \bar{Y}_{2d},\ s \in [0,1]; \\ \hat{D}^-(\mathbf{z}, s) \cdot [(1-s)t + 2sd] & \text{for } \mathbf{u} \in V_-,\ s \in [0,1]. \end{cases}$$

Clearly D is continuous on the closed subsets of $V_1 \times [0,1]$ obtained by taking the Cartesian product of each of $V_\pm$, $\bar{Y}_{2d}$, and P_1 with $[0,1]$, and it is easily verified that the piecewise definitions agree on any overlap. Thus D is continuous and by its definition is a strong deformation retraction of V_1 onto P_1.

In an analogous manner one can define a strong deformation retraction D^* of $W \setminus V_0$ onto P_1^* by a piecewise definition on the Cartesian product of sets $V_\pm^*$ and $\bar{Y}_{2d}^*$ with $[0,1]$ where $V_\pm^*$ and $\bar{Y}_{2d}^*$ are defined as follows:

$$V_-^* := Z^- \cdot \,]-d, 0]\,, \quad V_+^* := Z^+ \cdot [-2d, -d]\,,$$

and

$$\bar{Y}_{2d}^* := \bigcup \left\{\mathbf{u} \cdot [0, \tau_B(\mathbf{u}) - 2d] : \mathbf{u} \in \bar{Y}^-\right\}.$$

Then V_-^*, V_+^*, $\bar{Y}_{2d}^*$, and P_1^* are all closed relative to $W \setminus V_0$ and their union equals $W \setminus V_0$. Also for each $\mathbf{u} \in \bar{Y}_{2d}^*$, there exist unique $\mathbf{y} \equiv \mathbf{y}(\mathbf{u}) \in \bar{Y}^-$ and $r \equiv r(\mathbf{u}) \in [0,1]$ so that $\mathbf{u} = \mathbf{y} \cdot r(\tau_B(\mathbf{y}) - 2d)$, and for each $\mathbf{u} \in V_+^*$, there exist unique $\bar{\mathbf{u}} \in Z^+$ and $t \in [-2d, -d]$ so that $\mathbf{u} = \bar{\mathbf{u}} \cdot t$. Define D^* by

$$D^*(\mathbf{u}, s) := \begin{cases} \mathbf{u} & \text{for } \mathbf{u} \in P_1^*,\ s \in [0,1]; \\ D^-(\mathbf{u}, s) & \text{for } \mathbf{u} \in V_-^*,\ s \in [0,1]; \\ \hat{D}^-(\mathbf{y}, s) \cdot r\big[\tau_B(\hat{D}^-(\mathbf{y}, s)) - 2d\big] & \text{for } \mathbf{u} \in \bar{Y}_{2d}^*,\ s \in [0,1]; \\ \hat{D}^+(\bar{\mathbf{u}}, s) \cdot [(1-s)t - 2sd] & \text{for } \mathbf{u} \in V_+^*,\ s \in [0,1]. \end{cases}$$

It is easily verified that D^* is a strong deformation retraction of $W \setminus V_0$ onto P_1^*.

Presently, a weak deformation retraction of $Q \setminus V_0$ into $W \setminus V_0$ will be defined; a weak deformation retraction of $Q \setminus V_0^*$ into $W \setminus V_0^*$ can be defined analogously. The deformation will be defined in three stages. First, deform each maximal orbit segment in $Q \setminus V_0$ containing a point of the form $h(\mathbf{u}, s)$ where $\mathbf{u} \in c_B$ and $-1 < s < \frac{1}{2}$ onto the maximal orbit segment in Q containing the point $h(\mathbf{u}, \frac{1}{2})$ by using the bicollar to define the deformation and scaling time along the orbit appropriately. All other points are left fixed during the first stage. Each maximal orbit segment in Q through a point of the form $h(\mathbf{u}, \frac{1}{2})$ contains the subsegment $h(\mathbf{u}, \frac{1}{2}) \cdot [\tau_B(h(\mathbf{u}, \frac{1}{2})) - d, \tau_B(h(\mathbf{u}, \frac{1}{2})) + d[$ lying in $\mathrm{cl}_Q(V_0)$, and the second stage of the deformation contracts each such orbit segment to the point $h(\mathbf{u}, \frac{1}{2}) \cdot \tau_B(h(\mathbf{u}, \frac{1}{2})) - d$. All other points are left fixed during the second stage. Successive application of the two stages defined so far yields a strong deformation retraction of $Q \setminus V_0$ onto $\mathrm{cl}_Q(W \setminus V_0)$. Above, a strong deformation retraction of $W \setminus V_0$ onto P_1^* has been given and will easily be extended by the reader to a strong deformation retraction of $\mathrm{cl}_Q(W \setminus V_0)$ onto P_1^* that can then be regarded as a weak deformation retraction of $\mathrm{cl}_Q(W \setminus V_0)$ into V_1^* and taken as the third stage. Successive application of the three stages yields a weak deformation retraction of $Q \setminus V_0$ onto $W \setminus V_0$.

A strong deformation retraction of Q onto B can be described as follows. Over the time interval $0 \leq s \leq \frac{1}{2}$ use the collar of c_B to deform maximal orbit segments in Q that do not intersect B into an orbit segment that intersects c_B and note that orbit segments of the latter type lie in $c_B \cdot]-d, d[\subset B^- \cdot]-d, 0] \cup B^+ \cdot [0, d[$. Then over the time interval $\frac{1}{2} \leq s \leq 1$ use the flow to deform points of $B^- \cdot]-d, 0]$ into B^- and points of $B^+ \cdot [0, d[$ into B^+. Throughout the deformation leave points of B fixed. This yields a strong deformation retraction of Q onto B.

A strong deformation retraction of Q^+ onto B^+ will be described; a strong deformation retraction of Q^- onto B^- can be described similarly. Observe that Q^+ is the union of bounded, maximal orbit segments of two types: (1) those that do not intersect B and (2) those that intersect B^+ in a unique point. As with Q, over the time interval $0 \leq s \leq \frac{1}{2}$ use the bicollar of c_B to define a deformation of maximal orbit segments of Q^+ that do not intersect B into a maximal orbit segment contained in $c_B \cdot]-d, d[$ while leaving all other maximal orbit segments pointwise fixed. Then, over the time interval $\frac{1}{2} \leq s \leq 1$, use the flow to contract each maximal orbit segment intersecting B^+ into its unique point of intersection. This yields a strong deformation retraction of Q^+ onto B^+.

A weak deformation retraction of $Q \setminus V_1$ into $W \setminus V_1$ is described presently; an analogous definition yields a weak deformation retraction of $Q \setminus V_1^*$ into $W \setminus V_1^*$. First, observe that $Q \setminus V_1 \subset Q \setminus \tilde{P}_1 =: Q^-$ and that the strong deformation retraction of Q^- onto B^- restricts to a strong deformation retraction of $Q \setminus V_1$ onto B^- that deforms $W \setminus V_1$ within itself. Second, it is easy to use the collar to define a strong deformation retraction of B^- onto P_0^* that deforms $W \cap B^-$ within itself. The two taken successively yield a strong deformation retraction of $Q \setminus V_1$ onto P_0^* that deforms $W \setminus V_1$ within itself. Hence, as $P_0^* \subset W \setminus V_1$, the deformations taken successively yield a weak deformation retraction of $Q \setminus V_1$ into $W \setminus V_1$.

As $(B, B^{\pm})$ is a pair of compact manifolds with boundary, it is a pair of ENR's whence its Čech and singular cohomology modules are isomorphic; see [D, Proposition VIII.6.12]. For the pair (P_1, P_0), one might be able to verify that it too is a pair of compact manifolds with boundary; however, even if that is not the case, note that the construction of (V_1, V_0) and the definition of the strong deformation retraction of V_i onto P_i $(i = 0, 1)$ is readily modified to produce a cofinal family $\{U_j\}_{j\in\mathbf{N}}$ of open neighborhood pairs of (P_1, P_0) such that P_i is a strong deformation retract of $U_{i\,j}$, $i = 0, 1$, $j \in \mathbf{N}$. Thus, $H^*(U_{i\,j}) \simeq H^*(P_i)$, $i = 0, 1$, and the Five Lemma implies $H^*(U_{1\,j}, U_{0\,j}) \simeq H^*(P_1, P_0)$, $j \in \mathbf{N}$. Then

$$\check{H}^*(P_1, P_0) = \varinjlim H^*(U_{1\,j}, U_{0\,j}) \simeq H^*(P_1, P_0)\,.$$

A similar argument shows $\check{H}^*(P_1^*, P_0^*) \simeq H^*(P_1^*, P_0^*)$. $\square$

9.15 THEOREM (cf. [Mc2]). *If $\mathfrak{B}(S) \neq \emptyset$, then the Čech and singular cohomology modules (any coefficients) of the Conley indices of S are naturally isomorphic and the R-module homomorphism $D_k\colon \widetilde{H}_k\mathcal{C}(S) \to \widetilde{H}^{m-k}\mathcal{C}^*(S)$ is an isomorphism.*

Proof. For any $P \in \operatorname{ob}(\mathcal{AIP}(S))$, let $h^*_{(P_1^*, P_0^*)}$ denote the natural transformation from the Čech to the singular cohomology of the pair (P_1^*, P_0^*) and define a homomorphism $\bar{h}^*_{(P_1^*, P_0^*)}\colon \widetilde{\check{H}}{}^*\mathcal{C}^*(S) \to \widetilde{H}^*\mathcal{C}^*(S)$ by $\bar{h}^*_{(P_1^*, P_0^*)} := (\bar{p}^{-\,*})^{-1} \circ h^*_{(P_1^*, P_0^*)} \circ \bar{\bar{p}}^{-\,*}$. Similarly, define a homomorphism $\bar{h}^*_{(P_1, P_0)}\colon \widetilde{\check{H}}{}^*\mathcal{C}(S) \to \widetilde{H}^*\mathcal{C}(S)$. An argument analogous to that used to prove Proposition 9.12 shows that the definitions of $\bar{h}^*_{(P_1^*, P_0^*)}$ and $\bar{h}^*_{(P_1, P_0)}$ are independent of the choice of $P \in \operatorname{ob}(\mathcal{AIP}(S))$. In particular, Lemma 9.14 provides an AIP(regular) P so that both (P_1, P_0) and (P_1^*, P_0^*) have isomorphic Čech and singular cohomology. This and Proposition 4.2 yield that each factor in the composites defining $\bar{h}^*_{(P_1, P_0)}$ and $\bar{h}^*_{(P_1^*, P_0^*)}$ are isomorphisms. Thus, the Čech and singular modules of $\mathcal{C}^*(S)$ are naturally isomorphic as are those of $\mathcal{C}(S)$. We may therefore regard D_k as a homomorphism into $\widetilde{H}^{m-k}\mathcal{C}^*(S)$.

To see that D_k is an isomorphism, let P, V, W, and Q be as guaranteed by Lemma 9.14 and note that by definition of $\bar{D}_{P\,k}$ and Proposition 9.12 it will suffice to show that $D_{P\,k}$ is an isomorphism. To do this, let us examine the diagram

$$(9.2)\qquad \begin{array}{ccccc}
H_k(P_1, P_0) & \longrightarrow & H_k(Q \setminus P_0^*, Q \setminus P_1^*) & \xrightarrow[\simeq]{(\frown o^Q)^{-1}} & \check{H}^{m-k}(P_1^*, P_0^*) \\
& \searrow^{\simeq} & \uparrow & & \uparrow \simeq \\
& & H_k(V_1, V_0) & \xrightarrow[\simeq]{(\frown o^Q)^{-1}} & \check{H}^{m-k}(Q \setminus V_0, Q \setminus V_1)
\end{array}$$

where the unlabeled arrows are inclusion induced. The arrows labeled by $(\frown o^Q)^{-1}$ are the inverses of relative Poincaré-Alexander duality isomorphisms determined by the orientation of Q induced by that of M. Note that there is no need to use Čech cohomology with compact supports with regard to the duality isomorphism in the bottom row because $\operatorname{cl}_M(V_1 \setminus V_0)$ is a compact subset of Q by Lemma 9.14. The rectangle in the diagram commutes as a consequence of the naturality of Čech cap products (hence of the duality isomorphisms) with respect to inclusions. That the triangle commutes is obvious.

Each of the arrows in the diagram is an isomorphism. First, the vertical arrow that is the right edge of the diagram is an isomorphism. For $\check{H}^*(Q \setminus V_0, Q \setminus V_1) \simeq \check{H}^*(W \setminus V_0, W \setminus V_1)$ as a consequence of statement (8) of Lemma 9.14 and the Five Lemma, and $\check{H}^*(W \setminus V_0, W \setminus V_1) \simeq \check{H}^*(P_1^*, P_0^*)$ as a consequence of statement (7) of Lemma 9.14 and the Five Lemma. Second, the oblique arrow in the triangle is an isomorphism as follows from statement (6) of Lemma 9.14 and the Five Lemma. A trivial diagram chase shows that the remaining two unlabeled arrows in the diagram are isomorphisms.

Then $D_{P\,k}$ is an isomorphism. For Proposition 9.7 implies that in the top row the inclusion induced isomorphism into $H_k(Q \setminus P_0^*, Q \setminus P_1^*)$ and the duality isomorphism $(\frown o^Q)^{-1}$ can be replaced respectively by the inclusion induced homomorphism into $H_k(M \setminus P_0^*, M \setminus P_1^*)$ (an isomorphism because it factors via an excision isomorphism through $H_k(Q \setminus P_0^*, Q \setminus P_1^*)$) and the duality isomorphism $(\frown o^M)^{-1}$. □

9.16 LEMMA. *Let $B \in \mathfrak{B}(S)$ and let P, V, W, Q, and $Q^{\pm}$ be as guaranteed by Lemma 9.14; then $\frown o_B \colon H^{m-k}(B, B^-; G) \to H_k(B, B^+; G)$ is an isomorphism, the diagram (all coefficients are in G)*

$$\text{(9.3)} \qquad \begin{array}{ccccc}
 & \swarrow & H_k(P_1, P_0) & \overset{i_9}{\searrow} & \\
H_k(V_1, V_0) & \longrightarrow & H_k(Q \setminus P_0^*, Q \setminus P_1^*) \;\big\downarrow i_1 & & H_k(B, B^+) \\
\big\downarrow (\frown o^Q)^{-1} & & \big\downarrow (\frown o^Q)^{-1} & & \big\downarrow (\frown o_B)^{-1} \\
\check{H}^{m-k}(Q \setminus V_0, Q \setminus V_1) & \longrightarrow & \check{H}^{m-k}(P_1^*, P_0^*) & \overset{i_{15}}{\longleftarrow} & H^{m-k}(B, B^-) \\
 & \nwarrow & \big\uparrow & \nearrow & \\
 & & \check{H}^{m-k}(Q, Q \setminus V_1) & &
\end{array}$$

is commutative where the unlabeled arrows and those labeled by i_1, i_9, or i_{15} are inclusion induced, and each arrow in the diagram is an isomorphism. The analogous statement holds if the roles of (P_1, P_0) and (P_1^*, P_0^*), of (V_1, V_0) and (V_1^*, V_0^*), and of (B, B^+) and (B, B^-) are interchanged.

9.16.1 Remark. Note Lemma 9.14(12) ensures the existence of the inclusion induced homomorphism i_{15}.

9.16.2 Remark. An immediate consequence of Lemma 9.16 and the proof of Theorem 9.15 is that the duality isomorphism of Conley indices D_k is in the current context the inverse of the duality isomorphism given in [Mc2].

Proof of Lemma 9.16. That subdiagram of diagram (9.3) consisting of the triangle at the top left together with the rectangle directly below it commutes and has each of its arrows an isomorphism because it is congruent with diagram (9.2) from the proof of Theorem 9.15.

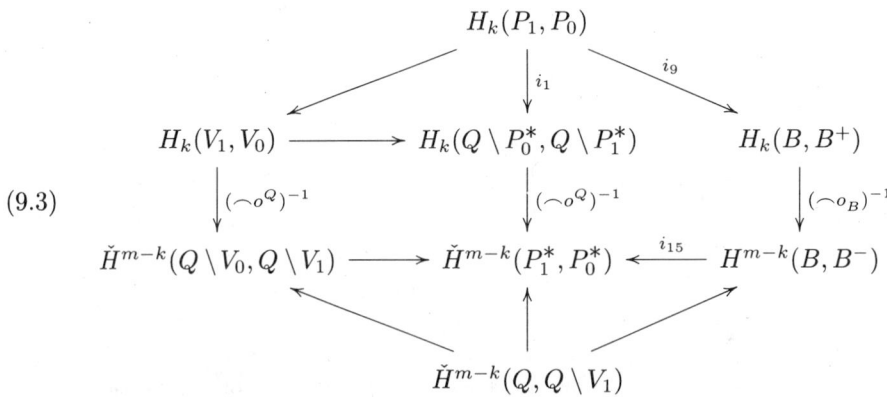

Each of the two triangles at the bottom of diagram (9.3) is obviously commutative, and each of their arrows is an isomorphism as follows. The oblique arrow in the triangle at the bottom right is an isomorphism as follows from statements (10) and (11) of Lemma 9.14 and the Five Lemma. The horizontal arrow of that triangle, labeled i_{15}, is an isomorphism because the inclusion inducing it is between regular index pairs. In such a situation the commutative rectangle of continuous maps consisting of the inclusion between regular index pairs at the top, the induced map on quotients at the bottom, and the collapse maps from the pairs to the quotients (as pointed spaces) on the sides is transformed by homology or cohomology to a commuting rectangle of module isomorphisms: the sides are isomorphisms by Proposition 4.2; the bottom because an inclusion induced map between index spaces in a Conley index has homotopy class the unique morphism in the index from the source to the target index space (see [K5, Proposition 3.1]) and is therefore a homotopy equivalence; hence, commutativity of the rectangle forces its top to be an isomorphism. Thus, commutativity of the triangle at the bottom right of diagram (9.3) implies its vertical arrow to be an isomorphism. Hence, two of the three arrows in the bottom left triangle are isomorphisms whence so is the third.

It remains to show $\frown o_B$ an isomorphism and to show the five-arrow subdiagram at the top right commutative and to consist of isomorphisms. This follows by analyzing the diagram

(9.4)

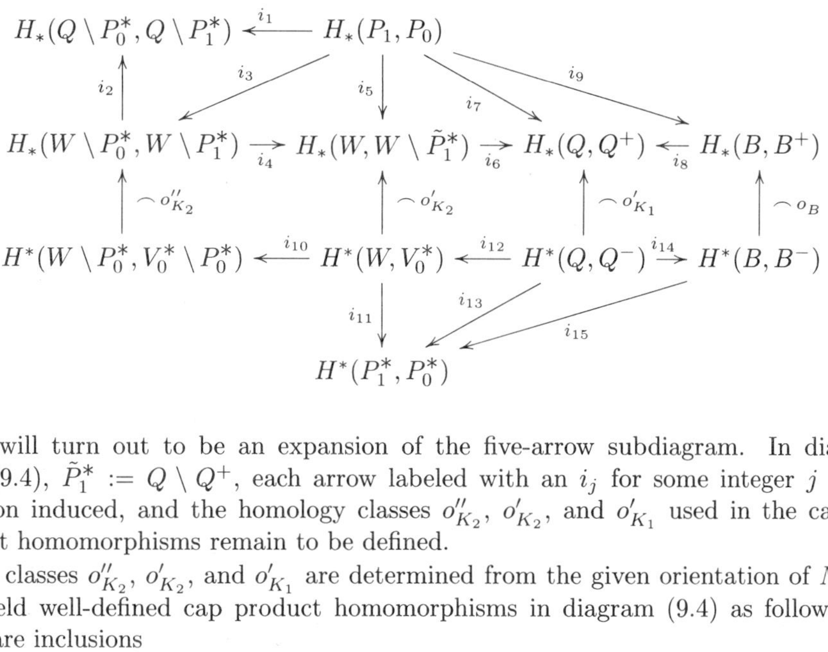

which will turn out to be an expansion of the five-arrow subdiagram. In diagram (9.4), $\tilde{P}_1^* := Q \setminus Q^+$, each arrow labeled with an i_j for some integer j is inclusion induced, and the homology classes o''_{K_2}, o'_{K_2}, and o'_{K_1} used in the cap product homomorphisms remain to be defined.

The classes o''_{K_2}, o'_{K_2}, and o'_{K_1} are determined from the given orientation of M and yield well-defined cap product homomorphisms in diagram (9.4) as follows. Their are inclusions

$$P_1 \cap P_1^* \subset K_1 := Q \setminus (Q^+ \cup Q^-) \subset P_1 \cap P_1^* \subset K_2 := P_1^* \cap W \setminus V_0^* \subset B^\circ$$

as a consequence of Lemma 9.14: the two leftmost follow from statement (4), the next from the inclusion implicit in statement (6), and the rightmost from statement (3). Thus, $K_1 = P_1 \cap P_1^*$ and K_1 and K_2 are compact subsets of B°; hence, also of W and Q. For $i = 1, 2$, let $o_{K_i} \in H_m(M, M \setminus K_i; R)$ be the fundamental class determined by the given orientation O^M of M. Then define o'_{K_1} to be the image of o_{K_1} under the excision isomorphism

$$H_m(M, M \setminus K_1; R) \simeq H_m(Q, Q \setminus K_1; R) = H_m(Q, Q^+ \cup Q^-; R)$$

and define o'_{K_2} to be the image of o_{K_2} under the excision isomorphism

$$H_m(M, M \setminus K_2; R) \simeq H_m(W, W \setminus K_2; R)\,.$$

It is immediate that the cap product homomorphism $\frown o'_{K_1}$ in diagram (9.4) is well-defined; in particular the excisiveness conditions are met because $Q^{\pm}$ is open. Too, the homomorphism $\frown o'_{K_2}$ in diagram (9.4) is well-defined because

$$W \setminus K_2 = V_0^* \cup (W \setminus P_1^*) = V_0^* \cup (W \setminus \tilde{P}_1^*)$$

where equality holds on the right as a consequence of Lemma 9.14(4); again, the excisiveness conditions are met because V_0^* and $W \setminus \tilde{P}_1^*$ are open. To define o''_{K_2} note that

$$W \setminus (K_2 \cup P_0^*) = (V_0^* \setminus P_0^*) \cup (W \setminus P_1^*)\,;$$

hence, define o''_{K_2} to be the image of o'_{K_2} under the excision isomorphism

$$H_m(W, W \setminus K_2) \simeq H_m(W \setminus P_0^*, (V_0^* \setminus P_0^*) \cup (W \setminus P_1^*))\,; \tag{9.5}$$

it follows that in diagram (9.4) the cap product homomorphism $\frown o''_{K_2}$ is well-defined; once again the excisiveness conditions are met because $V_0^* \setminus P_0^*$ and $W \setminus P_1^*$ are open.

Diagram (9.4) is commutative. Certainly each triangle of morphisms in the diagram is commutative since the arrows in each triangle are inclusion induced. It remains to show that the three rectangles across the middle part of the diagram are commutative. As o''_{K_2} is the image of o'_{K_2} under the excision isomorphism (9.5), it is immediate that the leftmost of the three rectangles commutes by naturality of cap products with respect to continuous maps; e.g., see [D, VII.12.6] or [Sp, 5.6.16]. Similarly, the other two rectangles will be known to commute once it is shown that the inclusion induced homomorphisms

$$H_m(W, W \setminus K_2; R) \to H_m(Q, Q \setminus K_1; R) \quad \text{and} \quad H_m(B, \partial B; R) \to H_m(Q, Q \setminus K_1; R)$$

respectively map o'_{K_2} to o'_{K_1} and o_B to o'_{K_1}.

That o'_{K_1} is the image of o'_{K_2} follows from the definition of o'_{K_i} $(i = 1, 2)$ and the commutativity of the diagram of inclusion induced homomorphisms

$$\begin{array}{ccc} H_m(W, W \setminus K_2; R) & \longrightarrow & H_m(Q, Q \setminus K_1; R) \\ \simeq\downarrow & & \downarrow\simeq \\ H_m(M, M \setminus K_2; R) & \longrightarrow & H_m(M, M \setminus K_1; R) \end{array}$$

because o_{K_2} is necessarily mapped to o_{K_1} by the bottom horizontal arrow since both are fundamental classes determined by the same orientation.

To prove that o'_{K_1} is the image of o_B proceed as follows. First, as $K_1 \subset B^\circ := \operatorname{int}(B)$, choose K_3 compact satisfying $K_1 \subset K_3 \subset B^\circ$ and $B \setminus K_3$ is a collar of ∂B. Next, consider the commutative diagram of inclusion induced homomorphisms

$$\begin{array}{ccccccc}
H_m(M, M\setminus K_3) & \xleftarrow{\simeq} & H_m(B^\circ, B^\circ \setminus K_3) & \xrightarrow{\simeq} & H_m(B, B\setminus K_3) & \xleftarrow{\simeq} & H_m(B, \partial B) \\
& \searrow & \downarrow & & \downarrow & \swarrow & \\
& & H_m(M, M\setminus K_1) & \xleftarrow{\simeq} & H_m(Q, Q\setminus K_1) & &
\end{array}$$

(coefficients are in R) and note that each arrow in the top row is an isomorphism: the leftmost by excision, the middle one because the inclusion of pairs inducing it is a weak deformation retract, the rightmost because ∂B is a strong deformation retract of the collar $B \setminus K_3$. Also, the bottom horizontal arrow is an excision isomorphism. Let o_{K_3} be the fundamental class along K_3 determined by O^M. Then the image in $H_m(B, \partial B)$ of o_{K_3} under the composite isomorphism that is the top row is by definition o_B and is independent of the particular choice of K_3 as is easily seen. Hence, as o_{K_1} is necessarily the image of o_{K_3} along the leftmost oblique arrow, the commutativity of the diagram and the definition of o'_{K_1} yield that o'_{K_1} is the image of o_B under the inclusion induced homomorphism. Thus, diagram (9.4) is commutative as claimed.

Each of the arrows $i_1, \dots, i_9$ in diagram (9.4) is an isomorphism. To prove the previous statement, diagram chasing shows it sufficient to prove that i_1, i_2, i_6, i_8, and i_9 are isomorphisms. The arrow i_1 is because it occurs as an arrow in diagram (9.2) in the proof of Theorem 9.15 where it is shown to be an isomorphism. The arrows i_2 and i_6 are excision isomorphisms. The arrows i_9 and i_8 are: i_9 by the general argument used above to show i_{15} an isomorphism; i_8 as a consequence of Lemma 9.14(10) and the Five-Lemma.

Each of the arrows $i_{10}, \dots, i_{15}$ in diagram (9.4) is an isomorphism. That i_{15} is one is already known, and i_{14} is by Lemma 9.14(10) and the Five Lemma, whence i_{13} is too. Next, statements (5) and (6) of Lemma 9.14 imply that $H^*(P_1^*) \simeq H^*(V_1^*) \simeq H^*(W)$ and that $H^*(P_0^*) \simeq H^*(V_0)$; the Five Lemma therefore implies i_{11} is an isomorphism. Therefore i_{12} is because i_{11} and i_{13} are. Finally, i_{10} is by excision.

The arrows $\frown o''_{K_2}$, $\frown o'_{K_2}$, $\frown o'_{K_1}$, and $\frown o_B$ in diagram (9.4) are isomorphisms, and up to natural isomorphism of the Čech and singular cohomology modules of the pair (P_1^*, P_0^*), the composite homomorphism $k := i_2 \circ (\frown o''_{K_2}) \circ i_{10} \circ i_{11}^{-1}$ equals the Poincaré-Alexander duality isomorphism

$$\frown o^Q \colon \check{H}^{m-k}(P_1^*, P_0^*; G) \to H_k(Q \setminus P_0^*, Q \setminus P_1^*; G).$$

For by definition of the Čech cap product given in [D], $\frown o^Q$ is obtained by passage to the direct limit over homomorphisms of the type $i_2 \circ (\frown o''_{K_2}) \circ i_{10}$ where the pair (W, V_0^*) is replaced by an arbitrary open neighborhood pair of (P_1^*, P_0^*) under the proviso that o''_{K_2} is the image of the fundamental class $o_{P_1^*}$ under a certain composite of inclusion induced homomorphisms to be described momentarily. With the proviso met, it follows that $\frown o^Q = k$ up to the isomorphism

$\check{H}^*(P_1^*, P_0^*) \simeq H^*(P_1^*, P_0^*)$ described for the proof of Lemma 9.14(12) by virtue of i_{11} being an isomorphism. The proviso is that o''_{K_2} must be the image of $o_{P_1^*}$ under the homomorphism that is the composite of the arrows in the top row of the commutative diagram of inclusion induced homomorphisms (all coefficients are taken in R and excision isomorphisms are marked by $\simeq$)

$$\begin{array}{ccccc} H_m(M, M \setminus P_1^*) & \to & H_m(M, M \setminus P_1^* \cup V_0^*) & \underset{\simeq}{\leftarrow} & H_m(W \setminus P_0^*, W \setminus P_1^* \cup V_0^* \setminus P_0^*) \\ \uparrow \simeq & & \uparrow \simeq & \swarrow \simeq & \\ H_m(W, W \setminus P_1^*) & \longrightarrow & H_m(W, W \setminus K_2) & & \end{array};$$

however, this follows immediately from the commutativity of the diagram and the definitions of o'_{K_2} and o''_{K_2} since o_{K_2} is necessarily the image of $o_{P_1^*}$ under the inclusion induced map. Thus, as $\frown o^Q$, i_{11}, and i_{10} are isomorphisms so too is $\frown o''_{K_2}$. Diagram chasing then shows $\frown o'_{K_2}$, $\frown o'_{K_1}$, and $\frown o_B$ to be isomorphisms.

It is immediate from the analysis of the last paragraph that diagram (9.4) is an expanded form of the five-arrow subdiagram in the top right corner of diagram (9.3). Hence, the subdiagram is commutative and has all its arrows isomorphisms since these properties hold for diagram (9.4). □

The next lemma is a statement about the diagrams in Figures 4 and 5. To aid in the printing of those diagrams all coefficient modules have been suppressed, but should be as follows: In the diagram of Figure 4 and in the top row of the diagram of Figure 5, each left-hand factor of a tensor product has coefficients in an R-module G and each right-hand factor has coefficients in an R-module G'; the cohomology and homology modules in the bottom two rows of the diagram of Figure 5 have coefficients in $G \otimes G'$.

9.17 LEMMA. *Let $B \in \mathfrak{B}(S)$, $S \neq \emptyset$, and let P, V, W, Q, Q^+, and Q^- be as guaranteed by Lemma 9.14. Then the diagrams in Figures 4 and 5 are commutative where all arrows not labeled as cup or cap products are inclusion induced. Further, all arrows in the diagram in Figure 4 are isomorphisms, and all the inclusion induced arrows in the diagram in Figure 5 are isomorphisms except possibly the two horizontal arrows in the rectangle at the bottom right corner of that diagram.*

Proof. The diagram in Figure 4 is obtained by tensoring each lattice point of diagram (9.3) with the corresponding lattice point of the diagram obtained from diagram (9.3) by interchanging the roles of (P_1, P_0) and (P_1^*, P_0^*), of (V_1, V_0) and (V_1^*, V_0^*), and of (B, B^+) and (B, B^-) and similarly tensoring the arrows. Thus, clearly the tensor product of the two diagrams is commutative since each factor is, and each arrow in the tensor product is an isomorphism since each arrow in each of the factors is.

To prove that the diagram in Figure 5 is commutative, let us first examine the definitions of the vertical arrows in that diagram. Each arrow marked as a cup product, but in particular the leftmost one, is an abbreviation for $\check{\Delta} \circ \times$, the composition of the map induced by the diagonal map with the cohomology cross-product. Where the source and target are Čech modules, the cross-product and map induced by the diagonal are those obtained from the singular theory by passing to the limit

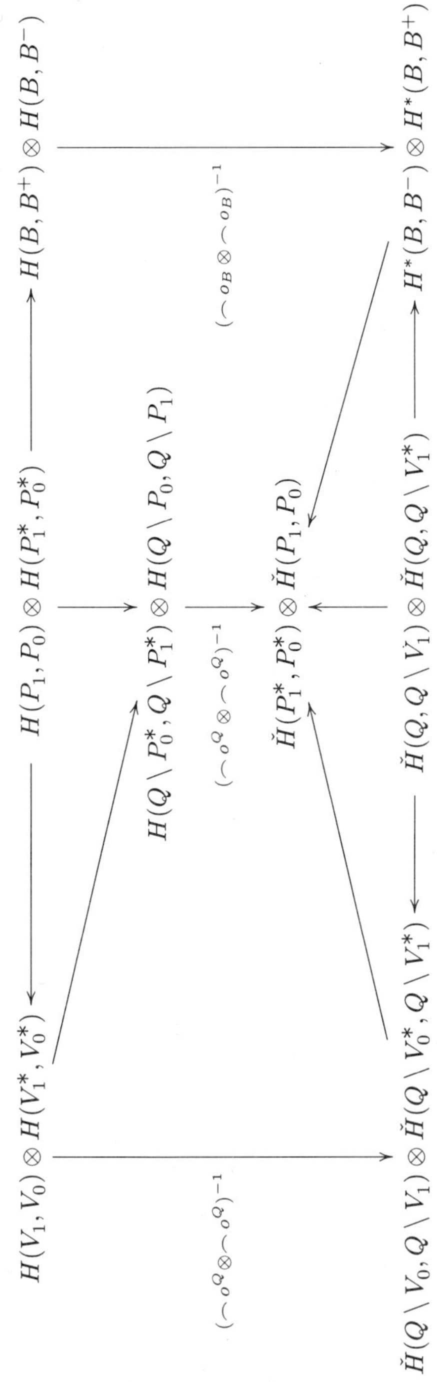

FIGURE 4

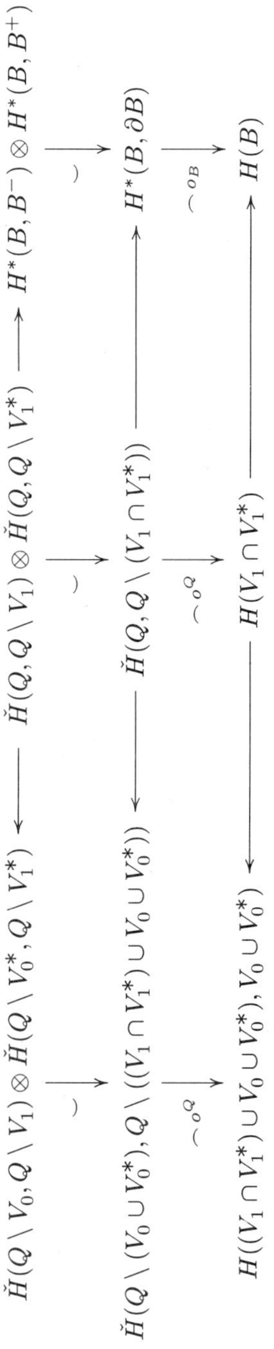

FIGURE 5

over neighborhood pairs, or alternatively, can be viewed as those obtained from the Alexander-Spanier theory. In either case, the functoriality of those constructions implies that the rectangle at the top left of the diagram in Figure 5 is commutative. Similarly, the rectangle at the top right is commutative using the fact that because B, ∂B, and $B^{\pm}$ are ENR's the Čech and singular cohomology of the pairs $(B, B^{\pm})$ and $(B, \partial B)$ are naturally isomorphic, and this is also used in defining the horizontal arrows in that rectangle. The cap product homomorphisms at the left and center of the diagram are Čech cap products and are well-defined, i.e., do not require the use of cohomology with compact supports, since $\mathrm{cl}_M\,(V_1 \setminus V_0)$, $\mathrm{cl}_M\,(V_1^* \setminus V_0^*)$, and $\mathrm{cl}_M\,(V_1 \cap V_1^*)$ are compact subsets of Q. Thus, by the naturality of Čech cap products with respect to inclusions (see formula VIII.(7.11) on page 296 of [D]) the rectangle at the bottom left of the diagram is commutative.

The proof that the rectangle at the bottom right of the diagram in Figure 5 is commutative is more difficult. First, as $V_1 \cap V_1^*$ is an open subset of B° recall that there is an inclusion induced homomorphism $\check{H}_c^*(V_1 \cap V_1^*) \to \check{H}_c^*(B^\circ)$. Second, as $(B^\circ, B^\circ \setminus K)$ is an open pair in M, it is cofinal in the set of all neighborhood pairs of itself contained in M. Thus, by definition its Čech and singular cohomology modules are equal. Consequently, where K_3 compact is chosen as in the proof of Lemma 9.16, i.e., $B \setminus K_3$ is an open collar of ∂B, the diagram of cap product and inclusion induced homomorphisms (all coefficients are in $G \otimes G'$)

(9.6)

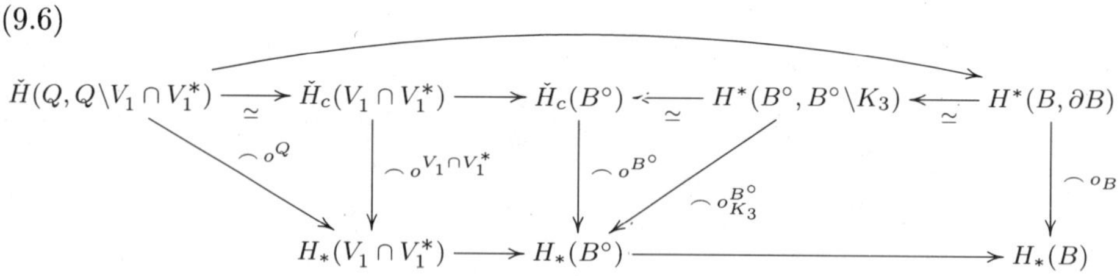

is a well-defined expansion of the rectangle in the lower right corner of the diagram in Figure 5. The class $o_{K_3}^{B^\circ} \in H_m(B^\circ, B^\circ \setminus K_3)$ is the image under the inverse of the excision isomorphism of the fundamental class along K_3 determined by O^M. The three non-horizontal arrows to the left of $\frown o_{K_3}^{B^\circ}$ are Poincaré duality isomorphisms; $\frown o_B$ of course is the Lefschetz duality isomorphism.

Diagram (9.6) is commutative: the subdiagram consisting of the curved arrow on top together with the horizontal arrows of the top row because each is inclusion induced; the trapezoid on the right by naturality of cap products in the singular theory; the triangle to the immediate left of the trapezoid because $\frown o^{B^\circ}$ is the direct limit of triangles of this type where K_3 is replaced by an arbitrary compact $K \subset B^\circ$. The commutativity of the rectangle follows immediately from Proposition 9.8. The commutativity of the triangle at the left of the diagram is an exercise in unwinding the definitions of the two duality isomorphisms that are its non-horizontal sides and utilizes that $\mathrm{cl}_M\,(V_1 \cap V_1^*)$ is a compact subset of Q so that $\check{H}(Q, Q \setminus V_1 \cap V_1^*) \simeq \check{H}_c(Q, Q \setminus V_1 \cap V_1^*) \simeq \check{H}_c(V_1 \cap V_1^*)$.

The sought for commutativity of the rectangle in the lower right corner of the diagram in Figure 5 follows from a simple diagram chase on diagram (9.6) once it is shown that the the two rightmost horizontal arrows in its top row are invertible.

That both are isomorphisms follows from an elementary diagram chase over the inclusion induced diagram (again coefficients are in $G \otimes G'$)

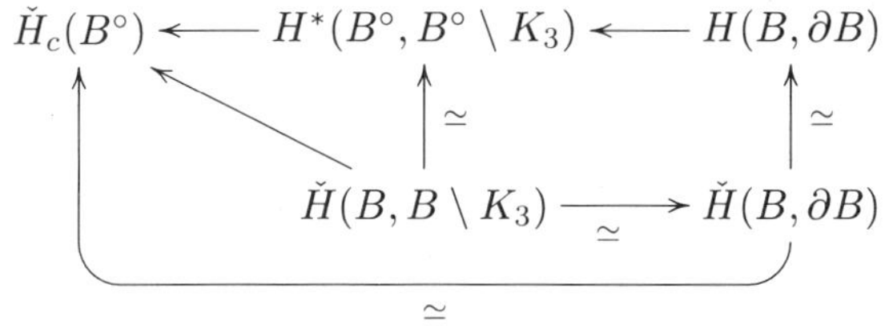

where the two vertical arrows were shown to be isomorphisms in the course of proving Lemma 9.16 and where the curved arrow on the bottom is the isomorphism $\check{H}(B, \partial B) = \check{H}_c(B, \partial B) \simeq \check{H}_c(B^\circ)$.

Finally, the left and right horizontal arrows in the top row of the diagram in Figure 5 are isomorphisms because they are the same respectively as the left and right horizontal arrows in the bottom of the diagram in Figure 4, and the two horizontal arrows in the rectangle at the bottom left of the diagram in Figure 5 are isomorphisms by excision. □

The last proposition needed for the proof of Theorem 9.4 gives precise interpretation to the formulae (2.5). It is inspired by [D, §VIII.13.30, Exercise 1, p. 345] which in the current context states that the composite homomorphism of (9.9) coincides with $\overset{\frown}{\bullet}^V$. In the course of the proof, the cup product of two Čech classes one of which has compact support must be factored through the map on cohomology induced by the diagonal embedding Δ. To carry out this factorization requires the introduction of a relative Čech cohomology module whose elements have compact support along the vertical, i.e., along the fiber of a bundle. As our sole interest is in product bundles, only the following situation need be considered. Let (K, L) be a closed pair in a manifold and define

$$\check{H}^*_{cv}\big((K, L) \times K\big) := \varinjlim \check{H}^*(K \times K, L \times K \cup K \times \omega)$$

where ω lies in the family of subsets of K that are locally compact and cobounded in K.

9.18 Proposition. *Let $V \in M^{(2)}_{\not\Delta o}$. Then where each unlabeled arrow is inclusion induced, each of the following composite homomorphisms equals $\mathfrak{L}_V$ upon identification of the singular and Čech homology modules of the open set $V_1 \cap V_1^*$:*

$$\begin{aligned}
(9.7)\quad & H_i(V_1, V_0; G) \otimes H_j(V_1^*, V_0^*; G') \xrightarrow{(\frown o^M)^{-1} \otimes (\frown o^M)^{-1}} \\
& \quad \check{H}_c^{m-i}(M \setminus V_0, M \setminus V_1; G) \otimes \check{H}_c^{m-j}(M \setminus V_0^*, M \setminus V_1^* G') \xrightarrow{\times} \\
& \quad \check{H}_c^{2m-i-j}\big((M \setminus V_0, M \setminus V_1) \times (M \setminus V_0^*, M \setminus V_1^*); G \otimes G'\big) \xrightarrow{\check{\Delta}^*} \\
& \quad \check{H}_c^{2m-i-j}\big(M \setminus (V_0 \cup V_0^*), M \setminus ((V_1 \cap V_1^*) \cup V_0 \cup V_0^*); G \otimes G'\big) \xrightarrow{\frown o^M} \\
& \quad H_{i+j-m}\big((V_1 \cap V_1^*) \cup V_0 \cup V_0^*, V_0 \cup V_0^*; G \otimes G'\big) \xrightarrow{(e_{1\,*})^{-1}} \\
& \qquad H_{i+j-m}(V_1 \cap V_1^*; G \otimes G');
\end{aligned}$$

$$(9.8)\quad H_i(V_1,V_0;G)\otimes H_j(V_1^*,V_0^*;G') \xrightarrow{(\frown o^M)^{-1}\otimes(\frown o^M)^{-1}}$$
$$\check{H}_c^{m-i}(M\setminus V_0, M\setminus V_1;G)\otimes \check{H}_c^{m-j}(M\setminus V_0^*, M\setminus V_1^*;G') \to$$
$$\check{H}^{m-i}(V_1^*\setminus V_0^*, V_1^*\setminus((V_1\cap V_1^*)\cup V_0^*);G)\otimes \check{H}_c^{m-j}(V_1^*\setminus V_0^*;G') \xrightarrow{\smile}$$
$$\check{H}_c^{2m-i-j}(V_1^*\setminus V_0^*, V_1^*\setminus((V_1\cap V_1^*)\cup V_0^*);G\otimes G') \xrightarrow{\frown o^{V_1^*}}$$
$$H_{i+j-m}((V_1\cap V_1^*)\cup V_0^*, V_0^*;G\otimes G') \xrightarrow{(e_{2*})^{-1}}$$
$$H_{i+j-m}(V_1\cap V_1^*;G\otimes G');$$

$$(9.9)\quad H_i(V_1,V_0;G)\otimes H_j(V_1^*,V_0^*;G') \xrightarrow{(\frown o^M)^{-1}\otimes 1_*}$$
$$\check{H}_c^{m-i}(M\setminus V_0, M\setminus V_1;G)\otimes H_j(V_1^*,V_0^*;G') \to$$
$$\check{H}^{m-i}(V_1^*\setminus V_0^*, V_1^*\setminus((V_1\cap V_1^*)\cup V_0^*);G)\otimes H_j(V_1^*,V_0^*;G') \xrightarrow{\frown}$$
$$H_{i+j-m}((V_1\cap V_1^*)\cup V_0^*, V_0^*;G\otimes G') \xrightarrow{(e_{2*})^{-1}}$$
$$H_{i+j-m}(V_1\cap V_1^*;G\otimes G').$$

Proof. The proof is a diagram chase on the diagram obtained by putting together the two diagrams of Figures 6 and 7 as follows: the two vertical arrows on the right edge of the diagram in Figure 6 coincide with the top two vertical arrows in the left edge of the diagram in Figure 7 and the two horizontal arrows in the bottom edge of the diagram of Figure 6 coincide with the bottom two vertical arrows in the left edge of the diagram of Figure 7. All unlabeled arrows are inclusion induced; those on Čech cohomology with compact support induced by an inclusion of the form $(K\setminus L,\emptyset)\subset(K,L)$ where (K,L) is a closed pair in a manifold are natural isomorphisms; see statement (6.24) on [D, p. 289]. Although all coefficient modules have been suppressed from the diagrams, they are easily determined from the statement of the proposition after reading the following paragraph.

The composition of the arrows of (9.9) yields the top row of the diagram in Figure 6 followed by the top row of the diagram in Figure 7, and $\frown\!\!\!\bullet^V$ is the left edge of the diagram in Figure 6 followed by the bottom edge of the diagram in Figure 7 followed by the identity homomorphism on $H_*(V_1\cap V_1^*)$ as represented by the right edge of the diagram in Figure 7 in order to carry out the chase. The composite homomorphisms of (9.7) and (9.8) are easily seen to lie "between" the two just given. Assuming the diagrams of Figures 6 and 8 are commutative, the reader can easily perform the necessary diagram chase to see that the composite homomorphisms of (9.7), (9.8), and (9.9) coincide with $\frown\!\!\!\bullet^V$.

The diagram of Figure 6 is seen to commute as follows. Where e_7^* and e_8^* are the excision induced isomorphisms (see [D, §VIII.6.22], in particular the isomorphism labeled (6.24), or see [Sp, Lemma 6.6.11])

$$\check{H}_c(M\times M\setminus(V_0\times V_1^*\cup V_1\times V_0^*), M\times M\setminus V_1\times V_1^*) \xrightarrow{e_7^*} \check{H}_c(V_1\times V_1^*\setminus(V_0\times V_1^*\cup V_1\times V_0^*))$$

and

$$\check{H}_c((M\setminus V_0, M\setminus V_1)\times(M\setminus V_0^*, M\setminus V_1^*)) \xrightarrow{e_8^*} \check{H}_c(V_1\times V_1^*\setminus(V_0\times V_1^*\cup V_1\times V_0^*)),$$

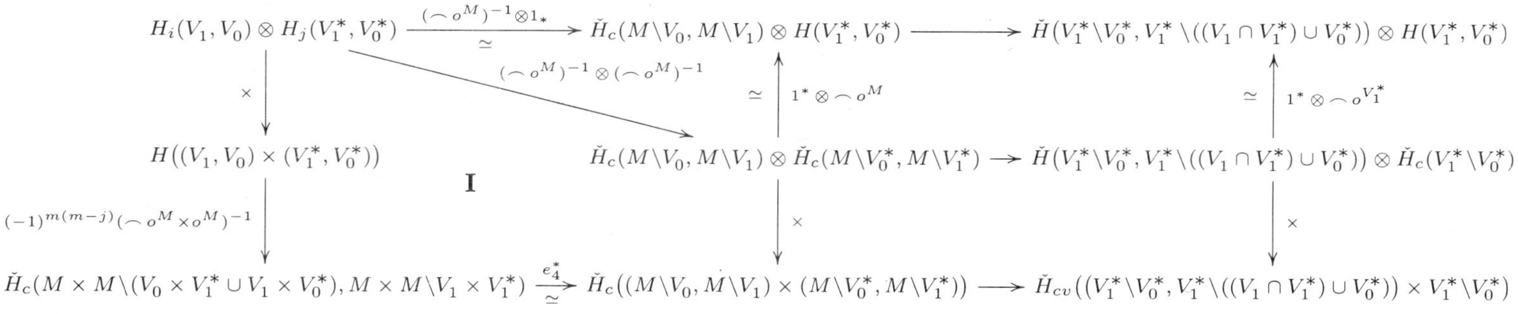

FIGURE 6

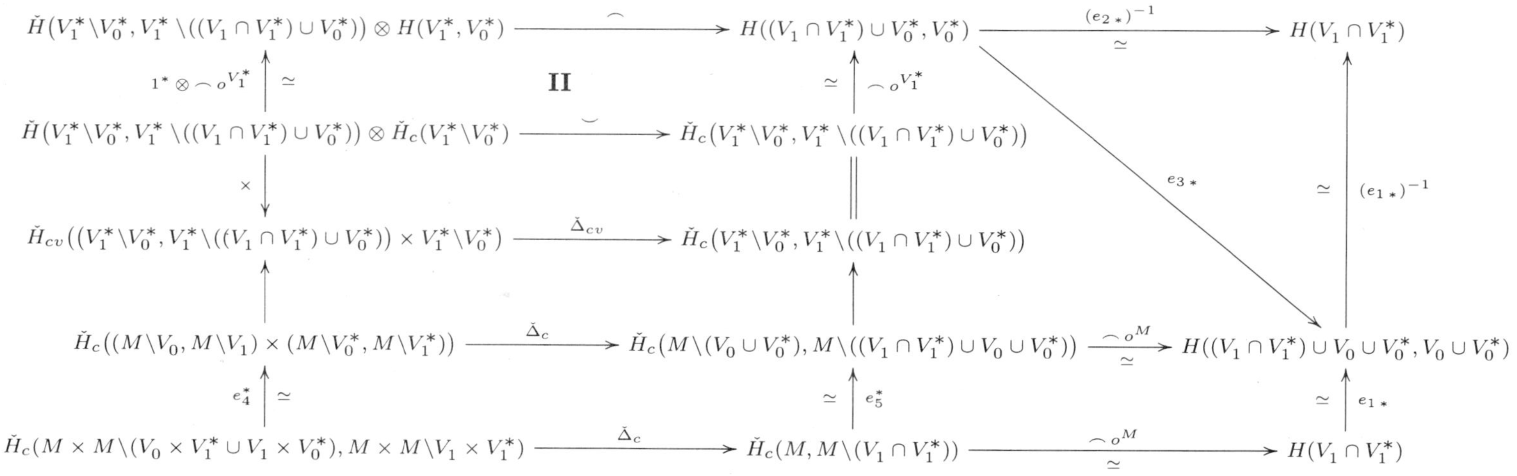

FIGURE 7

the commutativity of the trapezoid labeled **I** is a consequence of Proposition 9.10 and the fact that $e_7^* = e_8^* \circ e_4^*$. That the rectangle at the top right corner of the diagram in Figure 6 commutes follows from the definition of the duality isomorphism that is the right-hand factor of its left edge. Specifically, that duality isomorphism is by definition the composite of the isomorphism $\check{H}_c(M \setminus V_0^*, M \setminus V_1^*) \simeq \check{H}_c(V_1^* \setminus V_0^*)$ which is the right-hand factor in the bottom arrow of the rectangle with the duality isomorphism that is the right-hand factor of the right edge of the rectangle. It is evident that the remaining rectangle and the triangle in the diagram of Figure 6 commute.

The commutativity of the diagram in Figure 7 is seen as follows. The commutativity of the rectangle labeled **II** is an immediate consequence of Proposition 9.11. The rectangle below rectangle **II** commutes because the cup product that is its top horizontal arrow and $\check{\Delta}_{cv} \circ \times$ are both obtainable by passage to the limit over homomorphisms of the form $\Delta^* \circ \times$ on the tensor product of singular cohomology modules of open pairs. The rectangle at the bottom right corner of the diagram in Figure 7 and the triangle above it both commute as an immediate consequence of Proposition 9.9. The commutativity of the remaining two rectangles of the diagram and of the remaining triangle of excision isomorphisms is evident. □

9.19 The Proof of Theorem 9.4. The theorem follows from Lemma 9.17, Proposition 9.18, in particular, that $\mathfrak{L}_V$ can be identified with the composite homomorphism of the sequence in (9.7), and a diagram chase upon the composite diagram obtained by putting together the diagram

(9.10)

$$\begin{array}{ccc}
 & & H(V_1, V_0; G) \otimes H(V_1^*, V_0^*; G') \\
 & \swarrow{\scriptstyle (\frown o^M \otimes \frown o^M)^{-1}} & \downarrow{\scriptstyle (\frown o^Q \otimes \frown o^Q)^{-1}} \\
\check{H}(M\backslash V_0, M\backslash V_1) \otimes \check{H}(M\backslash V_0^*, M\backslash V_1^*) & \longrightarrow & \check{H}(Q\backslash V_0, Q\backslash V_1) \otimes \check{H}(Q\backslash V_0^*, Q\backslash V_1^*) \\
\downarrow{\scriptstyle \smile} & & \downarrow{\scriptstyle \smile} \\
\check{H}(M\backslash(V_0 \cup V_0^*), M\backslash((V_1 \cap V_1^*) \cup V_0 \cup V_0^*)) & \longrightarrow & \check{H}(Q\backslash(V_0 \cup V_0^*), Q\backslash((V_1 \cap V_1^*) \cup V_0 \cup V_0^*)) \\
 & \searrow{\scriptstyle \frown o^M} & \downarrow{\scriptstyle \frown o^Q} \\
 & & H((V_1 \cap V_1^*) \cup V_0 \cup V_0^*), V_0 \cup V_0^*)
\end{array}$$

and the diagrams of Figures 4 and 5. Specifically, given $B \in \mathfrak{B}(S)$, let P, V, Q, $Q^{\pm}$, and W be as guaranteed by Lemma 9.14, and join the diagrams of Figures 4 and 5 by identifying the bottom row of the former with the top row of the latter. Then combine this joined diagram with diagram (9.10) by identifying the left edge of the former with the right edge of the latter. Let us refer to the commutative diagram so obtained as the combined diagram. Note that in the sequence (9.7) each occurrence of Čech cohomology with compact support can be replaced by ordinary Čech cohomology because the sets $V_1 \setminus V_0$, $V_1^* \setminus V_0^*$, and $V_1 \cap V_1^*$ have compact closure in M.

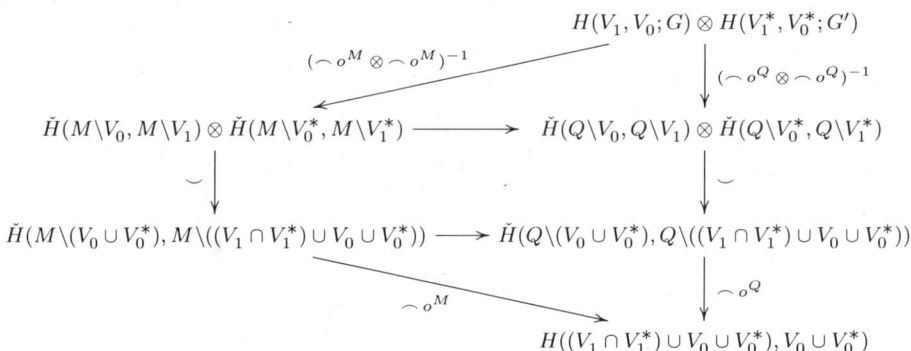

Then the composite homomorphism obtained by starting in the top row of the combined diagram at $H(P_1, P_0) \otimes H(P_1^*, P_0^*)$ and traveling to the left along the top row, then down the outside left edge of the combined diagram, and then along the bottom row of the combined diagram to $H(B)$ equals $\mathfrak{L}_B^P := i_B^{V_1 \cap V_1^*} \circ \mathfrak{L}_V^P$ as follows from Proposition 9.18 using the composite homomorphism from (9.7). A fairly straightforward diagram chase, but one that makes full use of the fact that many of the arrows in the combined diagram have been shown to be isomorphisms, shows that the composite, call it C_B^P, obtained by starting at $H(P_1, P_0) \otimes H(P_1^*, P_0^*)$ and traveling to the right along the top row of the combined diagram and then down its outer right edge to $H(B)$ equals $\mathfrak{L}_B^P$. Call the composite of the homomorphisms in the outer right edge of the combined diagram C_B^B. The equality $\mathfrak{L}_B^P = C_B^P$ together with the commutativity of the diagram

$$
\begin{array}{ccc}
 & \widetilde{H}_*\mathcal{C}(S;G) \otimes \widetilde{H}_*\mathcal{C}^*(S;G') & \\
 \swarrow {\scriptstyle (\bar{p}_*^+ \otimes \bar{p}_*^-)^{-1}} & & {\scriptstyle (\bar{b}_*^+ \otimes \bar{b}_*^-)^{-1}} \searrow \\
H_*(P_1, P_0; G) \otimes H_*(P_1^*, P_0^*; G') & \xrightarrow[\simeq]{} & H_*(B, B^+; G) \otimes H_*(B, B^-; G')
\end{array}
$$

which follows from the definitions yields via the definition of $\mathfrak{L}$ that

$$\pi_B^S \circ \mathfrak{L} = C_B^B \circ (\bar{b}_*^+ \otimes \bar{b}_*^-)^{-1} .$$

Application to $\boldsymbol{\alpha} \otimes \boldsymbol{\gamma} \in \widetilde{H}_*\mathcal{C}(S) \otimes \widetilde{H}_*\mathcal{C}^*(S)$ then yields

$$(\boldsymbol{\alpha} \mathbin{\widehat{\bullet}} \boldsymbol{\gamma})_B = (x_B \smile y_B) \frown o_B = x_B \frown \gamma_B$$

as desired where recall that x_B and y_B have been defined in the paragraph preceding the statement of Theorem 9.4. □

D. The case of $S \cap \partial M \neq \emptyset$. As mentioned earlier McCord's statement and proof of Poincaré duality of Conley indices is broader than that considered above. The sole goal of this section is to point out how his broader result allows one by analogy with Theorem 9.4 to associate two intersection pairings to an isolated invariant set S of a flow in a manifold with boundary M when

$$S \cap \partial M \neq \emptyset . \tag{9.11}$$

Note that by invariance of domain, the flow in M restricts to a flow in ∂M. It is then easy to see that the intersection of an isolated invariant set in M with ∂M yields an isolated invariant set of the flow in ∂M. Suppose (9.11) holds and for convenience, set

$$S_\partial := S \cap \partial M .$$

In this context, one can see from simple examples that the homotopy Conley index of S is often trivial either in forward or backward time although the corresponding index of S_∂ is non-trivial. In such situations, it is shown in [Mc1] that it is possible by dividing out what is happening in the boundary at the index space level to assign S a non-trivial relative Conley index. We refer the reader to [Mc1] for full generality, but here this index will only be described in terms of isolating blocks. For simplicity, assume M is C^r $(r \geq 1)$ and that the flow in M, hence also in ∂M, is generated by a continuous vectorfield. It is not difficult to modify the construction of isolating blocks in [WY] to obtain in each isolating neighborhood of S an isolating block B for S whose intersection with ∂M is an isolating block for S_∂. For convenience, set

$$B_\partial := B \cap \partial M .$$

In the modified construction B and B_∂ will be compact, C^0 manifolds with boundary. Also,

$$\partial B = B^+ \cup B^- \cup B_\partial \tag{9.12}$$

and $B_\partial^\pm = B^\pm \cap B_\partial$. Accordingly, the definition of $\mathfrak{B}(S)$ should be changed to reflect (9.12) whereas B_∂ will be in $\mathfrak{B}(S_\partial)$ as defined in §9.A. However, for the purposes of the discussion here, we can define $\mathfrak{B}(S)$ to consist of those isolating blocks B for S that are compact manifolds with boundary satisfying (9.12) with $B^\pm$ and B_∂ submanifolds with boundary of ∂B, either void or of codimension one in ∂B; also

$$\partial B_\partial = B_\partial^+ \cup B_\partial^-$$

and $B_\partial^\pm$ is a submanifold with boundary of ∂B_∂, either void or of codimension one.

Then as a connected simple system, the relative index of S in forward time, denote it $\mathcal{C}(S, S_\partial)$, includes in its set of objects those quotient spaces of the form $B/(B^+ \cup B_\partial)$. Similarly, the relative index of S in reverse time, denote it $\mathcal{C}^*(S, S_\partial)$, includes in its set of objects those quotient spaces of the form $B/(B^- \cup B_\partial)$. As in the absolute case one can assign homology modules to these relative indices which in keeping with the notation we have used in the absolute case we will denote by $\widetilde{H}_*\mathcal{C}(S, S_\partial)$ and $\widetilde{H}_*\mathcal{C}^*(S, S_\partial)$ where for simplicity coefficients are taken in R. Then note that there will be isomorphisms

$$\widetilde{H}_*\mathcal{C}(S, S_\partial) \simeq H_*(B, B^+ \cup B_\partial) \qquad \text{and} \qquad \widetilde{H}_*\mathcal{C}^*(S, S_\partial) \simeq H_*(B, B^- \cup B_\partial)$$

for any block $B \in \mathfrak{B}(S)$. Similar definitions and statements can be made for cohomology; their easy formulation is left to the reader. Then the full duality result of [Mc2] is the following:

9.20 THEOREM. *There are isomorphisms*

$$\widetilde{H}^k\mathcal{C}(S, S_\partial) \simeq \widetilde{H}_{m-k}\mathcal{C}^*(S) , \qquad \widetilde{H}^k\mathcal{C}^*(S) \simeq \widetilde{H}_{m-k}\mathcal{C}(S, S_\partial)$$

and

$$\widetilde{H}^k\mathcal{C}^*(S, S_\partial) \simeq \widetilde{H}_{m-k}\mathcal{C}(S) , \qquad \widetilde{H}^k\mathcal{C}(S) \simeq \widetilde{H}_{m-k}\mathcal{C}^*(S, S_\partial)$$

that in each block $B \in \mathfrak{B}(S)$ as above are given by cap product with the fundamental class $o_B \in H_m(B, \partial B)$.

9.21 DEFINITION. For B a block as above, for $\alpha \in H_{m-k}(B, B^+)$, let $x_\alpha \in H^k(B, B^- \cup B_\partial)$ be the unique element such that $\alpha = x_\alpha \frown o_B$. For such α and for $\gamma \in H_*(B, B^- \cup B_\partial)$, define a pairing

$$\mathfrak{L}_{\partial^-\ B}: H_*(B, B^+) \otimes H_*(B, B^- \cup B_\partial) \to H_*(B)$$

by

$$\mathfrak{L}_{\partial^-\ B}(\alpha \otimes \gamma) = x_\alpha \frown \gamma\,.$$

Similarly, one can define a pairing

$$\mathfrak{L}_{\partial^+\ B}: H_*(B, B^+ \cup B_\partial) \otimes H_*(B, B^-) \to H_*(B)\,.$$

Then define

$$\bar{\mathfrak{L}}_{\partial^-\ B} := \mathfrak{L}_{\partial^-\ B} \circ (\bar{b}^+ \otimes \bar{b}^-)^{-1}: \widetilde{H}_*\mathcal{C}(S) \otimes \widetilde{H}_*\mathcal{C}^*(S, S_\partial) \to H_*(B)$$

and similarly define $\bar{\mathfrak{L}}_{\partial^+\ B}$. It follows from the proof of Theorem 9.20, in particular [Mc2, Lemma 2.8], that if B and B' are two isolating blocks with $B' \subset B$, then

$$i_B^{B'} \circ \bar{\mathfrak{L}}_{\partial^-\ B'} = \bar{\mathfrak{L}}_{\partial^-\ B}\,.$$

Similarly,

$$i_B^{B'} \circ \bar{\mathfrak{L}}_{\partial^+\ B'} = \bar{\mathfrak{L}}_{\partial^+\ B}\,.$$

As $\mathfrak{B}(S)$ is cofinal in $\mathcal{N}(S)$ it follows that we can pass to the inverse limit to obtain pairings of degree $-m$

$$\mathfrak{L}_{\partial^-}: \widetilde{H}_*\mathcal{C}(S) \otimes \widetilde{H}_*\mathcal{C}^*(S, S_\partial) \to \check{H}_*(S)$$

and

$$\mathfrak{L}_{\partial^+}: \widetilde{H}_*\mathcal{C}(S, S_\partial) \otimes \widetilde{H}_*\mathcal{C}^*(S) \to \check{H}_*(S)\,.$$

Simple examples suggest that only one of these two pairings will be non-trivial for S satisfying (9.11) and $S = S_\partial$, e.g., hyperbolic rest points in the boundary, but the author has not investigated the matter further.

Also, it is certainly true that if one uses these pairings to define intersection number pairings in the same manner used to define $^\#\mathfrak{L}$, then the intersection number pairings so defined will be invariant under continuation and each of the full pairings has continuous lifts over a path of isolated invariant sets just as $\mathfrak{L}$ does. The proofs of these facts can either be obtained by suitably extending the notion of admissible index pairs with separate extensions required for each pairing and then mimicking the proofs for $\mathfrak{L}$, or at least for invariance of intersection numbers, by suitably adapting the proof in [Mc2] that the duality isomorphisms of Theorem 9.20 are invariant under continuation.

The cursory treatment here is not meant to imply a lack of importance of results that might be obtainable, but the present investigation draws to a close here. There are, however, numerous interesting differential equations modeling the dynamics of the relative population densities of n related entities where the natural phase space is the open n-simplex. Such systems often extend to the boundary of the simplex, but with singularities at the vertices, and the presence of critical points or more general isolated invariant sets in the $(n-1)$-faces certainly affects the dynamics in the open simplex.

APPENDIX A

INTERSECTION NUMBERS AND EXISTENCE RESULTS FOR TWO-POINT BOUNDARY VALUE PROBLEMS OF SINGULARLY PERTURBED SYSTEMS

The analysis of Example 3.5 above leads to the following existence result.

A.1 THEOREM. *Assume* $a\colon [0,1] \to \,]0,1[$ *is continuous, satisfies* $a(\lambda) \neq 1/2$ *for* $\lambda = 0, 1$, *and has the property that* $(a(x) - \frac{1}{2})$ *changes sign at least once over the interval* $[0,1]$. *Then there exists* $\varepsilon^* > 0$ *and a family of solutions* $\{v_\varepsilon\}_{0<\varepsilon\leq\varepsilon^*}$ *of the two-point boundary value problem*

$$\begin{gathered} \varepsilon^2 v'' + v(1-v)(v-a(x)) = 0 \\ v'(0) = 0 = v'(1) \end{gathered}$$

with the properties that for each $x \in [0,1]$, $0 < v_\varepsilon(x) < 1$ *and*

$$\lim_{\varepsilon\to 0} v_\varepsilon(x) = \begin{cases} 0, & \text{if } a(x) > \frac{1}{2} \\ 1, & \text{if } a(x) < \frac{1}{2}. \end{cases}$$

Remark. The hypothesis "$(a(x) - \frac{1}{2})$ changes sign at least once over $[0,1]$" ensures that the solutions found are not the constant solutions $v_\varepsilon \equiv 0$ or $v_\varepsilon \equiv 1$.

If the zero set of $(a(x) - \frac{1}{2})$ consists of an infinite discrete set at each point of which $(a(x) - \frac{1}{2})$ changes sign together with finitely many limit points (e.g., $a(\frac{1}{2}) = \frac{1}{2}$ and $a(x) = \frac{1}{2} + (x - \frac{1}{2})^2 \sin\left(\pi(2-4x)^{-1}\right)$ for $x \neq \frac{1}{2}$), then near the limit points the family exhibits "chattering"; i.e., in any given neighborhood of a limit point the number of oscillations from near zero to near one exhibited by v_ε increases as $\varepsilon \downarrow 0$.

If it is assumed that $(a(x) - \frac{1}{2})$ has only finitely many zeros at each of which $(a(x) - \frac{1}{2})$ changes sign, then the theorem implies that the family of solutions v_ε, $0 < \varepsilon \ll 1$, exhibits an interior transition layer at each such zero. It follows from the proof of Theorem A.1 given below that in these layers $(v_\varepsilon(x) - \frac{1}{2})$ changes sign in opposite direction from $(a(x) - \frac{1}{2})$. Hence, if in fact a is C^1 with non-zero derivative at each zero of $(a(x) - \frac{1}{2})$, then according to [AnMP] the non-constant stationary solution v_ε is a stable equilibrium of the Fisher equation with Neumann boundary conditions. Using the results of [K7] one can also find many unstable stationary states which together with the results of [K6] yield all the stationary states of the Fisher equation. Details will appear elsewhere.

The proof of Theorem A.1 is based on an existence theorem for singularly perturbed, vector two-point boundary value problem in $\mathbf{R}^m$ on the finite interval $[a,b]$ of the form

$$\varepsilon \mathbf{u}' = \mathbf{F}(\mathbf{u}, x, \varepsilon) \tag{A.1a}$$

$$\mathbf{u}(a) \in \mathbf{E}(a), \qquad \mathbf{u}(b) \in \mathbf{E}(b) \tag{A.1b}$$

where differentiation is with respect to $x \in \mathbf{R}$ and where $\mathbf{E}(\lambda)$ for $\lambda = a, b$ is the locus of points in $\mathbf{R}^m$ satisfying the endpoint condition at λ. The following hypotheses are made on $\mathbf{F}$ and on the endpoint sets:

HYPOTHESES BVP.

(1) $\mathbf{F}$ *is continuous on an open, connected subset of* $\mathbf{R}^{m+1} \times \mathbf{R}^+$, *and for* $\lambda \in W$, *an open interval containing* $[a,b]$, *the family of fast systems* $\dot{\mathbf{u}} = \mathbf{F}(\mathbf{u}, \lambda, 0)$ *is defined and generates a continuous family of flows* $\{\varphi_\lambda\}_{\lambda \in W}$ *in* $\mathbf{R}^m$ *and the family of extended systems* $\varepsilon \mathbf{u}' = \mathbf{F}(\mathbf{u}, x, \varepsilon)$, $x' = 1$ *is defined and generates a continuous family of flows* $\{\psi_\varepsilon\}_{0<\varepsilon \ll 1}$ *in* $\mathbf{R}^m \times \mathbf{R}$;

(2) *there exists a continuous section* $x \mapsto (S(x), x)$ *of* $\mathcal{S}(\varphi)$ *defined on an open interval containing* $[a,b]$;

(3) *there exists a nested index pair* (N_1, N_0) *for* $S(a)$ *relative to the forward time flow and a singular chain c with support in* $\mathbf{E}(a)$ *that is a relative cycle in* N_1 *modulo* N_0 *representing* $\boldsymbol{\alpha} \in \widetilde{H}_* \mathcal{C}(S)$;

(4) *there exists a nested index pair* (N_1^*, N_0^*) *for* $S(b)$ *relative to the reverse time flow and a singular chain* c^* *with support in* $\mathbf{E}(b)$ *that is a relative cycle in* N_1^* *modulo* N_0^* *representing* $\boldsymbol{\gamma} \in \widetilde{H}_* \mathcal{C}^*(S)$;

(5) $\mathcal{C}(\beta)_* \boldsymbol{\alpha} \# \boldsymbol{\gamma} \neq 0$ *where* β *is the path class in* $\mathcal{S}(\varphi)$ *from* $S(a)$ *to* $S(b)$ *defined by the section* $x \mapsto (S(x), x)$; *i.e.,* $\beta(t) = \big(S((1-t)a + tb), (1-t)a + tb\big)$.

Remark. In the context of hypotheses BVP, Corollary 6.3 and the non-singularity of the pairing $^{\#}\mathfrak{L}$ (hence of L) guaranteed by Corollary 9.6 (assuming coefficients taken in a field) imply that given a left endpoint set supporting non-zero homology in $\widetilde{H}_* \mathcal{C}(S(a))$, then there exist right endpoint sets so that hypotheses BVP are satisfied. More importantly, to satisfy hypotheses BVP, Corollary 9.6 implies that $\mathbf{E}(a)$ and $\mathbf{E}(b)$ must support dual classes relative to L modulo continuation along β.

A.2 THEOREM. *Under the assumption of hypotheses* BVP, *there exists a solution* $\mathbf{u}_\varepsilon$ *to the boundary value problem* (A.1) *for* $0 < \varepsilon \ll 1$. *Furthermore, if* $\mathcal{O}$ *is open in* $\mathbf{R}^m \times \mathbf{R}$, *if* $S(x) \times \{x\} \subset \mathcal{O}$ *for each* $x \in [a,b]$, *and if* $N_1 \times \{a\}$ *and* $N_1^* \times \{b\}$ *are subsets of* $\mathcal{O}$, *then* $\mathbf{u}_\varepsilon$ *can be chosen so that its graph, i.e., the point set* $\{(\mathbf{u}_\varepsilon(x), x) : x \in [a,b]\}$, *also lies in* $\mathcal{O}$.

Before giving the proof, it is convenient to make the following definitions and notational conventions: (1) A singular chain c'' is called a subchain of a singular chain c' if c'' is obtained from an iterated barycentric subdivision of c' by deleting some of the summands in the subdivision. (2) If c' is a singular chain in $\mathbf{R}^m$, then for $x \in \mathbf{R}$, let $c' \times \{x\}$ denote the singular chain in $\mathbf{R}^m \times \{x\}$ induced by the identification of $\mathbf{R}^m$ with $\mathbf{R}^m \times \{x\}$. (3) For $(\mathbf{u}, x, t)$ in the domain of ψ_ε, define

$$(\mathbf{u}, x) \overset{\varepsilon}{\cdot} t := \psi_\varepsilon(\mathbf{u}, x, t).$$

A.3 Proof of Theorem A.2. The proof is based on Theorem 3.1 of [K4] which takes hypotheses BVP(1)–(3) as hypotheses, but also requires that an index pair for $S(b)$ relative to the forward time flow be given. Let us give one by using Lemma 5.12 to choose $\langle N_1(b), N_0(b)\rangle$ to be an index pair for $S(b)$ relative to the fast flow φ_b with the property that $((N_1(b), \cap N_0(b)), (N_1^*, N_0^{*-s}))$ is an object of $\mathcal{AIP}(S(b))$ for some sufficiently large $s > 0$; see Remark 5.12.2.

By [K4, Theorem 3.1] there exists $\bar{\varepsilon} > 0$ with the property that if $0 < \varepsilon \leq \bar{\varepsilon}$, then there exists a subchain c_ε of c whose image under the chain map induced by the time $(b-a)$ flow map of ψ_ε mapping singular chains in $\mathbf{R}^m \times \{a\}$ to singular chains in $\mathbf{R}^m \times \{b\}$ is a relative cycle in $N_1(b)$ modulo $N_1(b) \cap N_0(b)$ representing $\mathcal{C}(\beta)_*\boldsymbol{\alpha}$. That is, where $\bar{c}_\varepsilon$ is defined by

$$\bar{c}_\varepsilon \times \{b\} := c_\varepsilon \times \{a\} \underset{\sharp}{\overset{\varepsilon}{\cdot}} (b-a),$$

$\bar{c}_\varepsilon$ is a relative cycle in $N_1(b)$ modulo $N_0(b)$ representing $\mathcal{C}(\beta)_*\boldsymbol{\alpha}$. It follows immediately from Proposition 2.7 and hypotheses BVP(4) and BVP(5) that there exists $(\mathbf{u}_{b\,\varepsilon}, b) \in (|\bar{c}_\varepsilon| \setminus N_0(b) \cap |c^*| \setminus N_0^{*-s}) \times \{b\}$. It follows that the curve $\mathbf{u}_\varepsilon$ defined by

$$(\mathbf{u}_\varepsilon(x), x) := (\mathbf{u}_{b\,\varepsilon}, b) \overset{\varepsilon}{\cdot} (x-b) \qquad \text{for } a \leq x \leq b$$

is a solution of the boundary value problem (A.1).

Furthermore, given $\mathcal{O}$ open in $\mathbf{R}^m \times \mathbf{R}$ with $S(x) \times \{x\} \subset \mathcal{O}$ for $x \in W$, Theorem 3.1 of [K4] also ensures that c_ε will satisfy $|c_\varepsilon| \times \{a\} \overset{\varepsilon}{\cdot} [0, b-a] \subset \mathcal{O}$ whence $\mathbf{u}_\varepsilon$ as defined will have graph contained in $\mathcal{O}$. □

A.4 Proof of Theorem A.1. Throughout the proof rather than work with the second-order equation, we instead work with the equivalent first-order non-autonomous system

$$(A.2_\varepsilon) \qquad \begin{aligned} \varepsilon v' &= w \\ \varepsilon w' &= -v(1-v)(v-a(x)). \end{aligned}$$

In fact because the argument is conveniently phrased in terms of invariant regions, we work with the flow in $\mathbf{R}^3$ generated by system $(A.2_\varepsilon)$ extended by the equation of the independent variable, $x' = 1$. Call the extended system, system $(A.2_\varepsilon^+)$.

Before entering into the details of the proofs of existence and the requisite limiting properties, let us first show that a non-constant solution v_ε of the boundary value problem must necessarily satisfy $0 < v_\varepsilon(x) < 1$ for each $x \in [0,1]$. We begin by partitioning $\mathbf{R}^3$ into five closed regions $W_I, W_{II}, \ldots, W_V$ as follows:

$(v, w, x) \in W_I$ if, and only if, $v \leq 0$ and $w \geq 0$;
$(v, w, x) \in W_{II}$ if, and only if, $v \leq 0$ and $w \leq 0$;
$(v, w, x) \in W_{III}$ if, and only if, $0 \leq v \leq 1$;
$(v, w, x) \in W_{IV}$ if, and only if, $v \geq 1$ and $w \geq 0$;
$(v, w, x) \in W_V$ if, and only if, $v \geq 1$ and $w \leq 0$.

These sets are illustrated in cross-section in Figure 8 where the arrows indicate the direction of flow across the boundary of these regions, but observe that the points $(0,0,x)$ and $(1,0,x)$ correspond to constant solutions of $\varepsilon^2 v'' + v(1-v)(v-a(x)) = 0$. In particular, as is easily checked by computing derivatives along trajectories at boundary points, regions W_{II} and W_{IV} are positively invariant relative to the flow of system $(A.2_\varepsilon^+)$ while regions W_I and W_V are negatively invariant.

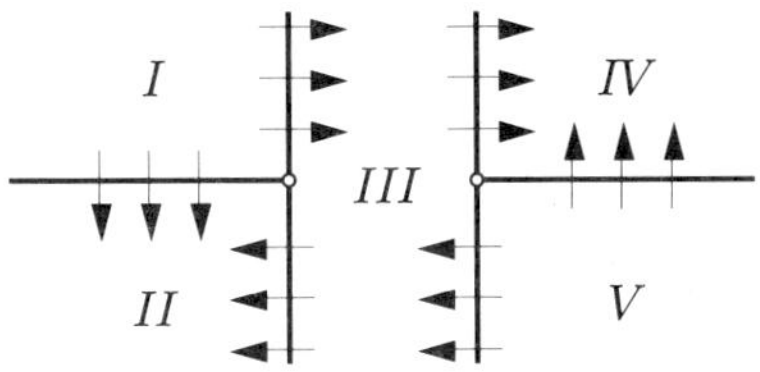

FIGURE 8. Cross-sections at $x \in \mathbf{R}$ of the closed regions W_I, W_{II}, W_{III}, W_{IV}, W_V.

Since $v_\varepsilon'(0) = 0$ and v_ε is not a constant solution, it is evident from Figure 8 that if $v_\varepsilon(0) < 0$ (respectively, $v_\varepsilon(0) > 1$), then $(v_\varepsilon(x), \varepsilon v_\varepsilon'(x), x)$ lies interior to W_{II} (respectively W_{IV}) for $0 < x \leq 1$ contradicting that $v_\varepsilon'(1) = 0$. Thus $0 < v_\varepsilon(0) < 1$. Similarly, interpreting Figure 8 in backwards time, it follows that $0 < v_\varepsilon(1) < 1$. Thus $z_\varepsilon(\lambda) := (v_\varepsilon(\lambda), \varepsilon v_\varepsilon'(\lambda), \lambda)$ lies in the interior of W_{III} for $\lambda = 0, 1$. Figure 8 again shows that the orbit segment through $z_\varepsilon(0)$ with $0 \leq x \leq 1$ cannot exit W_{III} and then reenter. It follows that $0 < v_\varepsilon(x) < 1$ for $x \in [0, 1]$.

For the existence and limit proofs, the notation established in Example 3.5 will be used. For definiteness, it will be assumed that $a(0) > \frac{1}{2} > a(1)$; the other possibilities are dealt with similarly.

First, let us see that the locus of points in the phase space of the fast system at λ satisfying the Neumann endpoint condition supports a relative 1-cycle in $B(\lambda)$ modulo its exit (entrance) set representing $\boldsymbol{\alpha}_0$ (respectively, $\boldsymbol{\gamma}_1$) when λ equals zero (respectively, one). As drawn in Figure 1, $B(\lambda)$ is a square with one pair of diametrically opposite vertices lying on the line $w = 0$. As these vertices are in the intersection of the exit and entrance sets of $B(\lambda)$, the line segment of $w = 0$ between these two vertices is an arc having one endpoint in each component of the exit set of $B(\lambda)$ and also having one endpoint in each component of the entrance set of $B(\lambda)$. Hence as observed in Example 3.5, this line segment when regarded as the homeomorphic image of the standard 1-simplex can on the one hand, when $\lambda = 0$, be taken as a geometric representative of the homology class $\boldsymbol{\alpha}_0 \in \widetilde{H}_1\mathcal{C}(S_a(0))$ and on the other, when $\lambda = 1$, can be regarded as a geometric representative of the homology class $\boldsymbol{\gamma}_1 \in \widetilde{H}_1\mathcal{C}^*(S_a(1))$.

These geometric representations of $\boldsymbol{\alpha}_0$ and $\boldsymbol{\gamma}_1$ and statement (3.7), viz.,

$$\mathcal{C}(\beta)_* \boldsymbol{\alpha}_0 \# \boldsymbol{\gamma}_1 \neq 0 ,$$

allow Theorem A.2 to be applied to system $(A.2_\varepsilon)$ yielding a family of solutions of the boundary value problem. However, for the family to possess the requisite limiting properties some care must be taken in applying Theorem A.2.

Accordingly, set $H := \{x \in [0, 1] : a(x) = \frac{1}{2}\}$ and set

$$\operatorname{gr}(\sigma_a) := \bigcup_{\lambda \in [0,1]} S_a(\lambda) \times \{\lambda\} \subset \mathbf{R}^2 \times [0, 1] .$$

Because the set map $\lambda \mapsto S_a(\lambda)$ varies upper semi-continuously with $\lambda \in [0, 1]$ as a consequence of the continuity of σ_a, it follows that $\operatorname{gr}(\sigma_a)$ is a compact subset of

$\mathbf{R}^3$. Hence, choose a sequence $\{\mathcal{O}_n\}_{n=1}^{\infty}$ of relatively compact open subsets of $\mathbf{R}^3$ satisfying

(1) if $a(\lambda) \neq \frac{1}{2}$ and $|\lambda - H| \geq (2n)^{-1}$, then $\mathcal{O}_n \cap \mathbf{R}^2 \times \{\lambda\}$ has two components each containing a component of $S_a(\lambda)$ $(n = 1, 2, \dots)$,
(2) $\mathrm{cl}\,(\mathcal{O}_{n+1}) \subset \mathcal{O}_n$ $(n = 1, 2, \dots)$,
(3) $\bigcap_{n=1}^{\infty} \mathcal{O}_n = \mathrm{gr}(\sigma_a)$.

For $\lambda = 0, 1$, the blocks $B(\lambda)$ and $C(\lambda)$ illustrated in Figure 1 depend implicitly on a parameter $r > 0$ where as r shrinks to zero $B(\lambda) \cup C(\lambda)$ shrinks to $S_a(\lambda)$. To see this, let us make this dependence on r explicit in the notation and write $B(\lambda, r)$ and $C(\lambda, r)$ from hereon instead of $B(\lambda)$ and $C(\lambda)$. Then regarded as subsets of $\mathbf{R}^2 \simeq \mathbf{R}^2 \times \{\lambda\}$, for $\lambda = 0, 1$,

$$B(\lambda, r) := \{(v, w) : |v - \lambda| + |w| \leq r\}$$

and

$$C(\lambda, r) := \{(v, w) : \big(|v - (1 - \lambda)| + |w| \leq r \text{ and } (-1)^{\lambda}(v + \lambda - 1) \geq 0\big) \\ \text{or } \big(H(v, w, \lambda) \leq H(1 - \lambda, r, \lambda) \text{ and } 0 \leq v \leq 1\big)\}$$

where r is chosen so small, say $0 < r < r_0$, that $B(\lambda, r)$ and $C(\lambda, r)$ are disjoint. From these descriptions and the definition of $S_a(\lambda)$, it follows easily that $B(\lambda, r) \cup C(\lambda, r)$ shrinks to $S_a(\lambda)$.

Then choose a sequence of positive reals $\{r_n\}_{n=1}^{\infty}$ within the open interval $]0, r_0[$ and strictly decreasing to zero so that for $\lambda = 0, 1$, $\big(B(\lambda, r_n) \cup C(\lambda, r_n)\big) \times \{\lambda\} \subset \mathcal{O}_n$. Also, let $\alpha_{n\,0}$ be the maximal segment of the line $w = 0$ contained in $B(0, r_n)$ and let $\gamma_{n\,1}$ be the maximal segment of $w = 0$ contained in $B(1, r_n)$. It follows that $\alpha_{n\,0}$ is a relative cycle in $B(0, r_n)$ modulo its exit set representing $\boldsymbol{\alpha}_0$ and that $\gamma_{n\,1}$ is a relative cycle in $B(1, r_n)$ modulo its entrance set representing $\boldsymbol{\gamma}_1$.

Application of Theorem A.2 to the family of systems $(A.2_\varepsilon)$ yields for each positive integer n an $\varepsilon_n > 0$ and a family of solutions $\{v_{\varepsilon\,n}\}_{0<\varepsilon\leq\varepsilon_n}$ of the boundary value problem with the property that

$$(v_{\varepsilon\,n}(x), \varepsilon v'_{\varepsilon\,n}(x), x) \in \mathcal{O}_n \qquad \text{for } x \in [0, 1].$$

We can, without loss of generality, assume that the sequence $\{\varepsilon_n\}_{n=1}^{\infty}$ strictly decreases to zero. Then set $\varepsilon^* := \varepsilon_1$ and for $\varepsilon \in\,]0, \varepsilon^*]$ if $\varepsilon_{n+1} < \varepsilon \leq \varepsilon_n$, set $v_\varepsilon := v_{\varepsilon\,n}$.

The family of solutions $\{v_\varepsilon\}_{0<\varepsilon\leq\varepsilon^*}$ has the desired limiting properties. This follows immediately from the choice of the sequence of sets $\mathcal{O}_n$ once it is realized that if $a(x) \neq \frac{1}{2}$ and if $\mathcal{O}_n \cap \mathbf{R}^2 \times \{x\}$ has two components each containing a component of $S_a(x) \times \{x\}$, then $(v_\varepsilon(x), \varepsilon v'_\varepsilon(x), x)$ always lies in the component of $\mathcal{O}_n \cap \mathbf{R}^2 \times \{x\}$, call it $C_n(x)$, that *does not* contain $(a(x), 0, x)$. Hence, as ε decreases and n therefore increases forcing $\mathcal{O}_n$ to squeeze down to $\mathrm{gr}(\sigma_a)$, the point $(v_\varepsilon(x), \varepsilon v'_\varepsilon(x), x)$ is forced to limit on $(0, 0, x)$ if $a(x) > \frac{1}{2}$, but on $(1, 0, x)$ if $a(x) < \frac{1}{2}$.

The placement of $(v_\varepsilon(x), \varepsilon v'_\varepsilon(x), x)$ in $C_n(x)$ is a consequence of the proof of [K4, Theorem 3.1] which is invoked to prove Theorem A.2. In particular, the component of $S_a(x) \times \{x\}$ containing $(a(x), 0, x)$ has, as observed in Example 3.5 above, trivial

Conley homotopy index in the forward time direction so that its Conley index has zero reduced homology in all dimensions. The singular chains selected in the proof of [K4, Theorem 4.1] always evolve so as to give the continuation of a non-zero homology class in the index so that at any point of the evolution any subchain lying in a component of $\mathcal{O}_n \cap \mathbf{R}^2 \times \{x\}$ whose closure isolates an invariant set with trivial index can be discarded as a consequence of the fact that the homology of the Conley index of an isolated invariant set that is the union of two disjoint isolated invariant sets is the direct sum of the homology modules of the Conley indices of the disjoint pieces which follows trivially from a Meyer-Vietoris sequence using an index pair for the whole isolated invariant set that is the union of disjoint index pairs for the pieces.

In more detail, at stage n when Theorem A.2 is applied to choose ε_n, Theorem 3.1 of [K4] is implicitly invoked. As applied to the current problem, the first step in the proof of the latter chooses a partition $\mathcal{P}$ of $[0,1]$ so that $\mathcal{C}\,(\beta) = F(\sigma_a, \mathcal{P})$ where $F(\sigma_a, \mathcal{P})$ is a composite of equivalences between Conley indices as in (5.4) above. It can be assumed that $\mathcal{P}$ contains sufficiently many partition points so that each partition interval of $\mathcal{P}$ satisfies at least one of two properties: (i) each point of the partition interval lies within distance n^{-1} of H; (ii) each point of the partition interval is at least distance $(2n)^{-1}$ from H. For a partition interval J satisfying the second, $\mathcal{O}_n \cap \mathbf{R}^2 \times J$ has components C_n and D_n where the labeling is such that $C_n(x) = C_n \cap \mathbf{R}^2 \times \{x\}$ for each $x \in J$. Thus, if $a(x) - \frac{1}{2} > 0$ for each $x \in J$, then $(0,0,x) \in C_n$ and $(1,0,x) \in D_n$ for each $x \in J$, but if $a(x) - \frac{1}{2} < 0$ for each $x \in J$, then $(1,0,x) \in C_n$ and $(0,0,x) \in D_n$ for each $x \in J$.

Let us consider one such partition interval J, write $J = [y,z]$, and note $J \subset [0,1]$. As applied to the family of systems $(A.2_\varepsilon)$, the proof of [K4, Theorem 3.1] is carried out by iteratively applying Sublemma(**) of Theorem 3.1 in [K4] to the family of autonomous stretched systems associated to the systems $(A.2_\varepsilon)$ restricted to successive partition intervals of $\mathcal{P}$. For present purposes, we need not specify these stretched systems, but instead rephrase the relevant parts of the proof of Sublemma(**) in terms of the flow of system $(A.2_\varepsilon^+)$.

As it applies in the present context and using the notation established in its proof, Sublemma(**) assumes given a singular chain c representing $\mathcal{C}\,(\beta^y)\,\boldsymbol{\alpha}_0$ relative to an index pair $\langle N_{1\,1}(y), N_{2\,1}(y)\rangle$ of $S_a(y)$ and various index pairs for $S_a(z)$—the only one of interest to us here is denoted there by $\langle N_{1\,3}(z), N_{2\,3}(z)\rangle$ and contains a perturbable index pair whence $(N_{1\,3}(z), \cap N_{2\,3}(z))$ has thick exit set. For sufficiently small $\varepsilon > 0$ (here sufficiently small means $\varepsilon \leq \varepsilon_n$) Sublemma(**) exhibits a subchain $\hat{c}_\varepsilon$ of c so that regarding $\hat{c}_\varepsilon$ as having support in $\mathbf{R}^2 \times \{y\}$, its image under the chain map induced by following the flow of system $(A.2_\varepsilon^+)$ for time $(z-y)$ is a chain $\bar{c}_\varepsilon$ with support in $\mathbf{R}^2 \times \{z\} \simeq \mathbf{R}^2$ and is a relative cycle in $(N_{1\,3}(z), \cap N_{2\,3}(z))$ representing $\mathcal{C}\,(\beta^z)_*\,\boldsymbol{\alpha}_0$. Further, for each time $x \in [0, z-y]$, the image of $\hat{c}_\varepsilon$ under the chain map induced by following the flow generated by $(A.2_\varepsilon^+)$ for time x has support in $\mathcal{O}_n \cap \mathbf{R}^2 \times \{y+x\}$. Thus, to complete the proof it will suffice to show that for $x \in J$, the image of $\hat{c}_\varepsilon$ obtained by following it under the flow for time x actually has support in $C_n(x)$.

The selection of $\hat{c}_\varepsilon$ is actually made by selecting $\bar{c}_\varepsilon$ as a subchain of the image of c under a certain chain map between index spaces using the barycentric subdivision operator on chains, but the details of this selection and how $\hat{c}_\varepsilon$ is produced from

$\bar{c}_\varepsilon$ are not relevant to us here. What is important here is that the selected chain $\bar{c}_\varepsilon$ is a relative cycle in $(N_{1\,3}(z), \cap N_{2\,3}(z))$ representing $\mathcal{C}\,(\beta^z)_*\,\boldsymbol{\alpha}_0$. Because $\mathcal{O}_n \cap \mathbf{R}^2 \times \{x\} = C_n(x) \cup D_n(x)$ we can, for $i = 1, 2$, write $N_{i\,3}(z)$ as the union of two disjoint sets $M_{i\,C}$ and $M_{i\,D}$, the former a subset of $C_n(z)$, the latter a subset of $D_n(z)$. It follows that $(M_{1\,C}(z), \cap M_{2\,C}(z))$ is an index pair for that part of $S_a(z)$ contained in $C_n(z)$ and that $(M_{1\,D}(Z), \cap M_{2\,D}(Z))$ is an index pair for that part of $S_a(z)$ contained in $D_n(z)$. Further, both these index pairs have thick exit set because $(N_{1\,3}(z), \cap N_{2\,3}(z))$ does. Then as that part of $S_a(z)$ contained in $D_n(z)$ has trivial Conley index, it follows that $H_*(M_{1\,D}(z), \cap M_{2\,D}(z)) \simeq 0$. Hence, as

$$H_*(M_{1\,C}(z), \cap M_{2\,C}(z)) \oplus H_*(M_{1\,D}(z), \cap M_{2\,D}(z)) \simeq H_*(N_{1\,3}(z), \cap N_{2\,2}(z))$$

with the isomorphism being given by the sum of inclusion induced homomorphisms, we can, without loss of generality, discard from the R-linear combination of simplexes giving $\bar{c}_\varepsilon$ those with support in $D_n(z)$ and assume that $\bar{c}_\varepsilon$ has support in $C_n(z)$. Because continuous maps preserve connectivity, it then follows by considering what happens to each simplex in the sum giving $\hat{c}_\varepsilon$ that the image of the support of $\hat{c}_\varepsilon$ over the time interval $[0, z - y]$ under the flow of $(A.2^+_\varepsilon)$ must lie entirely in C_n. Further, for each $x \in J$, the proof of Theorem A.2, and in particular, the conclusion of Theorem 3.1 of [K4], forces $(v_\varepsilon(x), \varepsilon v'_\varepsilon(x), x)$ to lie in the support of the image of the subchain $\hat{c}_\varepsilon$ followed for time x under the flow generated by system $(A.2^+_\varepsilon)$, i.e., it lies in $C_n(x)$ as desired. □

A stronger form of Theorem A.2 proved in [K7] allows after suitable modification of hypotheses (3) and (4) the endpoint sets in $(A.1b)$ to be replaced with endpoint sets $\mathbf{E}(\lambda, \varepsilon)$ that vary with ε and describes the limiting behavior of the solutions as ε decreases to zero. A key ingredient to this stronger result is Theorem 3.2. A brief discussion describing how results of the type of Theorem A.2 can be used to exhibit endpoint layers can be found in [K5, pp. 1129–1130]. For an explicit example exhibiting certain monotone equilibrium solutions of the above Fisher equation see [K6]; also see [K4, Example 1.4]. For a non-critical point section of the space of isolated invariant sets for a family of fast systems in $\mathbf{R}^3$ see [K4, Example 5.1]. Also examined in [K7] are situations where the family of fast systems associated to $(A.1a)$ exhibits the breakup of saddle-saddle connections, and these lead to families of solutions exhibiting interior transition layers. A detailed description of how solutions with interior transition layers are obtained is beyond the scope of the current work. The underlying argument is similar but requires additional algebraic machinery, namely, the exact sequence on homology of a repellor-attractor pair and the splitting map of such a sequence, in conjunction with [K4, Corollary 3.2]. Also, see [K4, Example 5.2]. It was the desire to exhibit solutions with interior transition layers that cannot be found via methods employing lower and upper solutions which provided the impetus for the results found in [K7].

Another potentially interesting situation where a theorem of the above type might apply and be of interest in the applications would be for a section of isolated invariant sets of the fast systems where at each point the isolated invariant set would have a Conley index equivalent to that of a suspension of a sub-shift of finite type. Invariant sets of this type have been realized for Smale flows on the three sphere [dR], but their Conley indices have not been computed and can be zero—the invariant set of the Smale horseshoe map is a subshift of finite type and the invariant set of the flow obtained by suspending this map has trivial Conley index. However, recent work of M. Mrozek [Mr] which defines a cohomological Conley type index (the Leray reduction of the Alexander-Spanier cohomology of an index pair suitably defined and an associated degree zero endomorphism) for each isolated invariant set of a homeomorphism of a locally compact Hausdorff space shows (see Theorem 2.9 and Example 8.1 of [Mr]) that there are suspensions of sub-shifts of finite type with non-trivial Conley index and with non-trivial rational homology. One should therefore be able to use Theorem A.2 and generalizations to discover solutions whose behavior over small sub-intervals would appear quite complicated, but was in fact mimicking a sub-shift of finite type.

Further, with the notice of acceptance for publication of this work, the referee informed the author that Mrozek, McCord, and Mischaikow have given an explicit relation between Mrozek's index for an isolated invariant set of a homeomorphism and the Conley index of the corresponding isolated invariant set of the suspended homeomorphism. The relation is expressed in terms of an exact sequence of cohomology modules; see [MMM].

APPENDIX B

PROOFS OF THE PROPOSITIONS IN §9.B

The proofs of Propositions 9.8–11 depend on the following statement found at the bottom of page 296 of [D, §VIII.7.12] which is elevated here to the status of a proposition. To state it, as in Chapter 9, let X denote an m-dimensional manifold oriented over a PID R and give every open submanifold of X the induced orientation. Also, coefficients of homology and cohomology modules are taken in some R-module G (unless specified otherwise) and the coefficient module is suppressed from the notation. Let K be a closed subset of X and set Ω_c^K to be the family of closed subsets of K that are cobounded in K. Note that Ω_c^K is cofinal in Ω^K, the family of locally compact subsets of K that are cobounded in K directed by containment. Thus, $\check{H}_c^*(K) = \varinjlim \check{H}^*(K, A)$ where $A \in \Omega_c^K$. Suppose $\check{x}_c \in \check{H}_c^i(K)$. By the definition of a direct limit, choose $A \in \Omega_c^K$ so that there exists $\check{x} \in \check{H}^i(K, A)$ whose image under the canonical map to the direct limit maps to $\check{x}_c$. As $\check{H}^*(X, A)$ is itself the direct limit of the singular cohomology modules of neighborhood pairs of (K, A), choose an open neighborhood pair (U, W) of (K, A) so that there exists $y \in H^i(U, W)$ whose image under the canonical map to the direct limit is $\check{x}$. Let $e_1\colon (U, U \setminus K) \hookrightarrow (X, X \setminus K)$ be the excision.

B.1 Proposition. *In $H_{m-i}(U, U \setminus K)$,*

$$\check{x}_c \frown o^U = y \frown o^U_{K \setminus W};$$

hence, in $H_{m-i}(X, X \setminus K)$,

$$\check{x}_c \frown o^X = e_{1*}(y \frown o^U_{K \setminus W}).$$

Proof. Let $C := \mathrm{cl}\,(K \setminus A)$, note that C is compact by choice of A, and observe that the following equalities between sets hold:

(1) $(U \setminus K) \cup W = U \setminus (K \setminus W)$;
(2) $(U \setminus K) \cup (W \setminus A) = (U \setminus A) \setminus (K \setminus W)$;
(3) $(U \setminus (K \cap C)) \cup (W \setminus (A \cap C)) = (U \setminus (A \cap C)) \setminus (K \setminus W)$.

Too, observe that the left-hand side of each equality is the union of open sets and that there are inclusions of pairs

(4) $(U \setminus A, (U \setminus A) \setminus (K \setminus W)) \subset (U, U \setminus (K \setminus W))$;
(5) $(U \setminus A, (U \setminus A) \setminus (K \setminus W)) \subset (U \setminus (A \cap C), (U \setminus (A \cap C)) \setminus (K \setminus W))$.

The proposition then follows from an analysis of the diagram in Figure 9 where the arrows will be defined as the analysis proceeds.

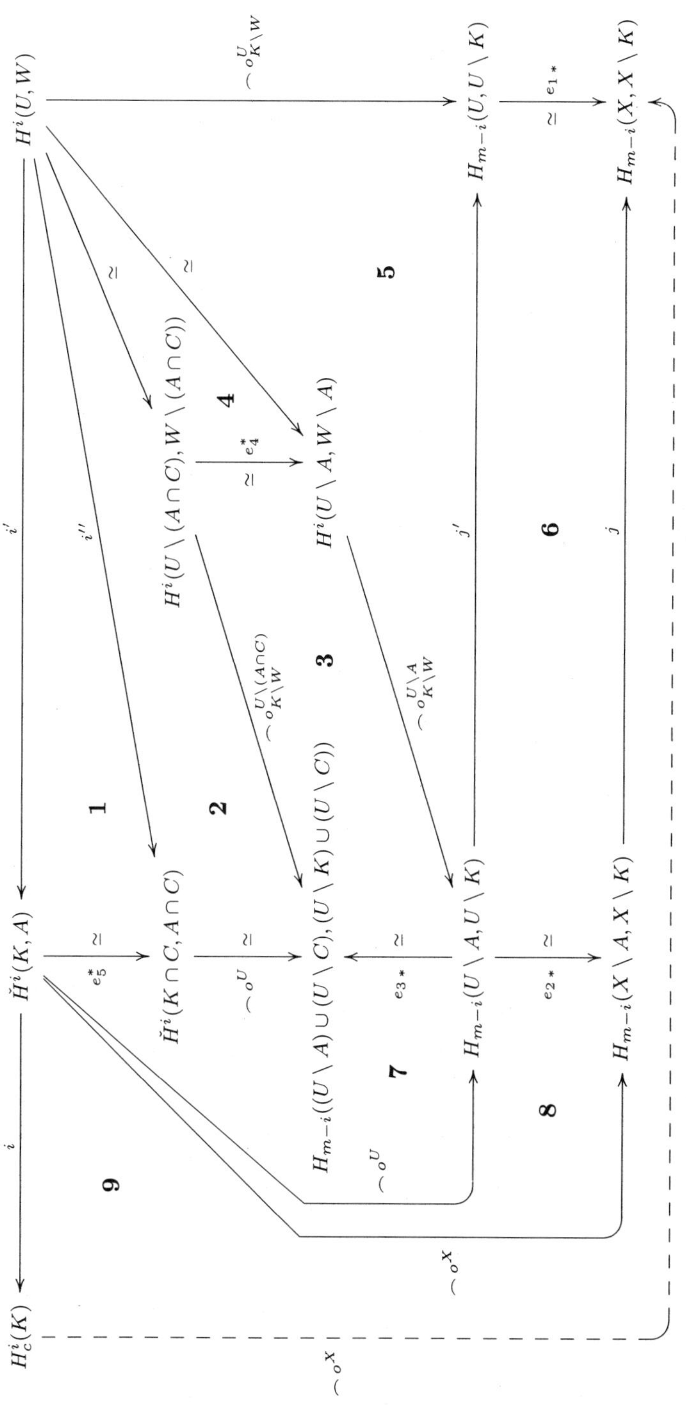
$H^i(U, W)$
$H_{m-i}(U, U \setminus K)$
$H_{m-i}(X, X \setminus K)$
$H^i(U \setminus (A \cap C), W \setminus (A \cap C))$
$H^i(U \setminus A, W \setminus A)$
$\check{H}^i(K, A)$
$\check{H}^i(K \cap C, A \cap C)$
$H_{m-i}((U \setminus A) \cup (U \setminus C), (U \setminus K) \cup (U \setminus C))$
$H_{m-i}(U \setminus A, U \setminus K)$
$H_{m-i}(X \setminus A, X \setminus K)$
$H^i_c(K)$
$\frown o^U_{K \setminus W}$
$\frown o^{U \setminus (A \cap C)}_{K \setminus W}$
$\frown o^{U \setminus A}_{K \setminus W}$
$\frown o^U$
$\frown o^X$
e_{1*}
e_{2*}
e_{3*}
e_4^*
e_5^*
i
i'
i''
j
j'
$\simeq$
1
2
3
4
5
6
7
8
9

FIGURE 9

The set equalities (1), (2), and (3) and the fact that the left-hand side of each is the union of open sets yield that in Figure 9 the cap product homomorphisms $\frown o^U_{K\setminus W}$, $\frown o^{U\setminus A}_{K\setminus W}$, and $\frown o^{U\setminus(A\cap C)}_{K\setminus W}$ determined by fundamental classes along the compact $K\setminus W$ are well-defined. Of course, the arrows labeled with $\frown o^X$ or $\frown o^U$ are Poincaré duality isomorphisms.

In Figure 9, the remaining arrows are defined and the diagram is shown commutative as follows. All unlabeled arrows are inclusion induced. All arrows labeled with an e_j^* or an e_{j*} for some integer j are excision isomorphisms. The arrow labeled i is the canonical map into the direct limit as are the arrows labeled i' and i'' in triangle **1**. It follows that triangle **1** commutes. The vertical arrow in quadrilateral **2**, labeled $\frown o^U$, is a relative Poincaré-Alexander duality isomorphism whence quadrilateral **2** commutes by the very definition of $\frown o^U$. Parallelogram **3** commutes by naturality of cap products since $o^{U\setminus A}_{K\setminus W}$ has image $o^{U\setminus(A\cap C)}_{K\setminus W}$ under the homomorphism induced by the inclusion in (5). Similarly, quadrilateral **5** commutes since $o^{U\setminus A}_{K\setminus W}$ has image $o^U_{K\setminus W}$ under the homomorphism induced by the inclusion in (4). Triangle **4** and rectangle **6** commute because all arrows in them are inclusion induced. Subdiagram **7** commutes because the curvilinear arrow of the subdiagram is by definition the composite of the vertical arrows in the subdiagram; similarly, by definition of its curvilinear arrow, subdiagram **8** commutes. Because $H_*(U,U\setminus K)=\varinjlim H_*(U\setminus A,U\setminus K)$ taken over $A\in\Omega_c^K$ (see [D, §VII.7.12, p. 296]) the inclusion induced homomorphism j' can be regarded as the canonical map into the direct limit, and analogous remarks apply when U is replaced by X and j' by j. Thus, passage to the direct limit over $A\in\Omega_c^K$ yields for $Z:=U$ or $Z:=X$ that the curvilinear arrow marked $\frown o^Z$ passes to the Poincaré duality isomorphism

$$(B.1)\qquad \frown o^Z\colon \check{H}^i_c(K)\to H_{m-i}(Z,Z\setminus K)\,.$$

With $Z:=X$, the last sentence immediately implies that

$$(B.2)\qquad \frown o^X\circ i=j\circ\frown o^X$$

where on the left $\frown o^X$ denotes the dashed curvilinear arrow of subdiagram **9** but on the right it denotes the solid curvilinear arrow so marked in subdiagrams **8** and **9**. Thus, equality $(B.2)$ states that subdiagram **9** is commutative. Analogously, one sees that with $Z:=U$

$$(B.3)\qquad \frown o^U\circ i=j'\circ\frown o^U$$

where on the left $\frown o^U$ is the arrow of $(B.1)$ which does not appear in Figure 9, but on the right, $\frown o^U$ is the curvilinear arrow so marked in subdiagram 7.

The proposition then follows easily from equality $(B.3)$ and a trivial chase on the diagram in Figure 9 starting at $y\in H^i(U,W)$. □

B.2 The Proof of Proposition 9.8. Suppose $\check{x}_c\in\check{H}^i_c(V)$. Then Proposition B.1 yields that for some open subset W of V, W is cobounded in V and where $K:=V\setminus W$, K is compact and for some $y\in H^i(V,W)=H^i(V,V\setminus K)$

$$\check{x}_c\frown o^V=y\frown o^V_K\,.$$

Next, note that the diagram

$$
\begin{array}{ccccccc}
\check{H}^i_c(V) & \xleftarrow{i^V_*} & H^i(V, V \setminus K) & \xleftarrow[\simeq]{e_{K*}} & H^i(X, X \setminus K) & \xrightarrow{i^X_*} & \check{H}^i_c(X) \\
& \searrow {\scriptstyle \frown o^V} & \downarrow {\scriptstyle \frown o^V_K} & & \downarrow {\scriptstyle \frown o^X_K} & \swarrow {\scriptstyle \frown o^X} & \\
& & H_{m-i}(V) & \longrightarrow & H_{m-i}(X) & &
\end{array}
$$

is commutative where i^V_* and i^X_* are the canonical maps into the direct limits: The triangles of morphisms at left and right are commutative because the oblique arrow in each triangle is obtained from the vertical arrow in the triangle by passage to the direct limit over the family of vertical maps obtained by replacing K with any compact subset of V (respectively X) in the triangle on the left (right). Also, as o^X_K is the image of o^V_K under the excision isomorphism induced by the inclusion of pairs $(V, V \setminus K) \subset (X, X \setminus K)$, it follows that the middle rectangle commutes by naturality of singular cap products. Since $\check{x}_c$ was an arbitrary element of $\check{H}^i_c(V)$, the proposition follows by chasing the class y around the diagram because by definition of the top horizontal arrow in the diagram in the statement of the proposition the image of $\check{x}_c$ under that horizontal arrow is $i^X_* \circ e^{-1}_{K*} y$. $\square$

B.3 The Proof of Proposition 9.9. First, note that the hypotheses imply

$$V \setminus K_1 = (V \setminus L_1) \setminus (K_1 \setminus L_1) \subset (X \setminus L_2) \setminus (K_2 \setminus L_2) = X \setminus K_2$$

whence there is an excision $(V \setminus L_1, V \setminus K_1) \subset (X \setminus L_2, X \setminus K_2)$. Thus, the bottom horizontal arrow of the diagram in the statement of the proposition is well-defined and is an isomorphism because the other three sides of the diagram are isomorphisms. Too, the reader can if desired easily verify that $(V \setminus L_1) \cup (X \setminus K_2) = X \setminus L_2$ and that $(V \setminus L_1) \cap (X \setminus K_2) = V \setminus K_1$ which together imply that the excision yields an isomorphism on homology as $V \setminus L_1$ and $X \setminus K_2$ are open in $X \setminus L_2$.

Next, set $K := K_1 \setminus L_1 = K_2 \setminus L_2$. Note that K is closed relative to both $V \setminus L_1$ and $X \setminus L_2$. Thus, as in the preamble to Proposition B.1, for any $\check{x}_c \in \check{H}^i_c(K)$ there exists $A \in \Omega^K_c$ and an open (relative to both $V \setminus L_1$ and $X \setminus L_2$) neighborhood pair (U, W) of (K, A) so that for some $\check{x} \in \check{H}^i(K, A)$ and for some $y \in H^i(U, W)$, under the canonical maps to the direct limits y has image $\check{x}$ and $\check{x}$ has image $\check{x}_c$. By Proposition B.1,

$$\check{x}_c \frown o^U = y \frown o^U_{K \setminus W} \in H_{m-i}(U, U \setminus K) .$$

Commutativity of the triangle of excision induced isomorphisms

$$
\begin{array}{ccc}
& H_{m-i}(U, U \setminus K) & \\
{\scriptstyle \simeq}\swarrow {\scriptstyle e_{2*}} & & {\scriptstyle e_{1*}} \searrow {\scriptstyle \simeq} \\
H_{m-i}(V \setminus L_1, V \setminus K_1) & \xrightarrow[\simeq]{e_{3*}} & H_{m-i}(X \setminus L_2, X \setminus K_2)
\end{array}
$$

yields by the second part of Proposition B.1 that

$$\check{x}_c \frown o^{X \setminus L_2} = e_{1*}(y \frown o^U_{K \setminus W}) = e_{3*} \circ e_{2*}(y \frown o^U_{K \setminus W}) = e_{3*}(\check{x}_c \frown o^{V \setminus L_1}) . \quad \square$$

B.4 The Proof of Proposition 9.10. Suppose $\check{x}_\ell \in \check{H}_c(K_\ell)$. Then Proposition B.1 yields that for some open pair (V_ℓ, W_ℓ) in M_ℓ where W_ℓ contains a cobounded subset A_ℓ of K_ℓ,

$$(B.4) \qquad \check{x}_\ell \frown o_\ell = y_\ell \frown o_{K_\ell \setminus W_\ell}$$

where $y_\ell \in H^{m_\ell - i_\ell}(V_\ell, W_\ell)$ and $o_{K_\ell \setminus W_\ell}$ is the fundamental class along the compact set $K_\ell \setminus W_\ell$. Then by the standard identity of singular theory relating cap and cross products (see [D, §VII.12.17])

$(B.5)$

$$(y_1 \frown o_{K_1 \setminus W_1}) \times (y_2 \frown o_{K_2 \setminus W_2}) = (-1)^{m_1(m_2 - i_2)}(y_1 \times y_2) \frown (o_{K_1 \setminus W_1} \times o_{K_2 \setminus W_2}).$$

Because

$$o_{K_1 \setminus W_1} \times o_{K_2 \setminus W_2} = (o_1 \times o_2)^{M_1 \times M_2}_{(K_1 \setminus W_1) \times (K_2 \setminus W_2)},$$

i.e., because the product on the left side is the generator of $H_{2m}(X \times X, X \times X \setminus (K_1 \setminus W_1 \times K_2 \setminus W_2))$ corresponding to the product orientation and because $\check{x}_1 \times \check{x}_2$ is represented in $\check{H}((K_1, A_1) \times (K_2, A_2))$ by the canonical image of $y_1 \times y_2$, the result follows from the equalities $(B.4)$ and $(B.5)$ and Proposition B.1. □

B.5 The Proof of Proposition 9.11. Let $\check{x} \in \check{H}^i(K, L)$ and let $\check{y}_c \in \check{H}^j_c(K)$. As in the preamble to Proposition B.1, for some $A \in \Omega^K_c$, for some open neighborhood pair (U, ω) of (K, A), some $z \in H^j(U, \omega)$ has image $\check{y}_c$ under the canonical map to the direct limit. By the properties of direct systems, we may without loss of generality assume that for some W, (U, W) is an open neighborhood pair of (K, L) and some $x \in H^i(U, W)$ has image $\check{x}$ under the canonical map to the direct limit. The proposition will follow from a diagram chase on the diagrams in Figures 10 and 11 once those diagrams are defined and shown commutative.

Toward that end, note that $K \setminus \omega$ and $K \setminus (W \cup \omega)$ are compact and observe that the following equalities between sets hold:

(1) $U \setminus (K \setminus \omega) = (U \setminus K) \cup \omega$;
(2) $U \setminus (K \setminus (W \cup \omega)) = (U \setminus K) \cup W \cup \omega$;
(3) $(U \setminus L) \setminus (K \setminus (W \cup \omega)) = (U \setminus K) \cup (W \setminus L) \cup (\omega \setminus L)$.

Observe too that there are inclusions of open pairs

(4) $(U, (U \setminus K) \cup \omega) \subset (U, (U \setminus K) \cup W \cup \omega)$;
(5) $(U \setminus L, (U \setminus K) \cup (W \setminus L) \cup (\omega \setminus L)) \subset (U, (U \setminus K) \cup W \cup \omega)$.

Also, as in earlier proofs, a homomorphism labeled by $e_{j\,*}$ or e^j_* for some integer j is an excision isomorphism, and one labeled by $i_{j\,*}$ or i^j_* for some integer j is inclusion induced.

Then the diagram in Figure 10 is well-defined and commutes: It follows from equalities (1), (2), and (3) and that the right-hand side of each is expressed as the union of open sets that each of the factor cap product homomorphisms in the diagram in Figure 10 labeled by $\frown o^U_{K \setminus W}$, $\frown o^U_{K \setminus (W \cup \omega)}$, or $\frown o^{U \setminus L}_{K \setminus (W \cup \omega)}$ is well-defined. It follows that the triangle of homomorphisms at the top left of the diagram in Figure 10 commutes by naturality of singular cap products because $o^U_{K \setminus (W \cup \omega)}$ is the image of $o^U_{K \setminus \omega}$ under the homomorphism induced by the inclusion

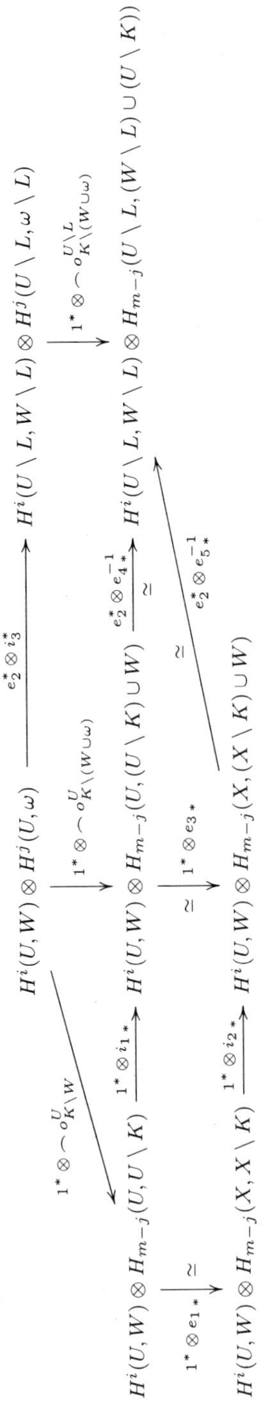

FIGURE 10

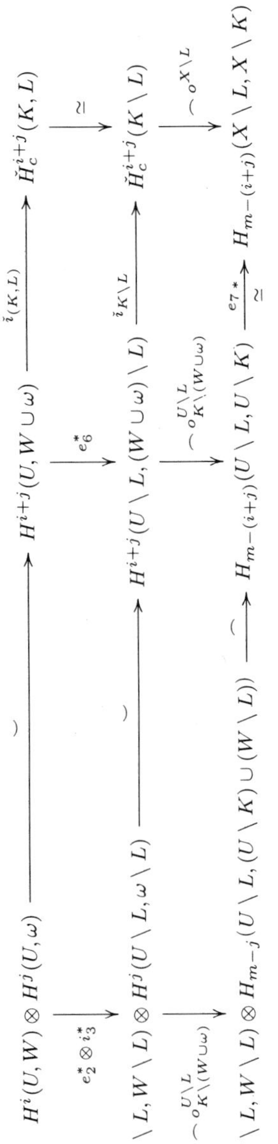

FIGURE 11

of (4). Similarly, the rectangle of homomorphisms at the top right of the diagram in Figure 10 commutes because $o^{U}_{K\setminus(W\cup\omega)}$ is the image of $o^{U\setminus L}_{K\setminus(W\cup\omega)}$ under the homomorphism induced by the inclusion of (5).

Also, the diagram in Figure 11 is well-defined and commutes: For the same reasons as in the diagram in Figure 10, in the diagram in Figure 11, the factor cap product homomorphisms labeled by $\frown o^{U\setminus L}_{K\setminus(W\cup\omega)}$ are well-defined. It follows that the lower left rectangle commutes as a consequence of the standard associative law of cup and cap products in singular theory; see [D, §VII.12.7]. The upper left rectangle commutes by naturality of cup products. The horizontal arrows $\check{\imath}^*_{(K,L)}$ and $\check{\imath}^*_{K\setminus L}$ are the canonical maps to the direct limits. It follows that the upper right rectangle commutes because the natural isomorphism that is the right edge is obtained by passage to the limit from e_6^* and analogous excisions obtained by replacing $W\cup\omega$ with V where (U,V) is a neighborhood pair of (K,L) and $K\setminus V$ is bounded. Finally, the rectangle in the lower right corner commutes as a consequence of the proof of Proposition B.1, in particular the commutativity of the diagram in Figure 9: For the composite $e_{7*}\circ \frown o^{U\setminus L}_{K\setminus(W\cup\omega)}$ corresponds to the composite of the two arrows in the right edge of the diagram in Figure 9, $i^*_{K\setminus L}$ corresponds to the composite of the arrows in the top edge of that diagram, and $\frown o^{X\setminus L}$ corresponds to the dashed curvilinear arrow in that diagram.

To begin the diagram chase, note that the diagrams in Figures 10 and 11 can be joined by identifying the top horizontal arrow of the diagram in Figure 10 with the top vertical arrow in the left edge of the diagram in Figure 11 and also by identifying the rightmost vertical arrow of the diagram in Figure 10 with the bottom vertical arrow in the left edge of the diagram in Figure 11. Next, note that $x\otimes z$ has image $\check{x}\smile\check{y}_c$ under the composite of the arrows in the top row of Figure 11 by definition of the cup product on $\check{H}^*(K,L)\otimes\check{H}^*_c(K)$. It follows that the image of $x\otimes z$ under the composition of the arrows in the top row and right edge of Figure 11 is $(\check{x}\smile\check{y}_c)\frown o^X$ where here $\frown o^X$ is just the composite of the arrows in the right edge. On the other hand, Proposition B.1 yields that the image of $x\otimes z$ under the composite obtained by starting at the top left hand corner of the diagram in Figure 10 and traveling down along its left edge to $H^i(U,W)\otimes H_{m-j}(X,X\setminus K)$ is $x\otimes(\check{y}_c\frown o^X)$. The commutativity of both diagrams shows that the image of $x\otimes(\check{y}_c\frown o^X)$ under the composite homomorphism $C^{(U,W)}$ obtained by first following the horizontal and oblique arrows in the bottom edge of the diagram in Figure 10 and then following the two bottom horizontal arrows in the diagram in Figure 11 is also $(\check{x}\smile\check{y}_c)\frown o^X$. Because the Čech cap product of elements in $\check{H}^*(K,L)\otimes H_*(X,X\setminus K)$ is obtained by passing to the direct limit from composites of the form $C^{(U,W)}$, it follows that

$$\check{x}\frown(\check{y}_c\frown o^X)=(\check{x}\smile\check{y}_c)\frown o^X$$

as desired. □

References

[Ab-R] R. Abraham and J. Robbin, *Transversal Mappings and Flows*, W.A. Benjamin, Inc., New York, 1967.

[AnMP] S. Angenent and J. Mallet-Paret, *Stable transition layers in a semilinear boundary value problem*, J. Differential Equations **67** (1987), 212–242.

[A] V.I. Arnold, *Ordinary Differential Equations*; translated from the Russian by R.A. Silverman, The MIT Press, Boston, 1978.

[Bg] Yu. P. Boglaev, *The two-point problem for a class of ordinary differential equations with a small parameter coefficient of the derivative*, USSR Comput. Math.-Math. Phys. **10** (1970), 190–204.

[Ch] R.C. Churchill, *Isolated invariant sets in compact metric spaces*, J. Differential Equations **12** (1972), 330–352.

[C] C.C. Conley, *Isolated Invariant Sets and the Morse Index*, CBMS Conference Proceedings, Amer. Math. Soc., Providence, R.I., 1978.

[C-E] C. Conley and R. Easton, *Isolated invariant sets and isolating blocks*, Trans. AMS **158** (1971), 35–61.

[D] A. Dold, *Lectures on Algebraic Topology*, Springer-Verlag, New York, 1972.

[F] R.A. Fisher, *The wave of advance of advantageous genes*, Ann. of Eugenics **7** (1937), 355–369.

[Ha] J.B.S. Haldane, *The theory of a cline*, J. Genetics **48** (1948), 277–284.

[H] M.W. Hirsch, *Differental Topology*, Springer-Verlag, New York, 1976.

[HPS] M.W. Hirsch, C.C. Pugh, and M. Shub, *Invariant Manifolds*, Lecture Notes in Mathematics, vol. 583, Springer-Verlag, New York, 1977.

[H-S] M.W. Hirsch and S. Smale, *Differential Equations, Dynamical Systems, and Linear Algebra*, Academic Press, New York, 1974.

[Hus] D. Husemoller, *Fibre Bundles*, second edition, Springer-Verlag, New York, 1975.

[K1] H.L. Kurland, *The Morse index of an isolated invariant set is a connected simple system*, J. Differential Equations **42** (1981), 234–259.

[K2] ———, *Homotopy invariants of a repeller-attractor pair,* I*: the Püppe sequence of an R-A pair*, J. Differential Equations **46** (1982), 1–31.

[K3] ———, *Homotopy invariants of a repeller-attractor pair,II: continuation of an R-A pair*, J. Differential Equations **49** (1983), 281–329.

[K4] ———, *Following homology in singularly perturbed systems*, J. Differential Equations **62** (1986), 1–72.

[K5] ———, *On the two definitions of the Conley index*, Proceedings of the AMS **106** (1989), 1117–1130.

[K6] ———, *Monotone and oscillatory solutions of a problem arising in population genetics*, Nonlinear Partial Differential Equations (J.A. Smoller, ed.), Contemporary Mathematics, vol. 17, AMS, Providence, R.I., 1983, pp. 323–342.

[K7] ———, *Layers in singularly perturbed systems via homology contintuation*, (unpublished typescript).

[Mc1] C.K. McCord, *Mappings and homological properties in the Conley index theory*, Ergod. Th. & Dynam. Sys. **8*** (1988), 175–198.

[Mc2] C.K. McCord, *Poincaré-Lefschetz duality for the homology Conley index*, Transactions of the AMS **329** (1992), 233–252.

[M] J.T. Montgomery, *Cohomology of isolated invariant sets under perturbation*, J. Differential Equations **13** (1973), 257–299.

[Mr] M. Mrozek, *Leray functor and cohomological Conley index for discrete dynamical systems*, Transactions of the AMS **318** (1990), 149–178.

[MMM] C.K. McCord, K. Mischaikow, and M. Mrozek, *Zeta functions, periodic trajectories, and the Conley index*, J. Differential Equations (to appear).

[N] M. Nagumo, *Uber die Differentialgleichung* $y'' = f(x, y, y')$, Proc. Phys. Math. Soc. Japan Ser. 3 **19** (1937), 861–866.

[Pa] J. Palis, *On Morse-Smale dynamical systems*, Topology **8** (1969), 385–405.

[PaS] J. Palis and S. Smale, *Structural stability theorems*, Global Analysis (S.-S. Chern and S. Smale, eds.), Proceedings of Symposia in Pure Mathematics, vol. XIV, AMS, Providence, R.I., 1970, pp. 223–231.

[PS] C. Pugh and M. Shub, *Linearization of Normally Hyperbolic Diffeomorphisms and Flows*, Inventiones math. **10** (1970), 187–198.

[dR] K. de Rezende, *Smale flows on the three sphere*, Trans. AMS **303** (1987), 283–310.

[R] C. Robinson, *Stable manifolds in hamiltonian systems*, Hamiltonian Dynamical Systems (K.R. Meyer and D.G. Saari, eds.), Contemporary Mathematics, vol. 81, AMS, Providence, R.I., 1988, pp. 77–97.

[Sl] D. Salamon, *Connected simple systems and the Conley index of isolated invariant sets*, Trans. AMS **291** (1985), 1–41.

[S] S. Smale, *Stable manifolds for differential equations and diffeomorphisms*, Annali della Scuola Norm. Sup. - Pisa **17** (1963), 97–116.

[Sm] J. Smoller, *Shock Waves and Reaction-Diffusion Equations*, Springer-Verlag, New York Heidelberg Berlin, 1983.

[Sp] E.H. Spanier, *Algebraic Topology*, McGraw-Hill, New York, 1966.

[St] N. Steenrod, *The Topology of Fibre Bundles*, Princeton University Press, Princeton, NJ, 1951.

[Sw] R. Switzer, *Algebraic Topology-Homotopy and Homology*, Springer-Verlag, New York, 1975.

[WY] F.W. Wilson, Jr. and J.A. Yorke, *Lyapunov functions and isolating blocks*, J. Differential Equations **13** (1973), 106–123.

10 Westchester Court, Clifton Park, NY 12065
E-mail address: hkurl@math.albany.edu

Editorial Information

To be published in the *Memoirs*, a paper must be correct, new, nontrivial, and significant. Further, it must be well written and of interest to a substantial number of mathematicians. Piecemeal results, such as an inconclusive step toward an unproved major theorem or a minor variation on a known result, are in general not acceptable for publication. *Transactions* Editors shall solicit and encourage publication of worthy papers. Papers appearing in *Memoirs* are generally longer than those appearing in *Transactions* with which it shares an editorial committee.

As of September 30, 1995, the backlog for this journal was approximately 6 volumes. This estimate is the result of dividing the number of manuscripts for this journal in the Providence office that have not yet gone to the printer on the above date by the average number of monographs per volume over the previous twelve months, reduced by the number of issues published in four months (the time necessary for preparing an issue for the printer). (There are 6 volumes per year, each containing at least 4 numbers.)

A Copyright Transfer Agreement is required before a paper will be published in this journal. By submitting a paper to this journal, authors certify that the manuscript has not been submitted to nor is it under consideration for publication by another journal, conference proceedings, or similar publication.

Information for Authors and Editors

Memoirs are printed by photo-offset from camera copy fully prepared by the author. This means that the finished book will look exactly like the copy submitted.

The paper must contain a *descriptive title* and an *abstract* that summarizes the article in language suitable for workers in the general field (algebra, analysis, etc.). The *descriptive title* should be short, but informative; useless or vague phrases such as "some remarks about" or "concerning" should be avoided. The *abstract* should be at least one complete sentence, and at most 300 words. Included with the footnotes to the paper, there should be the 1991 *Mathematics Subject Classification* representing the primary and secondary subjects of the article. This may be followed by a list of *key words and phrases* describing the subject matter of the article and taken from it. A list of the numbers may be found in the annual index of *Mathematical Reviews*, published with the December issue starting in 1990, as well as from the electronic service e-MATH [**telnet e-MATH.ams.org** (or **telnet 130.44.1.100**). Login and password are **e-math**]. For journal abbreviations used in bibliographies, see the list of serials in the latest *Mathematical Reviews* annual index. When the manuscript is submitted, authors should supply the editor with electronic addresses if available. These will be printed after the postal address at the end of each article.

Electronically prepared papers. The AMS encourages submission of electronically prepared papers in AMS-TEX or AMS-LATEX. The Society has prepared author packages for each AMS publication. Author packages include instructions for preparing electronic papers, the *AMS Author Handbook*, samples, and a style file that generates the particular design specifications of that publication series for both AMS-TEX and AMS-LATEX.

Authors with FTP access may retrieve an author package from the Society's Internet node e-MATH.ams.org (130.44.1.100). For those without FTP access, the author package can be obtained free of charge by sending e-mail to pub@math.ams.org (Internet) or from the Publication Division, American Mathematical Society, P.O. Box 6248, Providence, RI 02940-6248. When requesting an author package, please specify AMS-TEX or AMS-LATEX, Macintosh or IBM (3.5) format, and the publication in which your paper will appear. Please be sure to include your complete mailing address.

Submission of electronic files. At the time of submission, the source file(s) should be sent to the Providence office (this includes any TEX source file, any graphics files, and the DVI or PostScript file).

Before sending the source file, be sure you have proofread your paper carefully. The files you send must be the EXACT files used to generate the proof copy that was accepted for publication. For all publications, authors are required to send a printed copy of their paper, which exactly matches the copy approved for publication, along with any graphics that will appear in the paper.

TEX files may be submitted by email, FTP, or on diskette. The DVI file(s) and PostScript files should be submitted only by FTP or on diskette unless they are encoded properly to submit through e-mail. (DVI files are binary and PostScript files tend to be very large.)

Files sent by electronic mail should be addressed to the Internet address pub-submit@math.ams.org. The subject line of the message should include the publication code to identify it as a Memoir. TEX source files, DVI files, and PostScript files can be transferred over the Internet by FTP to the Internet node e-math.ams.org (130.44.1.100).

Electronic graphics. Figures may be submitted to the AMS in an electronic format. The AMS recommends that graphics created electronically be saved in Encapsulated PostScript (EPS) format. This includes graphics originated via a graphics application as well as scanned photographs or other computer-generated images.

If the graphics package used does not support EPS output, the graphics file should be saved in one of the standard graphics formats—such as TIFF, PICT, GIF, etc.—rather than in an application-dependent format. Graphics files submitted in an application-dependent format are not likely to be used. No matter what method was used to produce the graphic, it is necessary to provide a paper copy to the AMS.

Authors using graphics packages for the creation of electronic art should also avoid the use of any lines thinner than 0.5 points in width. Many graphics packages allow the user to specify a "hairline" for a very thin line. Hairlines often look acceptable when proofed on a typical laser printer. However, when produced on a high-resolution laser imagesetter, hairlines become nearly invisible and will be lost entirely in the final printing process.

Screens should be set to values between 15% and 85%. Screens which fall outside of this range are too light or too dark to print correctly.

Any inquiries concerning a paper that has been accepted for publication should be sent directly to the Editorial Department, American Mathematical Society, P. O. Box 6248, Providence, RI 02940-6248.

Editors

This journal is designed particularly for long research papers (and groups of cognate papers) in pure and applied mathematics. Papers intended for publication in the *Memoirs* should be addressed to one of the following editors:

Ordinary differential equations, partial differential equations, and applied mathematics to JOHN MALLET-PARET, Division of Applied Mathematics, Brown University, Providence, RI 02912-9000; e-mail: am438000@brownvm.brown.edu.

Harmonic analysis, representation theory, and Lie theory to ROBERT J. STANTON, Department of Mathematics, The Ohio State University, 231 West 18th Avenue, Columbus, OH 43210-1174; electronic mail: stanton@function.mps.ohio-state.edu.

Ergodic theory, dynamical systems, and abstract analysis to DANIEL J. RUDOLPH, Department of Mathematics, University of Maryland, College Park, MD 20742; e-mail: djr@math.umd.edu.

Real and harmonic analysis and elliptic partial differential equations to JILL C. PIPHER, Department of Mathematics, Brown University, Providence, RI 02910-9000; e-mail: jpipher@gauss.math.brown.edu.

Algebra and algebraic geometry to EFIM ZELMANOV, Department of Mathematics, University of Wisconsin, 480 Lincoln Drive, Madison, WI 53706-1388; e-mail: zelmanov@math.wisc.edu

Algebraic topology and cohomology of groups to STEWART PRIDDY, Department of Mathematics, Northwestern University, 2033 Sheridan Road, Evanston, IL 60208-2730; e-mail: s_priddy@math.nwu.edu.

Global analysis and differential geometry to ROBERT L. BRYANT, Department of Mathematics, Duke University, Durham, NC 27706-7706; e-mail: bryant@math.duke.edu.

Probability and statistics to RICHARD DURRETT, Department of Mathematics, Cornell University, White Hall, Ithaca, NY 14853-7901; e-mail: rtd@cornella.cit.cornell.edu.

Combinatorics and Lie theory to PHILIP J. HANLON, Department of Mathematics, University of Michigan, Ann Arbor, MI 48109-1003; e-mail: phil.hanlon@math.lsa.umich.edu.

Logic and universal algebra to GREGORY L. CHERLIN, Department of Mathematics, Rutgers University, Hill Center, Busch Campus, New Brunswick, NJ 08903; e-mail: cherlin@math.rutgers.edu.

Number theory and arithmetic algebraic geometry to ALICE SILVERBERG, Department of Mathematics, Ohio State University, Columbus, OH 43210-1174; e-mail: silver@math.ohio-state.edu.

Complex analysis and complex geometry to DANIEL M. BURNS, Department of Mathematics, University of Michigan, Ann Arbor, MI 48109-1003; e-mail: burns@gauss.stanford.edu.

Algebraic geometry and commutative algebra to LAWRENCE EIN, Department of Mathematics, University of Illinois, 851 S. Morgan (MIC 249), Chicago, IL 60607-7045; email: u22425@uicvm.uic.edu.

All other communications to the editors should be addressed to the Managing Editor, PETER SHALEN, Department of Mathematics, Statistics, and Computer Science, University of Illinois at Chicago, Chicago, IL 60680; e-mail: shalen@math.uic.edu.

Other Titles in This Series

(Continued from the front of this publication)

(See the AMS catalog for earlier titles)